AF601317

Reliability Theory Based on Uncertain Lifetimes

Ying Liu

Reliability Theory Based on Uncertain Lifetimes

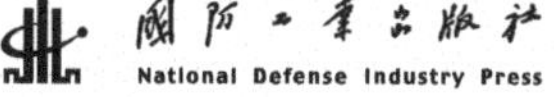

Ying Liu
Tianjin University of Science and Technology
Tianjin, China

ISBN 978-981-16-0994-7 ISBN 978-981-16-0995-4 (eBook)
https://doi.org/10.1007/978-981-16-0995-4

Jointly published with National Defense Industry Press
The print edition is not for sale in China Mainland. Customers from China Mainland please order the print book from: National Defense Industry Press.
ISBN of the Co-Publisher's edition: 978-7-118-11716-5

This Springer imprint is published by the registered company Springer Nature Singapore Pte Ltd.
The registered company address is: 152 Beach Road, #21-01/04 Gateway East, Singapore 189721, Singapore

Preface

Reliability system engineering is an interdisciplinary subject whose main research object is the life characteristics of products. It involves many fields of basic mathematics, engineering technology and management science. In recent years, many related books have appeared in the field of reliability. However, there are relatively few books on reliability mathematical models, or reliability mathematical models established only on the basis of probability theory or fuzzy theory. According to the different life characteristics of systems, this book uses three different mathematical tools: probability theory, fuzzy theory and random fuzzy theory to model and analyze the reliability of systems, which is also the feature of this book. The contents of this book can not only enrich and improve the traditional reliability theory, but also promote the development of related disciplines, which has important theoretical significance.

The book is divided into six chapters, among which Chaps. 3, 4 and 6 gather the original research achievements of the author for many years. The research was supported by the National Natural Science Foundation of China under Grant 11301382, the Humanities and Social Sciences Foundation of the Ministry of Education under Grant 18YJC630108, the Tianjin Research Program of Application Foundation and Advanced Technology under Grant 16JCYBJC18500, and the Enterprise Science and Technology Special Project under Grant 17JCTPJC54100. The contents have been reported many times in important academic conferences. Considering the applicability of this book and the wide range of readers, Chaps. 2 and 5 extract some contents from the "Introduction to Reliability Mathematics" coauthored by Professor Jinhua Cao and Professor Kan Cheng, so as to facilitate readers to compare the reliability mathematical models established by the three mathematical tools.

Chapter 1 introduces some basic concepts of uncertain mathematical theory. In Chaps. 2 and 5, the classical reliability mathematical models of nonrepairable systems and repairable systems are introduced, respectively. In Chap. 3, the reliability mathematical models of nonrepairable systems in fuzzy environment are established. In Chaps. 4 and 6, the reliability mathematical models of nonrepairable systems and repairable systems under random fuzzy environment are established, respectively.

This book can be used as a teaching material for senior undergraduates and postgraduates of science and engineering related majors, as well as a reference book for engineers and reliability practitioners to understand the basic theory of reliability. At the same time, readers are welcome to put forward valuable opinions and suggestions on this book.

As this book is to be published, I would like to express my heartfelt thanks to Prof. Wansheng Tang, Prof. Rui Kang, Prof. Xiaozhong Li, Prof. Ruiqing Zhao and Prof. Yanping Wang for their support and encouragement to publish this book. I would like to express my sincere thanks to my graduate students Yao Ma, Dongxue Liu, Liuying Ma, Lina Ha, Jinbang Shao, Zifeng Liu, Yunyun Yang and Shuqian Guan, who assisted the author to collate the literature! I would like to express my heartfelt thanks to editor Tianming Bai of National Defense Industry Press for editing and publishing the book!

Tianjin, China
August 2020

Ying Liu

Contents

Chapter 1
Uncertain Mathematical Foundation

1.1 Probability Theory

Probability theory is a branch of mathematics that studies the behavior of random phenomena. It is generally believed that the study of probability theory was started by Pascal and Fermat in the seventeenth century when they succeeded in deriving the exact probabilities for certain gambling problems. After that, probability theory was subsequently studied by many researchers. A great progress was achieved when von Mises [1] initialized the concept of sample space in 1931. A complete axiomatic foundation of probability theory was given by Kolmogorov [2] in 1933. Since then, probability theory has been developed steadily and widely applied in science and engineering.

1.1.1 Probability Measure

Let Ω be a nonempty set, and let $\mathcal{A}$ be a σ-algebra over Ω. Each element in $\mathcal{A}$ is called an event. In order to present an axiomatic definition of probability, the following three axioms are assumed:

Axiom 1 (Normality Axiom) $\Pr\{\Omega\} = 1$ for the universal set Ω.

Axiom 2 (Nonnegativity Axiom) $\Pr\{A\} \geq 0$ for any event A.

Axiom 3 (Additivity Axiom) For every countable sequence of mutually disjoint events $A_1, A_2, \ldots$, we have

$$\Pr\left\{\bigcup_{i=1}^{+\infty} A_i\right\} = \sum_{i=1}^{+\infty} \Pr\{A_i\}.$$

Definition 1.1.1 The set function Pr is called a probability measure if it satisfies the normality, nonnegativity and additivity axioms.

Y. Liu, *Reliability Theory Based on Uncertain Lifetimes*,
https://doi.org/10.1007/978-981-16-0995-4_1

Example 1.1.1 Let $\Omega = \{\omega_1, \omega_2, \ldots\}$, and let $\mathcal{A}$ be the power set of Ω. Assume that $p_1, p_2, \ldots$ are nonnegative numbers such that $p_1 + p_2 + \cdots = 1$. Define a set function on $\mathcal{A}$ as

$$\Pr\{A\} = \sum_{\omega_i \in A} p_i, \ A \in \mathcal{A}.$$

Then, Pr is a probability measure.

Example 1.1.2 Let ϕ be a nonnegative and integrable function on $\Re$ (the set of real numbers) such that

$$\int_{\Re} \phi(x)\mathrm{d}x = 1.$$

Define a set function on the Borel algebra as

$$\Pr\{A\} = \int_{A} \phi(x)\mathrm{d}x.$$

Then, Pr is a probability measure.

Definition 1.1.2 Let Ω be a nonempty set, let $\mathcal{A}$ be a σ-algebra over Ω, and let Pr be a probability measure. Then, the triplet $(\Omega, \mathcal{A}, \Pr)$ is called a probability space.

Example 1.1.3 Let $\Omega = \{\omega_1, \omega_2, \ldots\}$, let $\mathcal{A}$ be the power set of Ω, and let Pr be a probability measure defined in Example 1.1.1. Then, $(\Omega, \mathcal{A}, \Pr)$ is a probability space.

Theorem 1.1.1 *(Probability Continuity Theorem) Let* $(\Omega, \mathcal{A}, \Pr)$ *be a probability space. If* $A_1, A_2, \ldots \in \mathcal{A}$ *and* $\lim_{i \to +\infty} A_i$ *exists, then*

$$\lim_{i \to +\infty} \Pr\{A_i\} = \Pr\left\{\lim_{i \to +\infty} A_i\right\}.$$

Theorem 1.1.2 *(Product Probability Theorem) Let* $(\Omega_k, \mathcal{A}_k, \Pr_k), k = 1, 2, \ldots$ *be a sequence of probability spaces. Now we write*

$$\Omega = \Omega_1 \times \Omega_2 \times \cdots \text{ and } \mathcal{A} = \mathcal{A}_1 \times \mathcal{A}_2 \times \cdots$$

It has been proved that there is a unique probability measure Pr *on the product* σ*-algebra* $\mathcal{A}$ *such that*

$$\Pr\left\{\prod_{k=1}^{+\infty} A_k\right\} = \prod_{k=1}^{+\infty} \Pr_k\{A_k\},$$

where A_k *are arbitrarily chosen events from* $\mathcal{A}_k$ for $k = 1, 2, \ldots$, *respectively. This conclusion is called the product probability theorem. The measure* Pr, *is also a probability measure since*

$$\Pr\{\Omega\} = \prod_{k=1}^{+\infty} \Pr_k\{\Omega_k\} = 1.$$

Such a probability measure is called product probability measure, denoted by $\Pr = \Pr_1 \times \Pr_2 \times \cdots$

Definition 1.1.3 Assume $(\Omega_k, \mathcal{A}_k, \Pr_k)$ are probability space for $k = 1, 2, \ldots$ Let $\Omega = \Omega_1 \times \Omega_2 \times \cdots$, $\mathcal{A} = \mathcal{A}_1 \times \mathcal{A}_2 \times \cdots$ and $\Pr = \Pr_1 \times \Pr_2 \times \cdots$ Then, the triplet $(\Omega, \mathcal{A}, \Pr)$ is called a product probability space.

1.1.2 Random Variable

Definition 1.1.4 A random variable is a function ξ from a probability space $(\Omega, \mathcal{A}, \Pr)$ to the set of real numbers such that $\{\xi \in \mathrm{B}\}$ is an event for any Borel set B.

Example 1.1.4 Take $(\Omega, \mathcal{A}, \Pr)$ to be $\{\omega_1, \omega_2\}$ with $\Pr\{\omega_1\} = \Pr\{\omega_2\} = 0.5$. Then, the function

$$\xi(\omega) = \begin{cases} 0, \text{ if } \omega = \omega_1 \\ 1, \text{ if } \omega = \omega_2 \end{cases}$$

is a random variable.

Definition 1.1.5 The random variable ξ is said to be

(a) nonnegative if $\Pr\{\xi < 0\} = 0$;
(b) positive if $\Pr\{\xi \leq 0\} = 0$;
(c) continuous if $\Pr\{\xi = x\} = 0$ for each $x \in \Re$;
(d) simple if there exists a finite sequence $\{x_1, x_2, \ldots, x_m\}$ such that

$$\Pr\{\xi \neq x_1, \xi \neq x_2, \ldots, \xi \neq x_m\} = 0;$$

(e) discrete if there exists a countable sequence $\{x_1, x_2, \ldots\}$ such that

$$\Pr\{\xi \neq x_1, \xi \neq x_2, \ldots\} = 0.$$

Definition 1.1.6 Let ξ and η be random variables defined on the probability space $(\Omega, \mathcal{A}, \Pr)$. We say $\xi = \eta$ if and only if $\xi(\omega) = \eta(\omega)$ for almost all $\omega \in \Omega$.

Definition 1.1.7 An n-dimensional random vector is a measurable function from a probability space $(\Omega, \mathcal{A}, \Pr)$ to the set of n-dimensional real vectors.

Theorem 1.1.3 *The vector* $(\xi_1, \xi_2, \ldots, \xi_n)$ *is a random vector if and only if* $\xi_1, \xi_2, \ldots, \xi_n$ *are random variables.*

Definition 1.1.8 (Random Arithmetic on Single Probability Space) Let $f : \Re^n \to \Re$ be a measurable function, and let $\xi_1, \xi_2, \ldots, \xi_n$ be random variables defined on the probability space $(\Omega, \mathcal{A}, \Pr)$. Then, $\xi = f(\xi_1, \xi_2, \ldots, \xi_n)$ is a random variable defined by

$$\xi(\omega) = f(\xi_1(\omega), \xi_2(\omega), \ldots, \xi_n(\omega)), \ \forall \omega \in \Omega.$$

Example 1.1.5 Let ξ_1 and ξ_2 be random variables on the probability space $(\Omega, \mathcal{A}, \Pr)$. Then, their sum and product are defined by

$$(\xi_1 + \xi_2)(\omega) = \xi_1(\omega) + \xi_2(\omega)$$

and

$$(\xi_1 \times \xi_2)(\omega) = \xi_1(\omega) \times \xi_2(\omega)$$

for any $\omega \in \Omega$.

Definition 1.1.9 (Random Arithmetic on Different Probability Spaces) Let $f : \Re^n \to \Re$ be a measurable function, and let ξ_i be random variables defined on probability spaces $(\Omega_i, \mathcal{A}_i, \Pr_i)$, $i = 1, 2, \ldots, n$, respectively. Then, $\xi = f(\xi_1, \xi_2, \ldots, \xi_n)$ is a random variable on the product probability space $(\Omega, \mathcal{A}, \Pr)$, defined by

$$\xi(\omega_1, \omega_2, \ldots, \omega_n) = f(\xi_1(\omega_1), \xi_2(\omega_2), \ldots, \xi_n(\omega_n))$$

for all $\forall(\omega_1, \omega_2, \ldots, \omega_n) \in \Omega$.

Example 1.1.6 Let ξ_1 and ξ_2 be random variables on the probability spaces $(\Omega_1, \mathcal{A}_1, \Pr_1)$ and $(\Omega_2, \mathcal{A}_2, \Pr_2)$, respectively. Then, their sum and product are defined by

$$(\xi_1 + \xi_2)(\omega_1, \omega_2) = \xi_1(\omega_1) + \xi_2(\omega_2)$$

and

$$(\xi_1 \times \xi_2)(\omega_1, \omega_2) = \xi_1(\omega_1) \times \xi_2(\omega_2)$$

for any $(\omega_1, \omega_2) \in \Omega_1 \times \Omega_2$.

1.1.3 Probability Distribution

Definition 1.1.10 The probability distribution $\Phi : \Re \to [0, 1]$ of a random variable ξ is defined by

$$\Phi(x) = \Pr\{\xi \le x\}, x \in \Re.$$

Example 1.1.7 Take $(\Omega, \mathcal{A}, \Pr)$ to be $\{\omega_1, \omega_2\}$ with $\Pr\{\omega_1\} = \Pr\{\omega_2\} = 0.5$. We now define a random variable as follows,

$$\xi(\omega) = \begin{cases} 0, \text{ if } \omega = \omega_1 \\ 1, \text{ if } \omega = \omega_2 \end{cases}.$$

Then, ξ has a probability distribution

$$\Phi(x) = \begin{cases} 0, & \text{if } x < 0 \\ 0.5, & \text{if } 0 \le x < 1 \\ 1, & \text{if } x \ge 1. \end{cases}$$

Theorem 1.1.4 *(Sufficient and Necessary Condition for Probability Distribution) A function $\Phi : \Re \to [0, 1]$ is a probability distribution if and only if it is an increasing and right-continuous function with*

$$\lim_{x \to -\infty} \Phi(x) = 0; \quad \lim_{x \to +\infty} \Phi(x) = 1.$$

Theorem 1.1.5 *A random variable ξ with probability distribution function Φ is*

(a) nonnegative if and only if $\Phi(x) = 0$ for all $x < 0$;
(b) positive if and only if $\Phi(x) = 0$ for all $x \le 0$;
(c) simple if and only if Φ is a simple function;
(d) discrete if and only if Φ is a step function;
(e) continuous if and only if Φ is a continuous function.

Definition 1.1.11 The probability density function $\phi : \Re \to [0, +\infty)$ of a random variable ξ is a function such that

$$\Phi(x) = \int_{-\infty}^{x} \phi(y)\mathrm{d}y$$

holds for all $x \in \Re$, where Φ is the probability distribution of the random variable ξ.

Theorem 1.1.6 *Let ξ be a random variable whose probability density function ϕ exists. Then, for any Borel set* B *of real numbers, we have*

$$\Pr\{\xi \in \mathrm{B}\} = \int_B \phi(y)\mathrm{d}y.$$

Definition 1.1.12 (Uniform Distribution) A random variable ξ has a uniform distribution if its probability density function is

$$\phi(x) = \frac{1}{b-a}, \ a \le x \le b,$$

denoted by $U(a, b)$, where a and b are real numbers with $a < b$.

Definition 1.1.13 (Exponential Distribution) A random variable ξ has an exponential distribution if its probability density function is

$$\phi(x) = \frac{1}{\beta}\exp(-\frac{x}{\beta}), \ x \ge 0,$$

denoted by $\exp\left(\frac{1}{\beta}\right)$, where β is a positive number.

Definition 1.1.14 (Normal Distribution) A random variable ξ has a normal distribution if its probability density function is

$$\phi(x) = \frac{1}{\sigma\sqrt{2\pi}}\exp\left(-\frac{(x-\mu)^2}{2\sigma^2}\right), \quad -\infty < x < +\infty,$$

denoted by $N(\mu, \sigma^2)$, where μ and σ are real numbers with $\sigma > 0$.

Definition 1.1.15 (Lognormal Distribution) A random variable ξ has a lognormal distribution if its logarithm is normal distributed; i.e., its probability density function is

$$\phi(x) = \frac{1}{x\sigma\sqrt{2\pi}}\exp\left(-\frac{(\ln x-\mu)^2}{2\sigma^2}\right), \quad x > 0,$$

denoted by $LN(\mu, \sigma^2)$, where μ and σ are real numbers with $\sigma > 0$.

Definition 1.1.16 (Gamma Distribution) A random variable ξ has a Gamma distribution if its probability density function is

$$\phi(x) = \frac{\lambda(\lambda x)^{\alpha-1}}{\Gamma(\alpha)}e^{-\lambda x}, \quad x \ge 0; \quad \lambda, \alpha > 0,$$

denoted by $\Gamma(\alpha, \lambda; x)$, where $\Gamma(\alpha) = \int_0^{+\infty} t^{\alpha-1}e^{-t}\mathrm{d}t$, α is the shape parameter and λ is the scale parameter.

Definition 1.1.17 (Weibull Distribution) A random variable ξ has a Weibull distribution if its probability density function is

$$\phi(x) = \lambda\alpha(\lambda x)^{\alpha-1}e^{-(\lambda x)^{\alpha}}, \quad x \geq 0; \quad \alpha, \lambda > 0,$$

denoted by $\mathrm{W}(\alpha, \lambda;\ x)$, where α is the shape parameter and λ is the scale parameter.

Definition 1.1.18 (Two-Point Distribution) A random variable ξ has a two-point distribution with parameter p if

$$\Pr\{\xi = 1\} = p, \quad \Pr\{\xi = 0\} = 1 - p, \quad 0 < p < 1.$$

Definition 1.1.19 (Binomial Distribution) A random variable ξ has a binomial distribution with parameter (n, p) if

$$\Pr\{\xi = k\} = \binom{n}{k} p^k q^{n-k}, \quad k = 0, 1, \ldots, n,$$

where $p, q > 0$ and $p + q = 1$.

Definition 1.1.20 (Geometric Distribution) A random variable ξ has a geometric distribution with parameter p if

$$\Pr\{\xi = k\} = pq^k, \quad k = 0, 1, 2, \ldots$$

where $p, q > 0$ and $p + q = 1$.

Definition 1.1.21 (Negative Binomial Distribution) A nonnegative integer-valued random variable ξ has a negative binomial distribution with parameter (α, p) if

$$\Pr\{\xi = k\} = \binom{-\alpha}{k} p^{\alpha}(-q)^k, \quad k = 0, 1, 2, \ldots$$

where $\alpha, p, q > 0$ and $p + q = 1$.

Definition 1.1.22 (Poisson Distribution) A nonnegative integer-valued random variable ξ has a Poisson distribution with parameter λ if

$$\Pr\{\xi = k\} = \frac{\lambda^k}{k!}e^{-\lambda}, \quad k = 0, 1, 2, \ldots, \quad \lambda > 0.$$

Definition 1.1.23 The joint probability distribution $\Phi : \Re^n \to [0, 1]$ of a random vector $(\xi_1, \xi_2, \ldots, \xi_n)$ is defined by

$$\Phi(x_1, x_2, \ldots, x_n) = \Pr\{\omega \in \Omega | \xi_1(\omega) \leq x_1, \xi_2(\omega) \leq x_2, \ldots, \xi_n(\omega) \leq x_n\}.$$

Definition 1.1.24 The joint probability density distribution $\phi : \Re^n \to [0, +\infty)$ of a random vector $(\xi_1, \xi_2, \ldots, \xi_n)$ is a function such that

$$\Phi(x_1, x_2, \ldots, x_n) = \int_{-\infty}^{x_1}\int_{-\infty}^{x_2} \cdots \int_{-\infty}^{x_n} \phi(y_1, y_2, \ldots, y_n)\mathrm{d}y_1\mathrm{d}y_2 \cdots \mathrm{d}y_n,$$

holds for all $(x_1, x_2, \ldots, x_n) \in \Re^n$, where Φ is the probability distribution of the random vector $(\xi_1, \xi_2, \ldots, \xi_n)$.

1.1.4 Independence

Definition 1.1.25 The random variable $\xi_1, \xi_2, \ldots, \xi_m$ are said to be independent if

$$\Pr\left\{\bigcap_{i=1}^{m}\{\xi_i \in \mathrm{B}_i\}\right\} = \prod_{i=1}^{m}\Pr\{\xi_i \in \mathrm{B}_i\},$$

for any Borel sets $\mathrm{B}_1, \mathrm{B}_2, \ldots, \mathrm{B}_m$.

Example 1.1.8 Let $\xi_1(\omega_1)$ and $\xi_2(\omega_2)$ be random variables on the probability spaces $(\Omega_1, \mathcal{A}_1, \Pr_1)$ and $(\Omega_2, \mathcal{A}_2, \Pr_2)$, respectively. It is clear that they are also random variables on the product probability space $(\Omega_1, \mathcal{A}_1, \Pr_1) \times (\Omega_2, \mathcal{A}_2, \Pr_2)$. Then, for any Borel sets B_1 and B_2, we have

$$\begin{aligned}
&\Pr\{(\xi_1 \in \mathrm{B}_1) \cap (\xi_2 \in \mathrm{B}_2)\}\\
&= \Pr\{(\omega_1, \omega_2)|\xi_1(\omega_1) \in \mathrm{B}_1,\ \xi_2(\omega_2) \in \mathrm{B}_2\}\\
&= \Pr\{(\omega_1|\xi_1(\omega_1) \in \mathrm{B}_1) \times (\omega_2|\xi_2(\omega_2) \in \mathrm{B}_2)\}\\
&= \Pr_1\{\omega_1|\xi_1(\omega_1) \in \mathrm{B}_1\} \times \Pr_2\{\omega_2|\xi_2(\omega_2) \in \mathrm{B}_2\}\\
&= \Pr\{\xi_1 \in \mathrm{B}_1\} \times \Pr\{\xi_2 \in \mathrm{B}_2\}.
\end{aligned}$$

Thus, ξ_1 and ξ_2 are independent in the product probability space. In fact, it is true that random variables are always independent if they are defined on different probability spaces.

Theorem 1.1.7 *Let $\xi_1, \xi_2, \ldots, \xi_n$ be independent random variables, and let $f_1, f_2, \ldots, f_n$ be measurable functions. Then, $f_1(\xi_1), f_2(\xi_2), \ldots, f_n(\xi_n)$ are independent random variables.*

Theorem 1.1.8 *Let $\xi_1, \xi_2, \ldots, \xi_n$ be independent random variables with probability distributions $\Phi_1, \Phi_2, \ldots, \Phi_n$, respectively, and let $f : \Re^n \to \Re$ be a measurable function. Then, the random variable*

$$\xi = f(\xi_1, \xi_2, \ldots, \xi_n)$$

has a probability distribution

$$\Phi(x) = \int\limits_{f(x_1,x_2,\ldots,x_n)\le x} \mathrm{d}\Phi_1(x_1)\mathrm{d}\Phi_2(x_2)\cdots\mathrm{d}\Phi_n(x_n).$$

Remark 1.1.1 If $\xi_1, \xi_2, \ldots, \xi_n$ have probability density functions $\phi_1, \phi_2, \ldots, \phi_n$, respectively, then $\xi = f(\xi_1, \xi_2, \ldots, \xi_n)$ has a probability distribution

$$\Phi(x) = \int\limits_{f(x_1,x_2,\ldots,x_n)\le x} \phi_1(x_1)\phi_2(x_2)\cdots\phi_n(x_n)\mathrm{d}x_1\mathrm{d}x_2\cdots\mathrm{d}x_n$$

because $\mathrm{d}\Phi_i(x_i) = \phi_i(x_i)\mathrm{d}x_i$ for $i = 1, 2, \ldots, n$.

Example 1.1.9 Let $\xi_1, \xi_2, \ldots, \xi_n$ be independent random variables with probability distributions $\Phi_1, \Phi_2, \ldots, \Phi_n$, respectively. Show that the sum

$$\xi = \xi_1 + \xi_2 + \cdots + \xi_n,$$

has a probability distribution

$$\Phi(x) = \int\limits_{x_1+x_2+\cdots+x_n\le x} \mathrm{d}\Phi_1(x_1)\mathrm{d}\Phi_2(x_2)\cdots\mathrm{d}\Phi_n(x_n).$$

Especially, let ξ_1 and ξ_2 be independent random variables with probability distributions Φ_1 and Φ_2, respectively. Then, $\xi = \xi_1 + \xi_2$ has a probability distribution

$$\Phi(x) = \int_{-\infty}^{+\infty} \Phi_1(x-y)\mathrm{d}\Phi_2(y),$$

that is called the convolution of Φ_1 and Φ_2.

Example 1.1.10 Let $\xi_1, \xi_2, \ldots, \xi_n$ be independent random variables with probability distributions $\Phi_1, \Phi_2, \ldots, \Phi_n$, respectively. Show that the maximum

$$\xi = \xi_1 \vee \xi_2 \vee \cdots \vee \xi_n,$$

has a probability distribution

$$\Phi(x) = \Phi_1(x)\Phi_2(x)\cdots\Phi_n(x).$$

Example 1.1.11 Let $\xi_1, \xi_2, \ldots, \xi_n$ be independent random variables with probability distributions $\Phi_1, \Phi_2, \ldots, \Phi_n$, respectively. Show that the minimum

$$\xi = \xi_1 \wedge \xi_2 \wedge \cdots \wedge \xi_n$$

has a probability distribution

$$\Phi(x) = 1 - (1 - \Phi_1(x))(1 - \Phi_2(x)) \cdots (1 - \Phi_n(x)).$$

1.1.5 Identical Distribution

Definition 1.1.26 The random variables ξ and η are said to be identically distributed if

$$\Pr\{\xi \in B\} = \Pr\{\eta \in B\}.$$

for any Borel set B.

Theorem 1.1.9 *The random variables ξ and η are identically distributed if and only if they have the same probability distribution.*

Theorem 1.1.10 *Let ξ and η be two random variables whose probability density functions exist. Then, ξ and η are identically distributed if and only if they have the same probability density function.*

1.1.6 Expected Value

Definition 1.1.27 Let ξ be a random variable. Then, the expected value of ξ is defined by

$$E[\xi] = \int_0^{+\infty} \Pr\{\xi \geq x\} \mathrm{d}x - \int_{-\infty}^0 \Pr\{\xi \leq x\} \mathrm{d}x$$

provided that at least one of the two integrals is finite.

Example 1.1.12 Suppose that ξ is a discrete random variable taking values x_i with probabilities p_i, $i = 1, 2, \ldots, m$, respectively. Show that

$$E[\xi] = \sum_{i=1}^{m} p_i x_i.$$

Example 1.1.13 Let ξ be a uniformly distributed variable $U(a, b)$. Then, its expected value $E[\xi] = \frac{a+b}{2}$.

Example 1.1.14 Let ξ be an exponentially distributed variable $\exp\left(\frac{1}{\beta}\right)$. Then, its expected value $E[\xi] = \beta$.

Example 1.1.15 Let ξ be a normally distributed variable $N(\mu, \sigma^2)$. Then, its expected value $E[\xi] = \mu$.

Theorem 1.1.11 *Let ξ be a random variable with probability distribution Φ. Then,*

$$E[\xi] = \int_{0}^{+\infty} (1 - \Phi(x))\mathrm{d}x - \int_{-\infty}^{0} \Phi(x)\mathrm{d}x.$$

Theorem 1.1.12 *Let ξ be a random variable with probability distribution Φ. Then,*

$$E[\xi] = \int_{-\infty}^{+\infty} x\mathrm{d}\Phi(x).$$

Remark 1.1.2 Let $\phi(x)$ be the probability density function of ξ. Then, we immediately have

$$E[\xi] = \int_{-\infty}^{+\infty} x\phi(x)\mathrm{d}x$$

because $\mathrm{d}\Phi(x) = \phi(x)\mathrm{d}x$.

Theorem 1.1.13 *Let $\xi_1, \xi_2, \ldots, \xi_n$ be independent random variables with probability distributions $\Phi_1, \Phi_2, \ldots, \Phi_n$, respectively, and let $f : \Re^n \to \Re$ be a measurable function. Then, $\xi = f(\xi_1, \xi_2, \ldots, \xi_n)$ has an expected value*

$$E[\xi] = \int_{\Re^n} f(x_1, x_2, \ldots, x_n)\mathrm{d}\Phi_1(x_1)\mathrm{d}\Phi_2(x_2)\cdots\mathrm{d}\Phi_n(x_n).$$

Theorem 1.1.14 *Let $\xi_1, \xi_2, \ldots, \xi_n$ be independent random variables with probability density functions $\phi_1, \phi_2, \ldots, \phi_n$, respectively, and let $f : \Re^n \to \Re$ be a measurable function, then $\xi = f(\xi_1, \xi_2, \ldots, \xi_n)$ has an expected value*

$$E[\xi] = \int_{\Re^n} f(x_1, x_2, \ldots, x_n)\phi_1(x_1)\phi_2(x_2)\cdots\phi_n(x_n)\mathrm{d}x_1\mathrm{d}x_2\cdots\mathrm{d}x_n.$$

Theorem 1.1.15 *Let ξ and η be independent random variables with finite expected values. Then,*

$$E[\xi\eta] = E[\xi]E[\eta].$$

Theorem 1.1.16 *Let ξ and η be random variables with finite expected values. Then, for any numbers a and b, we have*

$$E[a\xi + b\eta] = aE[\xi] + bE[\eta].$$

Theorem 1.1.17 *Let ξ be a random variable, and let t be a positive number. If $E[|\xi|^t] < +\infty$, then*

$$\lim_{x\to+\infty} x^t \Pr\{|\xi| \geq x\} = 0.$$

Conversely, let ξ be a random variable satisfying $\lim_{x\to+\infty} x^t \Pr\{|\xi| \geq x\} = 0$ for some $t > 0$. Then, $E[|\xi|^s] < +\infty$ for any $0 \leq s < t$.

Example 1.1.16 The condition $\lim_{x\to+\infty} x^t \Pr\{|\xi| \geq x\} = 0$ in Theorem 1.1.17 does not ensure that $E[|\xi|^t] < +\infty$. We consider the positive random variable

$$\xi = \sqrt[t]{\frac{2^i}{i}} \text{ with probability } \frac{1}{2^i},\ i = 1, 2, \ldots$$

It is clear that

$$\lim_{x\to+\infty} x^t \Pr\{\xi \geq x\} = \lim_{n\to+\infty} \left(\sqrt[t]{\frac{2^n}{n}}\right)^t \sum_{i=n}^{+\infty} \frac{1}{2^i} = \lim_{n\to+\infty} \frac{2}{n} = 0.$$

However, the expected value of ξ^t is

$$E[\xi^t] = \sum_{i=1}^{+\infty} \left(\sqrt[t]{\frac{2^i}{i}}\right)^t \cdot \frac{1}{2^i} = \sum_{i=1}^{+\infty} \frac{1}{i} = +\infty.$$

Theorem 1.1.18 *Let ξ be a random variable, and let f be a nonnegative function. If f is even and increasing on $[0, +\infty)$, then, for any given number $t > 0$, we have*

$$\Pr\{|\xi| \geq t\} \leq \frac{E[f(\xi)]}{f(t)}.$$

Theorem 1.1.19 *(Markov Inequality) Let ξ be a random variable. Then, for any given numbers $t > 0$ and $p > 0$, we have*

$$\Pr\{|\xi| \geq t\} \leq \frac{E[|\xi|^p]}{t^p}.$$

1.1.7 Conditional Probability

We consider the probability of an event A after it has been learned that some other event B has occurred. This new probability is called the conditional probability A given B.

Definition 1.1.28 Let $(\Omega, \mathcal{A}, \Pr)$ be a probability space, and $A, B \in \mathcal{A}$. Then, the conditional probability of A given B is defined by

$$\Pr\{A|B\} = \frac{\Pr\{A \cap B\}}{\Pr\{B\}}$$

provided that $\Pr\{B\} > 0$.

Example 1.1.17 Let ξ be an exponentially distributed random variable with expected value β. Then, for any real numbers $a > 0$ and $x > 0$, the conditional probability of $\xi \geq a + x$ given $\xi \geq a$ is

$$\Pr\{\xi \geq a + x|\xi \geq a\} = \exp\{-x/\beta\} = \Pr\{\xi \geq x\},$$

which means that the conditional probability is identical to the original probability. This is the so-called memoryless property of exponential distribution. In other words, it is as good as new if it is functioning on inspection.

Theorem 1.1.20 *(Bayes Formula) Let the events $A_1, A_2, \ldots, A_n$ form a partition of the space Ω such that $\Pr\{A_i\} > 0$ for $i = 1, 2, \ldots, n$, and let B be an event with $\Pr\{B\} > 0$, then we have*

$$\Pr\{A_k|B\} = \frac{\Pr\{A_k\}\Pr\{B|A_k\}}{\sum_{i=1}^{n}\Pr\{A_i\}\Pr\{B|A_i\}}$$

for $k = 1, 2, \ldots, n$.

Remark 1.1.3 Especially, let A and B be two events with $\Pr\{A\} > 0$ and $\Pr\{B\} > 0$. Then, A and A^c form a partition of the space Ω, and the Bayes formula is

$$\Pr\{A|B\} = \frac{\Pr\{A\}\Pr\{B|A\}}{\Pr\{B\}} = \frac{\Pr\{A\}\Pr\{B|A\}}{\Pr\{A\}\Pr\{B|A\} + \Pr\{A^c\}\Pr\{B|A^c\}}.$$

Remark 1.1.4 In statistical applications, the events $A_1, A_2, \ldots, A_n$ are often called hypotheses. Furthermore, for each i, the $\Pr\{A_i\}$ is called a priori probability of A_i and $\Pr\{A_i|B\}$ is called a posteriori probability of A_i after the occurrence of event B.

Definition 1.1.29 Let $(\Omega, \mathcal{A}, \Pr)$ be a probability space. Then, the conditional probability distribution $\Phi : \Re \to [0, 1]$ of a random variable ξ given B is defined by

$$\Phi(x|B) = \Pr\{\xi \leq x|B\}$$

provided that $\Pr\{B\} > 0$.

Example 1.1.18 Let ξ and η be random variables. Then, the conditional probability distribution of ξ given $\eta = y$ is

$$\Phi(x|\eta = y) = \Pr\{\xi \leq x|\eta = y\} = \frac{\Pr\{\xi \leq x, \eta = y\}}{\Pr\{\eta = y\}}$$

provided that $\Pr\{\eta = y\} > 0$.

Definition 1.1.30 Let $(\Omega, \mathcal{A}, \Pr)$ be a probability space. Then, the conditional probability density function $\phi : \Re \to [0, +\infty)$ of a random variable ξ given B is a nonnegative function such that

$$\Phi(x|B) = \int_{-\infty}^{x} \phi(y|B)\mathrm{d}y$$

holds for all $\forall x \in \Re$, where $\Phi(x|B)$ is the conditional probability distribution of the random variable ξ given B provided that $\Pr\{B\} > 0$.

Example 1.1.19 Let (ξ, η) be a random vector with joint probability density function ψ. Then, the marginal probability density functions of ξ and η are

$$f(x) = \int_{-\infty}^{+\infty} \psi(x, y)\mathrm{d}y \quad \text{and} \quad g(y) = \int_{-\infty}^{+\infty} \psi(x, y)\mathrm{d}x,$$

respectively. Furthermore, we have

$$\Pr\{\xi \leq x, \eta \leq y\} = \int_{-\infty}^{x}\int_{-\infty}^{y} \psi(r, t)\mathrm{d}r\mathrm{d}t = \int_{-\infty}^{y}\left[\int_{-\infty}^{x} \frac{\psi(r, t)}{g(t)}\mathrm{d}r\right]g(t)\mathrm{d}t,$$

which implies that the conditional probability distribution of ξ given $\eta = y$ is

$$\Phi(x|\eta = y) = \int_{-\infty}^{x} \frac{\psi(r, y)}{g(y)}\mathrm{d}r, \quad a.s.$$

and the conditional probability density function of ξ given $\eta = y$ is

$$\phi(x|\eta = y) = \frac{\psi(x, y)}{g(y)} = \frac{\psi(x, y)}{\int_{-\infty}^{+\infty} \psi(x, y)\mathrm{d}x}, \quad a.s.$$

where $g(y) \neq 0$. In fact, the set $\{y|g(y) = 0\}$ has probability 0. Especially, if ξ and η are independent random variables, then $\psi(x, y) = f(x)g(y)$ and $\phi(x|\eta = y) = f(x)$.

1.1.8 Ranking Criteria

Let ξ_1 and ξ_2 be two random variables. Different from the situation of real numbers, there does not exist an ordership in a random world. Then, an important problem is how to rank random variables.

Definition 1.1.31 A collection of random variables $\mathcal{F}$ is said to be a totally ordered set with stochastic ordering if and only if for any given $\xi_1, \xi_2 \in \mathcal{F}$ and $r \in \Re$, either

$$\Pr\{\xi_1 \leq r\} \leq \Pr\{\xi_2 \leq r\} \quad \text{(denoted by } \xi_2 \preceq_d \xi_1)$$

or

$$\Pr\{\xi_1 \leq r\} \geq \Pr\{\xi_2 \leq r\} \quad \text{(denoted by } \xi_1 \preceq_d \xi_2).$$

Theorem 1.1.21 *For any given $\xi_1, \xi_2 \in \mathcal{F}$, we have*

$$E[\xi_1] \leq E[\xi_2] \Leftrightarrow \xi_1 \preceq_d \xi_2.$$

1.2 Credibility Theory

Credibility theory is a branch of mathematics that studies the behavior of fuzzy phenomena. The concept of fuzzy set was initiated by Zadeh [21] via membership function in 1965. In order to measure fuzzy events, Zadeh [22] proposed the concept of possibility measure in 1978. Although possibility measure has been widely used, it has no self-duality property. However, a self-dual measure is absolutely needed in both theory and practice. In order to define a self-dual measure, Liu and Liu [6] presented the concept of credibility measure in 2002.

1.2.1 Credibility Measure

Let Θ be a nonempty set, and $\mathcal{P}(\Theta)$ the power set of Θ. Each element in $\mathcal{P}(\Theta)$ is called an event. In order to present the axiomatic definition of credibility, it is necessary to assign to each event A a number $\text{Cr}\{A\}$ which indicates the credibility that A will occur. In order to ensure that the number $\text{Cr}\{A\}$ has certain mathematical properties which we intuitively expect a credibility to have, Liu [5] presented the following five axiom:

Axiom 1 $\text{Cr}\{\Theta\} = 1$.

Axiom 2 Cr is increasing, i.e., $\text{Cr}\{A\} \le \text{Cr}\{B\}$ whenever $A \subseteq B$.

Axiom 3 Cr is self-dual, i.e., $\text{Cr}\{A\} + \text{Cr}\{A^c\} = 1$ for any $A \in \mathcal{P}(\Theta)$.

Axiom 4 $\text{Cr}\{\cup_i A_i\} \wedge 0.5 = \sup_i \text{Cr}\{A_i\}$ for any $\{A_i\}$ with $\text{Cr}\{A_i\} \le 0.5$.

Axiom 5 Let Θ_k be nonempty sets on which Cr_k satisfies the first four axioms, $k = 1, 2, \ldots, n$, respectively, and $\Theta = \Theta_1 \times \Theta_2 \times \cdots \times \Theta_n$. Then,

$$\text{Cr}\{(\theta_1, \theta_2, \ldots, \theta_n)\} = \text{Cr}_1\{\theta_1\} \wedge \text{Cr}_2\{\theta_2\} \wedge \cdots \wedge \text{Cr}_n\{\theta_n\}$$

for each $(\theta_1, \theta_2, \ldots, \theta_n) \in \Theta$.

Definition 1.2.1 The set function Cr is called a credibility measure if it satisfies the first four axioms.

Theorem 1.2.1 *Let Θ be a nonempty set, $\mathcal{P}(\Theta)$ the power set of Θ and* Cr *the credibility measure. Then,* $\text{Cr}\{\theta\} = 0$ *and* $0 \le \text{Cr}\{A\} \le 1$ *for any $A \in \mathcal{P}(\Theta)$.*

Theorem 1.2.2 *(Credibility Subadditivity Theorem) The credibility measure is subadditive. That is,*

$$\text{Cr}\{A \cup B\} \le \text{Cr}\{A\} + \text{Cr}\{B\}$$

for any event $A, B \in \mathcal{P}(\Theta)$.

Definition 1.2.2 Let Θ be a nonempty set, $\mathcal{P}(\Theta)$ the power set of Θ, and Cr a credibility measure. Then, the triplet $(\Theta, \mathcal{P}(\Theta), \text{Cr})$ is called a credibility space.

Theorem 1.2.3 *(Product Credibility Theorem) Suppose that Θ_k are nonempty sets, Cr_k the credibility measures on $\mathcal{P}(\Theta_k)$, $k = 1, 2, \ldots, n$, respectively. Let $\Theta = \Theta_1 \times \Theta_2 \times \cdots \times \Theta_n$. Then, $\text{Cr} = \text{Cr}_1 \wedge \text{Cr}_2 \wedge \cdots \wedge \text{Cr}_n$ defined by Axiom 5 has a unique extension to a credibility measure on $\mathcal{P}(\Theta)$ as follows,*

$$\text{Cr}\{A\} = \begin{cases} \sup\limits_{(\theta_1,\theta_2,\ldots,\theta_n)\in A} \min\limits_{1\le k\le n} \text{Cr}_k\{\theta_k\}, & \text{if } \sup\limits_{(\theta_1,\theta_2,\ldots,\theta_n)\in A} \min\limits_{1\le k\le n} \text{Cr}_k\{\theta_k\} < 0.5 \\ 1 - \sup\limits_{(\theta_1,\theta_2,\ldots,\theta_n)\in A^c} \min\limits_{1\le k\le n} \text{Cr}_k\{\theta_k\}, & \text{if } \sup\limits_{(\theta_1,\theta_2,\ldots,\theta_n)\in A} \min\limits_{1\le k\le n} \text{Cr}_k\{\theta_k\} \ge 0.5 \end{cases}$$

for each $A \in \mathcal{P}(\Theta)$.

Definition 1.2.3 Let $(\Theta_k, \mathcal{P}(\Theta_k), \mathrm{Cr}_k),\ k = 1, 2, \ldots, n$ be credibility spaces, $\Theta = \Theta_1 \times \Theta_2 \times \cdots \times \Theta_n$ and $\mathrm{Cr} = \mathrm{Cr}_1 \wedge \mathrm{Cr}_2 \wedge \cdots \wedge \mathrm{Cr}_n$. Then, $(\Theta, \mathcal{P}(\Theta), \mathrm{Cr})$ is called the product credibility space of $(\Theta_k, \mathcal{P}(\Theta_k), \mathrm{Cr}_k),\ k = 1, 2, \ldots, n$.

Theorem 1.2.4 *(Infinite Product Credibility Theorem) Suppose that Θ_k are nonempty sets, Cr_k the credibility measure on $\mathcal{P}(\Theta_k), k = 1, 2, \ldots,$ respectively. Let $\Theta = \Theta_1 \times \Theta_2 \times \cdots$ Then,*

$$\mathrm{Cr}\{A\} = \begin{cases} \sup\limits_{(\theta_1,\theta_2,\ldots)\in A} \inf\limits_{1\le k\le+\infty} \mathrm{Cr}_k\{\theta_k\}, & \text{if } \sup\limits_{(\theta_1,\theta_2,\ldots)\in A} \inf\limits_{1\le k<+\infty} \mathrm{Cr}_k\{\theta_k\} < 0.5 \\ 1 - \sup\limits_{(\theta_1,\theta_2,\ldots)\in A^c} \inf\limits_{1\le k\le+\infty} \mathrm{Cr}_k\{\theta_k\}, & \text{if } \sup\limits_{(\theta_1,\theta_2,\ldots)\in A} \inf\limits_{1\le k<+\infty} \mathrm{Cr}_k\{\theta_k\} \ge 0.5 \end{cases}$$

is a credibility measure on $\mathcal{P}(\Theta)$.

Definition 1.2.4 Let $(\Theta_k, \mathcal{P}(\Theta_k), \mathrm{Cr}_k),\ k = 1, 2, \ldots$ be credibility spaces. Define $\Theta = \Theta_1 \times \Theta_2 \times \cdots$ and $\mathrm{Cr} = \mathrm{Cr}_1 \wedge \mathrm{Cr}_2 \wedge \cdots$ Then, $(\Theta, \mathcal{P}(\Theta), \mathrm{Cr})$ is called the infinite product credibility space of $(\Theta_k, \mathcal{P}(\Theta_k), \mathrm{Cr}_k),\ k = 1, 2, \ldots$

1.2.2 *Fuzzy Variable*

Definition 1.2.5 A fuzzy variable is defined as function from a credibility space $(\Theta, \mathcal{P}(\Theta), \mathrm{Cr})$ to the set of real numbers.

Example 1.2.1 Let $\Theta = \{\theta_1, \theta_2\}$ and $\mathrm{Cr}\{\theta_1\} = \mathrm{Cr}\{\theta_2\} = 0.5$. Then, $(\Theta, \mathcal{P}(\Theta), \mathrm{Cr})$ is a credibility space, and the function

$$\xi(\theta) = \begin{cases} 0, & \text{if } \theta = \theta_1 \\ 1, & \text{if } \theta = \theta_2 \end{cases}$$

is a fuzzy variable.

Example 1.2.2 Let $\Theta = [0, 1]$, and $\mathrm{Cr}\{\theta\} = \theta/2$ for each $\theta \in \Theta$. Then, $(\Theta, \mathcal{P}(\Theta), \mathrm{Cr})$ is a credibility space, and the identity function $\xi(\theta) = \theta$ is a fuzzy variable.

Example 1.2.3 A crisp number c may be regarded as a special fuzzy variable. In fact, it is the constant function $\xi(\theta) \equiv c$ on the credibility space $(\Theta, \mathcal{P}(\Theta), \mathrm{Cr})$.

Definition 1.2.6 Let ξ be a fuzzy variable defined on the credibility space $(\Theta, \mathcal{P}(\Theta), \mathrm{Cr})$. Then, the set

$$\xi_\alpha = \{\xi(\theta) | \theta \in \Theta, \mathrm{Cr}\{\theta\} \ge \alpha/2\}$$

is called the α-level set of ξ.

Definition 1.2.7 A fuzzy variable ξ is said to be

(a) nonnegative if $\text{Cr}\{\xi < 0\} = 0$;
(b) positive if $\text{Cr}\{\xi \leq 0\} = 0$;
(c) continuous if $\text{Cr}\{\xi = x\}$ is a continuous function of x;
(d) simple if there exists a finite sequence $\{x_1, x_2, \ldots, x_m\}$ such that

$$\text{Cr}\{\xi \neq x_1, \xi \neq x_2, \ldots, \xi \neq x_m\} = 0;$$

(e) discrete if there exists a countable sequence $\{x_1, x_2, \ldots\}$ such that

$$\text{Cr}\{\xi \neq x_1, \xi \neq x_2, \ldots\} = 0.$$

Definition 1.2.8 Let ξ and η be fuzzy variables defined on the credibility space $(\Theta, \mathcal{P}(\Theta), \text{Cr})$. Then, $\xi = \eta$ if and only if $\xi(\theta) = \eta(\theta)$ for almost all $\theta \in \Theta$.

Definition 1.2.9 An n-dimensional fuzzy vector is defined as a function from a credibility space $(\Theta, \mathcal{P}(\Theta), \text{Cr})$ to the set of n-dimensional real vectors.

Theorem 1.2.5 *The vector $(\xi_1, \xi_2, \ldots, \xi_n)$ is a fuzzy vector if and only if $\xi_1, \xi_2, \ldots, \xi_n$ are fuzzy variables.*

Definition 1.2.10 (Fuzzy Arithmetic on Single Credibility Space) Let $f : \Re^n \to \Re$ be a function, and $\xi_1, \xi_2, \ldots, \xi_n$ fuzzy variables on the credibility space $(\Theta, \mathcal{P}(\Theta), \text{Cr})$. Then, $\xi = f(\xi_1, \xi_2, \ldots, \xi_n)$ is a fuzzy variable defined as

$$\xi(\theta) = f(\xi_1(\theta), \xi_2(\theta), \ldots, \xi_n(\theta))$$

for any $\theta \in \Theta$.

Example 1.2.4 Let ξ_1 and ξ_2 be fuzzy variables on the credibility space $(\Theta, \mathcal{P}(\Theta), \text{Cr})$. Then, the sum of ξ_1 and ξ_2 is

$$(\xi_1 + \xi_2)(\theta) = \xi_1(\theta) + \xi_2(\theta), \quad \forall\theta \in \Theta,$$

the product of ξ_1 and ξ_2 is

$$(\xi_1 \times \xi_2)(\theta) = \xi_1(\theta) \times \xi_2(\theta), \quad \forall\theta \in \Theta.$$

Definition 1.2.11 (Fuzzy Arithmetic on Different Credibility Spaces) Let $f : \Re^n \to \Re$ be a function, and ξ_i be fuzzy variables defined on credibility spaces $(\Theta_i, \mathcal{P}(\Theta_i), \text{Cr}_i)$, $i = 1, 2, \ldots, n$, respectively. Then, $\xi = f(\xi_1, \xi_2, \ldots, \xi_n)$ is a fuzzy variable defined on the product credibility space $(\Theta, \mathcal{P}(\Theta), \text{Cr})$ as

$$\xi(\theta_1, \theta_2, \ldots, \theta_n) = f(\xi_1(\theta_1), \xi_2(\theta_2), \ldots, \xi_n(\theta_n))$$

for any $(\theta_1, \theta_2, \ldots, \theta_n) \in \Theta$.

Example 1.2.5 Let ξ_1 and ξ_2 be fuzzy variables on the credibility spaces $(\Theta_1, \mathcal{P}(\Theta_1), \text{Cr}_1)$ and $(\Theta_2, \mathcal{P}(\Theta_2), \text{Cr}_2)$, respectively. Then, the sum of ξ_1 and ξ_2 is

$$(\xi_1 + \xi_2)(\theta_1, \theta_2) = \xi_1(\theta_1) + \xi_2(\theta_2), \ \forall (\theta_1, \theta_2) \in \Theta_1 \times \Theta_2,$$

the product of ξ_1 and ξ_2 is

$$(\xi_1 \times \xi_2)(\theta_1, \theta_2) = \xi_1(\theta_1) \times \xi_2(\theta_2), \ \forall (\theta_1, \theta_2) \in \Theta_1 \times \Theta_2.$$

1.2.3 Membership Function

Definition 1.2.12 Let ξ be a fuzzy variable defined on the credibility space $(\Theta, \mathcal{P}(\Theta), \text{Cr})$. Then, its membership function is derived from the credibility measure by

$$\mu(x) = (2\text{Cr}\{\xi = x\}) \wedge 1, \quad x \in \Re.$$

Definition 1.2.13 Let ξ be a fuzzy variable and $\alpha \in (0, 1]$. Then,

$$\xi_\alpha^L = \inf\{x | \mu(x) \geq \alpha\} \quad \text{and} \quad \xi_\alpha^U = \sup\{x | \mu(x) \geq \alpha\}$$

are called the α-pessimistic value and the α-optimistic value of ξ, respectively.

Theorem 1.2.6 *(Credibility Inversion Theorem) Let ξ be a fuzzy variable with membership function μ. Then, for any set* B *of real numbers, we have*

$$\text{Cr}\{\xi \in \text{B}\} = \frac{1}{2}\left(\sup_{x \in \text{B}} \mu(x) + 1 - \sup_{x \in \text{B}^c} \mu(x)\right).$$

Example 1.2.6 Let ξ be a fuzzy variable with membership function μ. Then, the following equations follow immediately from Theorem1.2.6

$$\text{Cr}\{\xi = x\} = \frac{1}{2}\left(\mu(x) + 1 - \sup_{y \neq x} \mu(y)\right), \quad \forall x \in \Re;$$

$$\text{Cr}\{\xi \leq x\} = \frac{1}{2}\left(\sup_{y \leq x} \mu(y) + 1 - \sup_{y > x} \mu(y)\right), \quad \forall x \in \Re;$$

$$\text{Cr}\{\xi \geq x\} = \frac{1}{2}\left(\sup_{y \geq x} \mu(y) + 1 - \sup_{y < x} \mu(y)\right), \quad \forall x \in \Re.$$

Especially, if μ is a continuous function, then $\mathrm{Cr}\{\xi = x\} = \mu(x)/2$.

Theorem 1.2.7 *(Sufficient and Necessary Condition for Membership Function) A function $\mu : \Re \to [0, 1]$ is a membership function if and only if* $\sup_{x\in\Re} \mu(x) = 1$.

Theorem 1.2.8 *Let ξ be a fuzzy variable with membership function μ. Then, its α-level set is*

$$\xi_\alpha = \{x | x \in \Re,\ \mu(x) \ge \alpha\}.$$

Theorem 1.2.9 *A fuzzy variable ξ with membership function μ is*

(a) nonnegative if and only if $\mu(x) = 0$ for all $x < 0$;
(b) positive if and only if $\mu(x) = 0$ for all $x \le 0$;
(c) simple if and only if μ takes nonzero values at a finite number of points;
(d) discrete if and only if μ takes nonzero values at a countable set of points;
(e) continuous if and only if μ is a continuous function.

Definition 1.2.14 A fuzzy variable is a triangular fuzzy variable, denoted by the triplet (a, b, c) of crisp numbers with $a < b < c$, if its membership function is

$$\mu(x) = \begin{cases} \frac{x-a}{b-a}, & \text{if } a \le x \le b \\ \frac{x-c}{b-c}, & \text{if } b \le x \le c \\ 0, & \text{otherwise.} \end{cases}$$

Definition 1.2.15 A fuzzy variable is a trapezoidal fuzzy variable, denoted by the quadruplet (a, b, c, d) of crisp numbers with $a < b < c < d$, if its membership function is

$$\mu(x) = \begin{cases} \frac{x-a}{b-a}, & \text{if } a \le x \le b \\ 1, & \text{if } b \le x \le c \\ \frac{x-d}{c-d}, & \text{if } c \le x \le d \\ 0, & \text{otherwise.} \end{cases}$$

1.2.4 Credibility Distribution

Definition 1.2.16 The credibility distribution $\Phi : \Re \to [0, 1]$ of a fuzzy variable ξ is defined by

$$\Phi(x) = \mathrm{Cr}\{\theta \in \Theta | \xi(\theta) \le x\}.$$

That is, $\Phi(x)$ is the credibility that the fuzzy variable ξ takes a value less than or equal to x. Generally speaking, the credibility distribution Φ is neither left-continuous nor right-continuous.

Theorem 1.2.10 *Let ξ be a fuzzy variable with membership function μ. Then, its credibility distribution is*

$$\Phi(x) = \frac{1}{2}\left(\sup_{y \le x} \mu(y) + 1 - \sup_{y > x} \mu(y)\right), \quad \forall x \in \Re.$$

Theorem 1.2.11 *(Sufficient and Necessary Condition for Credibility Distribution) A function $\Phi : \Re \to [0, 1]$ is a credibility distribution if and only if it is an increasing function with*

$$\lim_{x \to -\infty} \Phi(x) \le 0.5 \le \lim_{x \to +\infty} \Phi(x),$$

$$\lim_{y \downarrow x} \Phi(y) = \Phi(x) \quad \text{if} \lim_{y \downarrow x} \Phi(y) > 0.5 \text{ or } \Phi(x) \ge 0.5.$$

Example 1.2.6 Let a and b be two numbers with $0 \le a \le 0.5 \le b \le 1$. We define a fuzzy variable by the following membership function,

$$\mu(x) = \begin{cases} 2a, & \text{if } x < 0 \\ 1, & \text{if } x = 0 \\ 2 - 2b, & \text{if } x > 0. \end{cases}$$

Then, its credibility distribution is

$$\Phi(x) = \begin{cases} a, & \text{if } x < 0 \\ b, & \text{if } x \ge 0. \end{cases}$$

Thus, we have

$$\lim_{x \to -\infty} \Phi(x) = a, \qquad \lim_{x \to +\infty} \Phi(x) = b.$$

Theorem 1.2.12 *A fuzzy variable with credibility distribution Φ is*

(a) nonnegative if and only if $\Phi(x) = 0$ for all $x < 0$;
(b) positive if and only if $\Phi(x) = 0$ for all $x \le 0$.

Theorem 1.2.13 *Let ξ be a fuzzy variable. Then, we have*

(a) if ξ is simple, then its credibility distribution is a simple function;
(b) if ξ is discrete, then its credibility distribution is a step function;
(c) if ξ is continuous, then its credibility distribution is a continuous function.

Example 1.2.7 However, the inverse of Theorem 1.2.13 is not true. For example, let ξ be a fuzzy variable whose membership function is

$$\mu(x) = \begin{cases} x, & \text{if } 0 \leq x \leq 1 \\ 1, & \text{otherwise.} \end{cases}$$

Then, its credibility distribution is $\Phi(x) \equiv 0.5$. It is clear that $\Phi(x)$ is simple and continuous. But the fuzzy variable ξ is neither simple nor continuous.

Definition 1.2.17 The credibility density function $\phi : \Re \to [0, +\infty)$ of a fuzzy variable ξ is a function such that

$$\Phi(x) = \int_{-\infty}^{x} \phi(y)\mathrm{d}y, \quad \forall x \in \Re,$$
$$\int_{-\infty}^{+\infty} \phi(y)\mathrm{d}y = 1,$$

where Φ is the credibility distribution of the fuzzy variable ξ.

Example 1.2.8 The credibility distribution of a triangular fuzzy variable (a, b, c) is

$$\Phi(x) = \begin{cases} 0, & \text{if } x \leq a \\ \frac{x-a}{2(b-a)}, & \text{if } a \leq x \leq b \\ \frac{x+c-2b}{2(c-b)}, & \text{if } b \leq x \leq c \\ 1, & \text{if } x \geq c \end{cases}$$

and its credibility density function is

$$\phi(x) = \begin{cases} \frac{1}{2(b-a)}, & \text{if } a \leq x \leq b \\ \frac{1}{2(c-b)}, & \text{if } b \leq x \leq c \\ 0, & \text{otherwise.} \end{cases}$$

Example 1.2.9 The credibility distribution of a trapezoidal fuzzy variable (a, b, c, d) is

$$\Phi(x) = \begin{cases} 0, & \text{if } x \leq a \\ \frac{x-a}{2(b-a)}, & \text{if } a \leq x \leq b \\ \frac{1}{2}, & \text{if } b \leq x \leq c \\ \frac{x+d-2c}{2(d-c)}, & \text{if } c \leq x \leq d \\ 1, & \text{if } x \geq d \end{cases}$$

and its credibility density function is

$$\phi(x) = \begin{cases} \frac{1}{2(b-a)}, & \text{if } a \le x \le b \\ \frac{1}{2(d-c)}, & \text{if } c \le x \le d \\ 0, & \text{otherwise.} \end{cases}$$

Theorem 1.2.14 *Let ξ be a fuzzy variable whose credibility density function ϕ exists. Then, we have*

$$\int_{-\infty}^{+\infty} \phi(y)\mathrm{d}y = 1,$$

$$\mathrm{Cr}\{\xi \le x\} = \int_{-\infty}^{x} \phi(y)\mathrm{d}y$$

and

$$\mathrm{Cr}\{\xi \ge x\} = \int_{x}^{+\infty} \phi(y)\mathrm{d}y.$$

Example 1.2.10 Different from the random case, generally speaking,

$$\mathrm{Cr}\{a \le \xi \le b\} \ne \int_{a}^{b} \phi(y)\mathrm{d}y.$$

Consider the trapezoidal fuzzy variable $\xi = (1, 2, 3, 4)$. Then, $\mathrm{Cr}\{2 \le \xi \le 3\} = 0.5$. However, it is obvious that $\phi(x) = 0$ when $2 \le x \le 3$ and

$$\int_{2}^{3} \phi(y)\mathrm{d}y = 0 \ne 0.5 = \mathrm{Cr}\{2 \le \xi \le 3\}.$$

Definition 1.2.18 Let $(\xi_1, \xi_2, \ldots, \xi_n)$ be a fuzzy vector. Then, the joint credibility distribution $\Phi : \Re^n \to [0, 1]$ is defined by

$$\Phi(x_1, x_2, \ldots, x_n) = \mathrm{Cr}\{\theta \in \Theta | \xi_1(\theta) \le x_1, \xi_2(\theta) \le x_2, \ldots, \xi_n(\theta) \le x_n\}.$$

Definition 1.2.19 The joint credibility density function $\phi : \Re^n \to [0, +\infty)$ of a fuzzy vector $(\xi_1, \xi_2, \ldots, \xi_n)$ is a function such that

$$\Phi(x_1, x_2, \ldots, x_n) = \int_{-\infty}^{x_1} \int_{-\infty}^{x_2} \cdots \int_{-\infty}^{x_n} \phi(y_1, y_2, \ldots, y_n)\mathrm{d}y_1\mathrm{d}y_2 \ldots \mathrm{d}y_n$$

holds for all $(x_1, x_2, \ldots, x_n) \in \Re^n$, and

$$\int_{-\infty}^{+\infty}\int_{-\infty}^{+\infty}\cdots\int_{-\infty}^{+\infty} \phi(y_1, y_2, \ldots, y_n)\mathrm{d}y_1\mathrm{d}y_2\cdots\mathrm{d}y_n = 1,$$

where Φ is the joint credibility distribution of the fuzzy vector $(\xi_1, \xi_2, \ldots, \xi_n)$.

1.2.5 Independence

Definition 1.2.20 The fuzzy variables $\xi_1, \xi_2, \ldots, \xi_m$ are said to be independent if

$$\mathrm{Cr}\left\{\bigcap_{i=1}^{m}\{\xi_i \in \mathrm{B}_i\}\right\} = \min_{1\le i\le m} \mathrm{Cr}\{\xi_i \in \mathrm{B}_i\}$$

for any sets $\mathrm{B}_1, \mathrm{B}_2, \ldots, \mathrm{B}_m$ of $\Re$.

Theorem 1.2.15 *The fuzzy variables $\xi_1, \xi_2, \ldots, \xi_m$ are independent if and only if*

$$\mathrm{Cr}\left\{\bigcup_{i=1}^{m}\{\xi_i \in \mathrm{B}_i\}\right\} = \max_{1\le i\le m} \mathrm{Cr}\{\xi_i \in \mathrm{B}_i\}$$

for any sets $\mathrm{B}_1, \mathrm{B}_2, \ldots, \mathrm{B}_m$ of $\Re$.

Theorem 1.2.16 *Let $\xi_1, \xi_2, \ldots, \xi_m$ be independent fuzzy variables, and let $f_i : \Re \to \Re$ be functions for $i = 1, 2, \ldots, m$. Then, $f_1(\xi_1), f_2(\xi_2), \ldots, f_m(\xi_m)$ are independent fuzzy variables.*

Theorem 1.2.17 *Let ξ and η be independent fuzzy variables. Then,*

(a) for any $\alpha \in (0, 1]$, $(\xi+\eta)_\alpha^L = \xi_\alpha^L+\eta_\alpha^L$;
(b) for any $\alpha \in (0, 1]$, $(\xi+\eta)_\alpha^U = \xi_\alpha^U+\eta_\alpha^U$.

Furthermore, if ξ and η are positive, then

(c) for any $\alpha \in (0, 1]$, $(\xi \cdot \eta)_\alpha^L = \xi_\alpha^L \cdot \eta_\alpha^L$;
(d) for any $\alpha \in (0, 1]$, $(\xi \cdot \eta)_\alpha^U = \xi_\alpha^U \cdot \eta_\alpha^U$.

Theorem 1.2.18 *(Expansion Principle of Zadeh) Let $\xi_1, \xi_2, \ldots, \xi_n$ be independent fuzzy variables with membership functions $\mu_1, \mu_2, \ldots, \mu_n$, respectively, and $f : \Re^n \to \Re$ a function. Then, the membership function μ of $\xi = f(\xi_1, \xi_2, \ldots, \xi_n)$ is derived from the membership functions $\mu_1, \mu_2, \ldots, \mu_n$ by*

$$\mu(x) = \sup_{x=f(x_1,x_2,\ldots,x_n)} \min_{1\le i\le n} \mu_i(x_i)$$

for any $x \in \Re$. *Here we set* $\mu(x) = 0$ *if there are not real numbers* $x_1, x_2, \ldots, x_n$ *such that* $x = f(x_1, x_2, \ldots, x_n)$.

Example 1.2.11 Let ξ_1 and ξ_2 be independent fuzzy variables with membership functions μ_1 and μ_2, respectively. Then, $\xi_1 + \xi_2$ has the membership function

$$\mu(x) = \sup_{y \in \Re} \mu_1(x - y) \wedge \mu_2(y).$$

Example 1.2.12 By using Theorem 1.2.18, we may verify that the sum of independent trapezoidal fuzzy variables $\xi = (a_1, a_2, a_3, a_4)$ and $\eta = (b_1, b_2, b_3, b_4)$ is also a trapezoidal fuzzy variable, and $\xi + \eta = (a_1 + b_1, a_2 + b_2, a_3 + b_3, a_4 + b_4)$. The product of a trapezoidal fuzzy variable $\xi = (a_1, a_2, a_3, a_4)$ and a scalar number λ is

$$\lambda \cdot \xi = \begin{cases} (\lambda a_1, \lambda a_2, \lambda a_3, \lambda a_4), & \text{if } \lambda \geq 0 \\ (\lambda a_4, \lambda a_3, \lambda a_2, \lambda a_1), & \text{if } \lambda < 0. \end{cases}$$

That is, the product of a trapezoidal fuzzy variable and a scalar number is also a trapezoidal fuzzy variable.

Theorem 1.2.19 *(Credibility Inversion Theorem) Let* $\xi_1, \xi_2, \ldots, \xi_n$ *be independent fuzzy variables with membership functions* $\mu_1, \mu_2, \ldots, \mu_n$, *respectively, and* $f : \Re^n \to \Re^m$ *a function. Then, for any set* B *of* $\Re^m$, *the credibility*

$$\text{Cr}\{f(\xi_1, \xi_2, \ldots, \xi_n) \in \text{B}\} = \frac{1}{2}\left(\sup_{f(x_1, x_2, \ldots, x_n) \in \text{B}} \min_{1 \leq i \leq n} \mu_i(x_i) \right.$$
$$\left. + 1 - \sup_{f(x_1, x_2, \ldots, x_n) \in \text{B}^c} \min_{1 \leq i \leq n} \mu_i(x_i) \right).$$

1.2.6 *Identical Distribution*

Definition 1.2.21 The fuzzy variables ξ and η are said to be identically distributed if

$$\text{Cr}\{\xi \in \text{B}\} = \text{Cr}\{\eta \in \text{B}\}$$

for any set B of $\Re$.

Theorem 1.2.20 *The fuzzy variables* ξ *and* η *are identically distributed if and only if* ξ *and* η *have the same membership function.*

Theorem 1.2.21 *If* ξ *and* η *are identically distributed fuzzy variables, then* ξ *and* η *have the same credibility distribution.*

Example 1.2.13 The inverse of Theorem 1.2.21 is not true. We consider two fuzzy variables ξ and η with the following membership functions

$$\mu_1(x) = \begin{cases} 1.0, & \text{if } x = 0 \\ 0.6, & \text{if } x = 1 \\ 0.8, & \text{if } x = 2 \end{cases} \quad \text{and} \quad \mu_2(x) = \begin{cases} 1.0, & \text{if } x = 0 \\ 0.7, & \text{if } x = 1 \\ 0.8, & \text{if } x = 2 \end{cases}$$

respectively. It is easy to verify that ξ and η have the same credibility distribution

$$\Phi(x) = \begin{cases} 0, & \text{if } x < 0 \\ 0.6, & \text{if } 0 \le x \le 2 \\ 1, & \text{if } x \ge 2. \end{cases}$$

However, they are not identically distributed fuzzy variables.

Theorem 1.2.22 *Let ξ and η be two fuzzy variables whose credibility density functions exist. If ξ and η are identically distributed, then they have the same credibility density function.*

1.2.7 Expected Value

Definition 1.2.22 Let ξ be a fuzzy variable. Then, the expected value of ξ is defined by

$$E[\xi] = \int_0^{+\infty} \mathrm{Cr}\{\xi \ge r\}\mathrm{d}r - \int_{-\infty}^{0} \mathrm{Cr}\{\xi \le r\}\mathrm{d}r$$

provided that at least one of the two integrals is finite. In particular, if ξ is a nonnegative fuzzy variable, then

$$E[\xi] = \int_0^{+\infty} \mathrm{Cr}\{\xi \ge r\}\mathrm{d}r.$$

Theorem 1.2.23 *Let ξ be a fuzzy variable with finite expected value $E[\xi]$, then*

$$E[\xi] = \frac{1}{2}\int_0^1 [\xi_\alpha^L + \xi_\alpha^U]\mathrm{d}\alpha,$$

where ξ_α^L and ξ_α^U are the α-pessimistic value and the α-optimistic value of ξ, respectively.

Example 1.2.14 The triangular fuzzy variable $\xi = (a, b, c)$ has an expected value

$$E[\xi] = \frac{1}{4}(a + 2b + c).$$

Example 1.2.15 The trapezoidal fuzzy variable $\xi = (a, b, c, d)$ has an expected value

$$E[\xi] = \frac{1}{4}(a + b + c + d).$$

Example 1.2.16 The definition of expected value operator is also applicable to discrete case. Suppose that ξ is a simple fuzzy variable whose membership function is given by

$$\mu(x) = \begin{cases} \mu_1, & \text{if} x = a_1 \\ \mu_2, & \text{if} x = a_2 \\ \vdots & \\ \mu_m, & \text{if} x = a_m. \end{cases}$$

Without loss of generality, suppose that $a_1, a_2, \ldots, a_m$ are distinct. Definition 1.2.22 implies that the expected value of ξ is

$$E[\xi] = \sum_{i=1}^{m} \omega_i a_i,$$

where the weights ω_i, $i = 1, 2, \ldots, m$ are given by

$$\omega_i = \frac{1}{2}\left(\max_{1 \le k \le m} \{\mu_k | a_k \le a_i\} - \max_{1 \le k \le m} \{\mu_k | a_k < a_i\} \right.$$
$$\left. + \max_{1 \le k \le m} \{\mu_k | a_k \ge a_i\} - \max_{1 \le k \le m} \{\mu_k | a_k > a_i\} \right)$$

for $i = 1, 2, \ldots, m$. It is easy to verify that all $\omega_i \ge 0$ and the sum of all weights is just 1.

Example 1.2.17 Let ξ be a fuzzy variable with membership function

$$\mu(x) = \begin{cases} 0, & \text{if} x < 0 \\ x, & \text{if} 0 \le x \le 1 \\ 1, & \text{if} x > 1. \end{cases}$$

Then, its expected value is $+\infty$. If ξ is a fuzzy variable with membership function

$$\mu(x) = \begin{cases} 1, & \text{if } x < 0 \\ 1 - x, & \text{if } 0 \le x \le 1 \\ 0, & \text{if } x > 1. \end{cases}$$

Then, its expected value is $-\infty$.

Example 1.2.18 The expected value may not exist for some fuzzy variables. For example, the fuzzy variable ξ with membership function $\mu(x) = 1/(1 + |x|)$ does not have expected value because both of the integrals

$$\int_0^{+\infty} \mathrm{Cr}\{\xi \ge r\}\mathrm{d}r \text{ and } \int_{-\infty}^{0} \mathrm{Cr}\{\xi \le r\}\mathrm{d}r$$

are infinite.

Theorem 1.2.24 *Let ξ be a fuzzy variable whose credibility density function ϕ exists. If the Lebesgue integral*

$$\int_{-\infty}^{+\infty} x\phi(x)\mathrm{d}x$$

is finite, then we have

$$E[\xi] = \int_{-\infty}^{+\infty} x\phi(x)\mathrm{d}x.$$

Example 1.2.19 Let ξ be a fuzzy variable with credibility distribution Φ. Generally speaking,

$$E[\xi] \ne \int_{-\infty}^{+\infty} x\mathrm{d}\Phi(x).$$

For example, let ξ be a fuzzy variable with membership function

$$\mu(x) = \begin{cases} 0, & \text{if } x < 0 \\ x, & \text{if } 0 \le x \le 1 \\ 1, & \text{if } x > 1. \end{cases}$$

Then, $E[\xi] = +\infty$. However

$$\int_{-\infty}^{+\infty} x \mathrm{d}\Phi(x) = \frac{1}{4} \neq +\infty.$$

Theorem 1.2.25 *Let ξ be a fuzzy variable with credibility distribution Φ. If*

$$\lim_{x \to -\infty} \Phi(x) = 0, \quad \lim_{x \to +\infty} \Phi(x) = 1$$

and the Lebesgue-Stieltjes integral

$$\int_{-\infty}^{+\infty} x \mathrm{d}\Phi(x)$$

is finite, then we have

$$E[\xi] = \int_{-\infty}^{+\infty} x \mathrm{d}\Phi(x).$$

Theorem 1.2.26 *Let ξ and η be independent fuzzy variables with finite expected values. Then, for any numbers a and b, we have*

$$E[a\xi + b\eta] = aE[\xi] + bE[\eta].$$

Example 1.2.20 Theorem 1.2.6 does not hold if ξ_1 and ξ_2 are not independent. For example, $\Theta = \{\theta_1, \theta_2, \theta_3\}$, $\mathrm{Cr}\{\theta_1\} = 0.7$, $\mathrm{Cr}\{\theta_2\} = 0.3$, $\mathrm{Cr}\{\theta_3\} = 0.2$ and the fuzzy variables are defined by

$$\xi_1(\theta) = \begin{cases} 1, & \text{if } \theta = \theta_1 \\ 0, & \text{if } \theta = \theta_2 \\ 2, & \text{if } \theta = \theta_3, \end{cases} \quad \xi_2(\theta) = \begin{cases} 0, & \text{if} \theta = \theta_1 \\ 2, & \text{if} \theta = \theta_2 \\ 3, & \text{if} \theta = \theta_3. \end{cases}$$

Then, we have

$$(\xi_1 + \xi_2)(\theta) = \begin{cases} 1, & \text{if} \theta = \theta_1 \\ 2, & \text{if} \theta = \theta_2 \\ 5, & \text{if} \theta = \theta_3. \end{cases}$$

Thus, $E[\xi_1] = 0.9,\ E[\xi_2] = 0.8$ and $E[\xi_1 + \xi_2] = 1.9$. This fact implies that

$$E[\xi_1 + \xi_2] > E[\xi_1] + E[\xi_2].$$

If the fuzzy variables are defined by

$$\xi_1(\theta)=\begin{cases}0, & \text{if}\theta=\theta_1\\ 1, & \text{if}\theta=\theta_2\\ 2, & \text{if}\theta=\theta_3,\end{cases}\qquad \xi_2(\theta)=\begin{cases}0, & \text{if}\theta=\theta_1\\ 3, & \text{if}\theta=\theta_2\\ 1, & \text{if}\theta=\theta_3.\end{cases}$$

Then, we have

$$(\xi_1+\xi_2)(\theta)=\begin{cases}0, & \text{if}\theta=\theta_1\\ 4, & \text{if}\theta=\theta_2\\ 3, & \text{if}\theta=\theta_3.\end{cases}$$

Thus, $E[\xi_1]=0.5,\ E[\xi_2]=0.9$ and $E[\xi_1+\xi_2]=1.2$. This fact implies that

$$E[\xi_1+\xi_2]<E[\xi_1]+E[\xi_2].$$

Definition 1.2.23 Let ξ be a fuzzy variable, and $f:\Re\to\Re$ a function. Then, the expected value of $f(\xi)$ is

$$E[f(\xi)]=\int_0^{+\infty}\mathrm{Cr}\{f(\xi)\geq r\}\mathrm{d}r-\int_{-\infty}^{0}\mathrm{Cr}\{f(\xi)\leq r\}\mathrm{d}r.$$

1.3 Random Fuzzy Theory

Random fuzzy theory is a branch of mathematics that studies the behavior of random fuzzy phenomena. Random fuzzy variable was defined by Liu [3] as a function from a credibility space to the set of random variables. In other words, a random fuzzy variable is a fuzzy variable taking "random variable" values.

1.3.1 Random Fuzzy Variable

Definition 1.3.1 A random fuzzy variable is a function from the credibility space $(\Theta,\mathcal{P}(\Theta),\mathrm{Cr})$ to the set of random variables.

Example 1.3.1 Let $\eta_1,\eta_2,\ldots,\eta_m$ be random variables, and $\mu_1,\mu_2,\ldots,\mu_m$ real numbers in $[0,1]$ such that $\mu_1\vee\mu_2\vee\cdots\vee\mu_{\mathrm{m}}=1$. Then,

$$\xi = \begin{cases} \eta_1, & \text{with membership degree } \mu_1 \\ \eta_2, & \text{with membership degree } \mu_2 \\ \vdots & \\ \eta_m, & \text{with membership degree } \mu_m \end{cases}$$

is clearly a random fuzzy variable.

Example 1.3.2 If η is a random variable and $\tilde{a}$ is a fuzzy variable defined on the credibility space $(\Theta, \mathcal{P}(\Theta), \text{Cr})$, then the sum $\xi = \eta + \tilde{a}$ is a random fuzzy variable defined by

$$\xi(\theta) = \eta + \tilde{a}(\theta), \quad \forall\theta \in \Theta.$$

The product $\xi = \eta \cdot \tilde{a}$ is also a random fuzzy variable defined by

$$\xi(\theta) = \eta \cdot \tilde{a}(\theta), \quad \forall\theta \in \Theta.$$

Example 1.3.3 Let $\xi \sim N(\rho, 1)$, where ρ is a fuzzy variable with membership function $\mu_\rho(x) = [1 - |x - 2|] \vee 0$. Then, ξ is a random fuzzy variable taking "normally distributed variable $N(\rho, 1)$" values.

Example 1.3.4 In many statistics problems, the probability distribution is completely known except for the values of one or more parameters. For example, it might be known that the lifetime ξ of a modern engine is an exponentially distributed variable with an unknown expected value β and has the following form of probability density function

$$\phi(x) = \begin{cases} \frac{1}{\beta}\exp\left(-\frac{x}{\beta}\right), & \text{if } x \geq 0 \\ 0, & \text{if } x < 0. \end{cases}$$

Usually, there is some relevant information in practice. It is thus possible to specify an interval in which the value of β is likely to lie or to give an approximate estimate of the value of β. It is typically not possible to determine the value of β exactly. If the value of β is provided as a fuzzy variable, then ξ is a random fuzzy variable, denoted by $\xi \sim \mathcal{EXP}\left(\frac{1}{\beta}\right)$.

Definition 1.3.2 A random fuzzy variable ξ defined on the credibility space $(\Theta, \mathcal{P}(\Theta), \text{Cr})$ is a said to be nonnegative if and only if $\Pr\{\xi(\theta) < 0\} = 0$ for each $\theta \in \Theta$ with $\mu_{\xi(\theta)}(x) > 0$.

Theorem 1.3.1 *Let ξ be a random fuzzy variable on the credibility space* $(\Theta, \mathcal{P}(\Theta), \text{Cr})$. *Then, for* $\forall\theta \in \Theta$,

(a) $\Pr\{\xi(\theta) \in \text{B}\}$ is a fuzzy variable for any Borel set B on $\Re$;
(b) $E[\xi(\theta)]$ is a fuzzy variable provided that $E[\xi(\theta)]$ is finite.

Definition 1.3.3 Let ξ and η be random fuzzy variables defined on the credibility space $(\Theta, \mathcal{P}(\Theta), \mathrm{Cr})$. Then, $\xi = \eta$ if and only if $\xi(\theta) = \eta(\theta)$ for almost all $\theta \in \Theta$.

Definition 1.3.4 An n-dimensional random fuzzy vector is a function from the credibility space $(\Theta, \mathcal{P}(\Theta), \mathrm{Cr})$ to the set of n-dimensional random vectors.

Theorem 1.3.2 *The vector $(\xi_1, \xi_2, \ldots, \xi_n)$ is a random fuzzy vector if and only if $\xi_1, \xi_2, \ldots, \xi_n$ are random fuzzy variables.*

Theorem 1.3.3 *Let ξ be an n-dimensional random fuzzy vector, and $f : \Re^n \to \Re$ a measurable function. Then, $f(\xi)$ is a random fuzzy variable.*

Definition 1.3.5 (Random Fuzzy Arithmetic on Single Space) Let $f : \Re^n \to \Re$ be a measurable function, and $\xi_1, \xi_2, \ldots, \xi_n$ random fuzzy variables on the credibility space $(\Theta, \mathcal{P}(\Theta), \mathrm{Cr})$. Then, $\xi = f(\xi_1, \xi_2, \ldots, \xi_n)$ is a random fuzzy variable defined as

$$\xi(\theta) = f(\xi_1(\theta), \xi_2(\theta), \ldots, \xi_n(\theta)), \quad \forall \theta \in \Theta.$$

Example 1.3.5 Let ξ_1 and ξ_2 be two random fuzzy variables defined on the credibility space $(\Theta, \mathcal{P}(\Theta), \mathrm{Cr})$. Then, the sum $\xi = \xi_1 + \xi_2$ is a random fuzzy variable defined by

$$\xi(\theta) = \xi_1(\theta) + \xi_2(\theta), \quad \forall \theta \in \Theta.$$

The product $\xi = \xi_1 \cdot \xi_2$ is also a random fuzzy variable defined by

$$\xi(\theta) = \xi_1(\theta) \cdot \xi_2(\theta), \quad \forall \theta \in \Theta.$$

Definition 1.3.6 (Random Fuzzy Arithmetic on Different Spaces) Let $f : \Re^n \to \Re$ be a measurable function, and ξ_i random fuzzy variables on the credibility spaces $(\Theta_i, \mathcal{P}(\Theta_i), \mathrm{Cr}_i), i = 1, 2, \ldots, n$, respectively. Then, $\xi = f(\xi_1, \xi_2, \ldots, \xi_n)$ is a random fuzzy variable on the product credibility space $(\Theta, \mathcal{P}(\Theta), \mathrm{Cr})$, defined as

$$\xi(\theta_1, \theta_2, \ldots, \theta_n) = f(\xi_1(\theta_1), \xi_2(\theta_2), \ldots, \xi_n(\theta_n))$$

for all $(\theta_1, \theta_2, \ldots, \theta_n) \in \Theta$.

Example 1.3.6 Let ξ_1 and ξ_2 be two random fuzzy variables defined on the credibility spaces $(\Theta_1, \mathcal{P}(\Theta_1), \mathrm{Cr}_1)$ and $(\Theta_2, \mathcal{P}(\Theta_2), \mathrm{Cr}_2)$, respectively. Then, the sum $\xi = \xi_1 + \xi_2$ is a random fuzzy variable on the credibility space $(\Theta_1 \times \Theta_2, \mathcal{P}(\Theta_1 \times \Theta_2), \mathrm{Cr}_1 \wedge \mathrm{Cr}_2)$ defined by

$$\xi(\theta_1, \theta_2) = \xi_1(\theta_1) + \xi_2(\theta_2), \ \forall (\theta_1, \theta_2) \in \Theta_1 \times \Theta_2.$$

The product $\xi = \xi_1 \cdot \xi_2$ is a random fuzzy variable defined on the credibility space $(\Theta_1 \times \Theta_2, \mathcal{P}(\Theta_1 \times \Theta_2), \mathrm{Cr}_1 \wedge \mathrm{Cr}_2)$ as

$$\xi(\theta_1, \theta_2) = \xi_1(\theta_1) \cdot \xi_2(\theta_2),\ \forall(\theta_1, \theta_2) \in \Theta_1 \times \Theta_2.$$

Example 1.3.7 Let ξ_1 and ξ_2 be two random fuzzy variables defined as follows,

$$\xi_1 \sim \begin{cases} \mathcal{N}(\mu_1, \sigma_1^2) & \text{with membership degree } 0.7 \\ \mathcal{N}(\mu_2, \sigma_2^2) & \text{with membership degree } 1.0, \end{cases}$$

$$\xi_2 \sim \begin{cases} \mathcal{N}(\mu_3, \sigma_3^2) & \text{with membership degree } 1.0 \\ \mathcal{N}(\mu_4, \sigma_4^2) & \text{with membership degree } 0.8. \end{cases}$$

Then, the sum of the two random fuzzy variables is also a random fuzzy variable,

$$\xi \sim \begin{cases} \mathcal{N}(\mu_1 + \mu_3, \sigma_1^2 + \sigma_3^2) & \text{with membership degree } 0.7 \\ \mathcal{N}(\mu_1 + \mu_4, \sigma_1^2 + \sigma_4^2) & \text{with membership degree } 0.7 \\ \mathcal{N}(\mu_2 + \mu_3, \sigma_2^2 + \sigma_3^2) & \text{with membership degree } 1.0 \\ \mathcal{N}(\mu_2 + \mu_4, \sigma_2^2 + \sigma_4^2) & \text{with membership degree } 0.8. \end{cases}$$

1.3.2 Chance Measure

Definition 1.3.7 Let ξ be a random fuzzy variable on the credibility space $(\Theta, \mathcal{P}(\Theta), \mathrm{Cr})$. Then, the average chance, denoted by Ch, of a random fuzzy event characterized by $\{\xi \in \mathrm{B}\}$ is defined as

$$\mathrm{Ch}\{\xi \in \mathrm{B}\} = \int_0^1 \mathrm{Cr}\{\theta \in \Theta | \Pr\{\xi(\theta) \in \mathrm{B}\} \geq p\} \mathrm{d}p.$$

Remark 1.3.1 If ξ degenerates to a random variable, then $\mathrm{Ch}\{\xi \in \mathrm{B}\}$ degenerates to $\Pr\{\xi \in \mathrm{B}\}$, which is just the probability of the random event $\{\xi \in \mathrm{B}\}$. If ξ degenerates to a fuzzy variable, then $\mathrm{Ch}\{\xi \in \mathrm{B}\}$ degenerates to $\mathrm{Cr}\{\xi \in \mathrm{B}\}$, which is just the credibility of the random event $\{\xi \in \mathrm{B}\}$.

1.3.3 Independence

Definition 1.3.8 The random fuzzy variables $\xi_1, \xi_2, \ldots, \xi_n$ are said to be independent if

(a) $\xi_1(\theta), \xi_2(\theta), \ldots, \xi_n(\theta)$ are independent random variables for each $\theta \in \Theta$;
(b) $E[\xi_1(\cdot)], E[\xi_2(\cdot)], \ldots, E[\xi_n(\cdot)]$ are independent fuzzy variables.

Remark 1.3.2 If $\xi_1, \xi_2, \ldots, \xi_n$ are independent random fuzzy variables, then $a_1\xi_1 + b_1, a_2\xi_2 + b_2, \ldots, a_n\xi_n + b_n$ are independent for any real numbers a_i and $b_i, i = 1, 2, \ldots, n$.

1.3.4 Expected Value

Definition 1.3.9 Let ξ be a random fuzzy variable. Then, the expected value of ξ is defined by

$$E[\xi] = \int_0^{+\infty} \mathrm{Cr}\{\theta \in \Theta | E[\xi(\theta) \geq r]\}\mathrm{d}r - \int_{-\infty}^{0} \mathrm{Cr}\{\theta \in \Theta | E[\xi(\theta) \leq r]\}\mathrm{d}r$$

provided that at least one of the two integrals is finite. Especially, if ξ is a nonnegative random fuzzy variable, we have

$$E[\xi] = \int_0^{+\infty} \mathrm{Cr}\{\theta \in \Theta | E[\xi(\theta) \geq r]\}\mathrm{d}r.$$

Remark 1.3.3 If the random fuzzy variable ξ degenerates to a random variable, then the expected value becomes

$$E[\xi] = \int_0^{+\infty} \Pr\{\xi \geq r\}\mathrm{d}r - \int_{-\infty}^{0} \Pr\{\xi \leq r\}\mathrm{d}r,$$

which is just the conventional mathematical expectation of random variable ξ. If the random fuzzy variable ξ degenerates to a fuzzy variable, then the expected value becomes

$$E[\xi] = \int_0^{+\infty} \mathrm{Cr}\{\xi \geq r\}\mathrm{d}r - \int_{-\infty}^{0} \mathrm{Cr}\{\xi \leq r\}\mathrm{d}r,$$

which is just the expected value of fuzzy variable ξ.

Example 1.3.8 Suppose that ξ is a random fuzzy variable defined as

$$\xi \sim \mathcal{U}(\rho, \rho + 2) \text{ with } \rho = (0, 1, 2).$$

Without loss of generality, we assume that ρ is defined on the credibility space $(\Theta, \mathcal{P}(\Theta), \mathrm{Cr})$. Then, for each $\theta \in \Theta$, $\xi(\theta)$ is a random variable and $E[\xi(\theta)] =$

$\rho(\theta) + 1$. Thus, the expected value of ξ is

$$E[\xi] = E[\rho] + 1 = 2.$$

Theorem 1.3.4 *Assume ξ and η are independent random fuzzy variables with finite expected values. Then, for any real numbers a and b, we have*

$$E[a\xi + b\eta] = aE[\xi] + bE[\eta].$$

Chapter 2
Nonrepairable Systems with Stochastic Lifetimes

This chapter mainly recalls the nonrepairable system in stochastic case. The so-called nonrepairable system means that no repair is carried out on the failed component. There are many reasons for not performing any maintenance: Some cannot be repaired due to technical reasons; some are not worth repairing due to economic reasons; some systems themselves are repairable, but for convenience of analysis, they are considered as a nonrepairable system for research. Therefore, it has great practical significance to study nonrepairable systems. This chapter will extract reliability mathematical models of nonrepairable systems in the book [24], including series systems, parallel systems, series–parallel systems, parallel–series systems, cold standby systems, warm standby systems, shock models and the bivariate exponential distribution. It is convenient for readers to compare the contents of this chapter with the corresponding results in Chaps. 3 and 4.

The main reliability indices of nonrepairable systems are reliability (denoted by $R(t)$) and mean time to failure (denoted by MTTF), which can describe the reliability characteristics of nonrepairable systems. We usually use a nonnegative random variable X to describe the lifetime of a system, and the corresponding distribution function is

$$F(t) = \Pr\{X \leq t\},\ t \geq 0.$$

Based on the lifetime distribution $F(t)$, we can arrive at the probability that the system is normal at time t; that is, the survival probability of the nonrepairable system at time t is

$$R(t) = \Pr\{X > t\} = 1 - F(t) = \overline{F}(t).$$

$R(t)$ is called the reliability function or reliability of the system. Therefore, reliability can also be defined as the probability that the system completes the specified function in the specified time under the specified conditions.

The MTTF of the system is defined as

Y. Liu, *Reliability Theory Based on Uncertain Lifetimes*,
https://doi.org/10.1007/978-981-16-0995-4_2

$$\text{MTTF} = EX = \int_0^{+\infty} t\text{d}F(t).$$

If the system lifetime X is a nonnegative continuous random variable, its distribution function is $F(t)$ and the density function is $f(t)$, the failure rate function of random variable X is defined as

$$r(t) = \frac{f(t)}{\overline{F}(t)} \quad \text{for } t \in \{t : F(t) < 1\},$$

it is called failure rate for short. $r(t)$ has the probability explanations that if the system is normal at time t, the failure probability of the system in $(t, t + \Delta t]$ is

$$\Pr\{X \leq t + \Delta t | X > t\} = \frac{F(t + \Delta t) - F(t)}{1 - F(t)} \sim \frac{f(t)\Delta t}{\overline{F}(t)} = r(t)\Delta t.$$

Therefore, when Δt is very small, $r(t)\Delta t$ represents the failure probability in $(t, t + \Delta t]$ under the condition that the system is normal before time t.

2.1 Series and Parallel Systems

2.1.1 *The Series System*

Consider the series system consisting of n components and the failure of any component will cause the failure of system. Figure 2.1 shows the reliability block diagram of the series system composed of n components. Let X_i be the lifetime of component i, $i = 1, 2, \ldots, n$. The reliability of component i is $R_i(t) = \Pr\{X_i > t\}$, $i = 1, 2, \ldots, n$. Assume $X_1, X_2, \ldots, X_n$ are independent with each other. At the initial moment, all components are new and start working at the same time. Obviously, the lifetime of the series system is $X = \min\{X_1, X_2, \ldots, X_n\}$.

Then, the reliability of the series system is

$$\begin{aligned} R(t) &= \Pr\{\min\{X_1, X_2, \ldots, X_n\} > t\} \\ &= \Pr\{X_1 > t, X_2 > t, \ldots, X_n > t\} \\ &= \prod_{i=1}^{n} \Pr\{X_i > t\} = \prod_{i=1}^{n} R_i(t). \end{aligned} \tag{2.1.1}$$

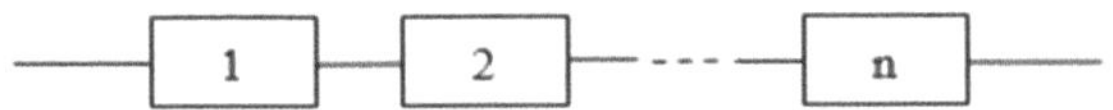

Fig. 2.1 Series system

When the failure rate of components i is $\lambda_i(t), i = 1, 2, \ldots, n,$ the reliability of the series system is

$$R(t) = \prod_{i=1}^{n} \exp\left\{ -\int_0^t \lambda_i(u)\mathrm{d}u \right\} = \exp\left\{ -\int_0^t \sum_{i=1}^{n} \lambda_i(u)\mathrm{d}u \right\}. \tag{2.1.2}$$

Then, the failure rate of the series system is

$$\lambda(t) = \frac{-R'(t)}{R(t)} = \sum_{i=1}^{n} \lambda_i(t).$$

It is easy to see that the failure rate of the series system composed of independent components is the sum of the failure rates of all components.

The MTTF of the series system is

$$\mathrm{MTTF} = \int_0^{+\infty} R(t)\mathrm{d}t = \int_0^{+\infty} \exp\left\{ -\int_0^t \lambda(u)\mathrm{d}u \right\} \mathrm{d}t. \tag{2.1.3}$$

If $R_i(t) = \exp\{-\lambda_i t\}, \ i = 1, 2, \ldots, n,$ that is, if the lifetime of component i follows an exponential distribution with parameter $\lambda_i, \ i = 1, 2, \ldots, n,$ the reliability and MTTF of the series system are

$$\begin{cases} R(t) = \exp\left\{ -\sum\limits_{i=1}^{n} \lambda_i t \right\} \\ \mathrm{MTTF} = \frac{1}{\sum\limits_{i=1}^{n} \lambda_i}. \end{cases} \tag{2.1.4}$$

In particular, if $R_i(t) = \mathrm{e}^{-\lambda t}, i = 1, 2, \ldots, n,$ we have

$$\begin{cases} R(t) = \mathrm{e}^{-n\lambda t} \\ \mathrm{MTTF} = \frac{1}{n\lambda}. \end{cases} \tag{2.1.5}$$

2.1.2 The Parallel System

Consider the parallel system consisting of n components, and the system fails when all components fail. Figure 2.2 shows the reliability block diagram of the parallel system composed of n components. Let X_i be the lifetime of component $i, \ i = 1, 2, \ldots, n.$ The reliability of component i is $R_i(t) = \Pr\{X_i > t\}, \ i = 1, 2, \ldots, n.$ Assume that $X_1, X_2, \ldots, X_n$ are independent with each other. At the initial moment, all components

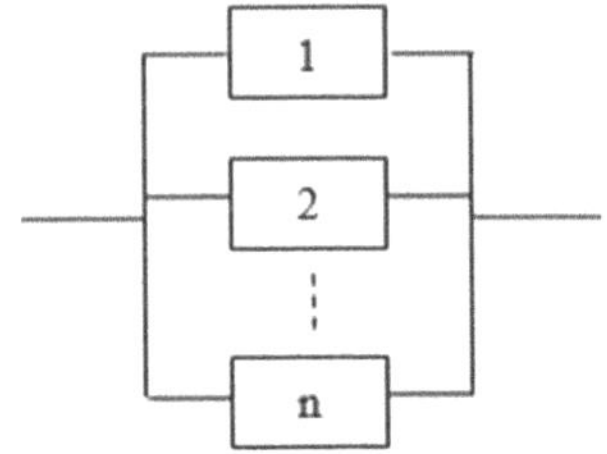

Fig. 2.2 Parallel system

are new and start working at the same time. The lifetime of the parallel system is $X = \max\{X_1, X_2, \ldots, X_n\}$.

Then the reliability of the parallel system is

$$\begin{aligned} R(t) &= \Pr\{\max\{X_1, X_2, \ldots, X_n\} > t\} \\ &= 1 - \Pr\{\max\{X_1, X_2, \ldots, X_n\} \le t\} \\ &= 1 - \Pr\{X_1 \le t, X_2 \le t, \ldots, X_n \le t\} \\ &= 1 - \prod_{i=1}^{n} [1 - R_i(t)]. \end{aligned} \tag{2.1.6}$$

If $R_i(t) = \mathrm{e}^{-\lambda_i t}, i = 1, 2, \ldots, n$, we have

$$R(t) = 1 - \prod_{i=1}^{n} \left[1 - \mathrm{e}^{-\lambda_i t}\right]. \tag{2.1.7}$$

Equation (2.1.7) can also be written as

$$\begin{aligned} R(t) = &\sum_{i=1}^{n} \mathrm{e}^{-\lambda_i t} - \sum_{1 \le i < j \le n} \mathrm{e}^{-(\lambda_i + \lambda_j)t} + \cdots + (-1)^{i-1} \sum_{1 \le j_1 < \cdots < j_i \le n} \mathrm{e}^{-(\lambda_{j_1} + \lambda_{j_2} + \cdots + \lambda_{j_i})t} \\ &+ \cdots + (-1)^{n-1} \mathrm{e}^{-(\lambda_1 + \cdots + \lambda_n)t}. \end{aligned}$$

The MTTF of the parallel system is

$$\begin{aligned} \text{MTTF} &= \int_0^{+\infty} R(t)\mathrm{d}t \\ &= \sum_{i=1}^{n} \frac{1}{\lambda_i} - \sum_{1 \le i < j \le n} \frac{1}{\lambda_i + \lambda_j} + \cdots + (-1)^{n-1} \frac{1}{\lambda_1 + \lambda_2 + \cdots + \lambda_n}. \end{aligned} \tag{2.1.8}$$

In particular, if $n = 2$, we have

$$\begin{cases} R(t) = \mathrm{e}^{-\lambda_1 t} + \mathrm{e}^{-\lambda_2 t} - \mathrm{e}^{-(\lambda_1+\lambda_2)t} \\ \mathrm{MTTF} = \frac{1}{\lambda_1} + \frac{1}{\lambda_2} - \frac{1}{\lambda_1+\lambda_2}. \end{cases} \tag{2.1.9}$$

If $R_i(t) = \mathrm{e}^{-\lambda t}, i = 1, 2, \ldots, n$, we have

$$\begin{cases} R(t) = 1 - (1 - \mathrm{e}^{-\lambda t})^n \\ \mathrm{MTTF} = \int_0^{+\infty} \left[1 - (1 - \mathrm{e}^{-\lambda t})^n\right]\mathrm{d}t = \sum_{i=1}^{n} \frac{1}{i\lambda}. \end{cases} \tag{2.1.10}$$

2.1.3 The Series–Parallel System

The system shown in Fig. 2.3 is called the series–parallel system. If the reliability of component is $R_{ij}(t),\ i = 1, 2, \ldots, n,\ j = 1, 2, \ldots, m_i$, the lifetimes of all components are independent of each other.

According to Eqs. (2.1.1) and (2.1.6), the reliability of the series–parallel system is

$$R(t) = \prod_{i=1}^{n} \left\{ 1 - \prod_{j=1}^{m_i} \left[1 - R_{ij}(t)\right] \right\}. \tag{2.1.11}$$

If $R_{ij}(t) = R_0(t),\ m_i = m$, we have

$$R(t) = \left\{1 - [1 - R_0(t)]^m\right\}^n. \tag{2.1.12}$$

In particular, if $R_0(t) = \mathrm{e}^{-\lambda t}$, we can arrive at

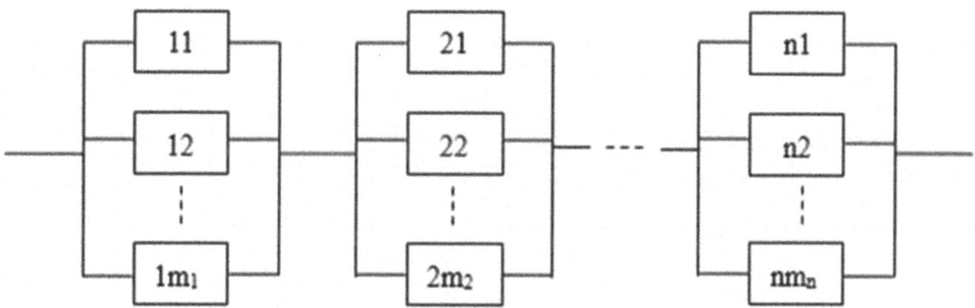

Fig. 2.3 Series–parallel system

$$\begin{cases} R(t) = \left\{1 - \left[1 - e^{-\lambda t}\right]^m\right\}^n \\ \text{MTTF} = \frac{1}{\lambda}\sum_{j=1}^{n}(-1)^j\binom{n}{j}\sum_{k=1}^{m_j}(-1)^k\binom{m_j}{k}\frac{1}{k}. \end{cases} \quad (2.1.13)$$

2.1.4 The Parallel–Series System

The system shown in Fig. 2.4 is called the parallel–series system. If the reliability of component is $R_{ij}(t),\ i = 1, 2, \ldots, n,\ j = 1, 2, \ldots, m_i$, the lifetimes of all components are independent of each other.

According to Eqs. (2.1.1) and (2.1.6), the reliability of the parallel–series system is

$$R(t) = 1 - \prod_{i=1}^{n}\left[1 - \prod_{j=1}^{m_j} R_{ij}(t)\right]. \quad (2.1.14)$$

If $R_{ij}(t) = R_0(t),\ m_i = m$, we have

$$R(t) = 1 - \left[1 - R_0^m(t)\right]^n. \quad (2.1.15)$$

In particular, if $R_0(t) = e^{-\lambda t}$, we can arrive at

$$\begin{cases} R(t) = 1 - \left[1 - e^{-m\lambda t}\right]^n \\ \text{MTTF} = \frac{1}{m\lambda}\sum_{i=1}^{n}\frac{1}{i}. \end{cases} \quad (2.1.16)$$

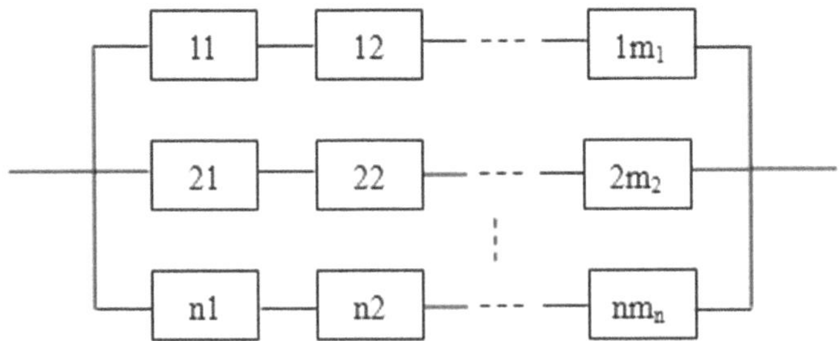

Fig. 2.4 Parallel–series system

2.2 Cold Standby Systems

2.2.1 The Perfect Conversion Switch Case

Consider the cold standby system consisting of n components. At the initial moment, one component starts to work and the other $n-1$ components are in cold standby. When the functioning component fails, the components in standby are replaced one by one until all components fail and, the cold standby system fails. The cold standby system means that the components in standby do not fail or deteriorate, and the standby period has no effect on the lifetimes of components in future use. Figure 2.5 shows the reliability block diagram of the cold standby system. When the failed component is replaced by the standby component, the conversion switch is completely reliable and the transfer is instantaneous.

Let $X_1, X_2, \ldots, X_n$ be the lifetimes of the n components. We also assume $X_1, X_2, \ldots, X_n$ are independent. Then, the lifetime of the cold standby system is

$$X = X_1 + X_2 + \cdots + X_n.$$

The distribution of the lifetime of the cold standby system is

$$F(t) = \Pr\{ X_1 + \cdots + X_n \leq t\} = F_1(t) * F_2(t) * \cdots * F_n(t),$$

where $F_i(t)$ is the distribution of the lifetime of component i, $i = 1, 2, \ldots, n$, and $*$ represents the convolution operator. Therefore, the reliability of the cold standby system is

$$R(t) = 1 - F_1(t) * F_2(t) * \cdots * F_n(t). \tag{2.2.1}$$

The MTTF of the cold standby system is

$$\text{MTTF} = E[X_1 + X_2 + \cdots + X_n] = \sum_{i=1}^{n} EX_i. \tag{2.2.2}$$

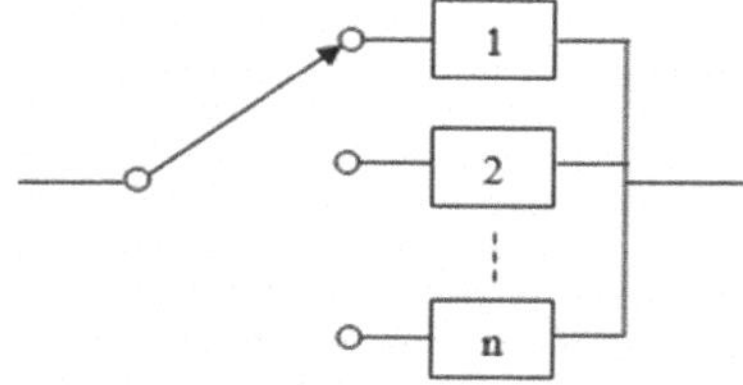

Fig. 2.5 Cold standby system

If $F_i(t) = 1 - e^{-\lambda t}$, $i = 1, 2, \ldots, n$, the lifetime of the cold standby system is the sum of the n independent and identically distributed random variables. Therefore, the reliability and MTTF of the cold standby system are

$$\begin{cases} R(t) = e^{-\lambda t} \sum\limits_{k=0}^{n-1} \frac{(\lambda t)^k}{k!} \\ \text{MTTF} = \frac{n}{\lambda} \end{cases}. \tag{2.2.3}$$

If $F_i(t) = 1 - e^{-\lambda_i t}$, $i = 1, 2, \ldots, n$ and $\lambda_1, \lambda_2, \ldots, \lambda_n$ are not equal to each other, denote the Laplace-Stieltjes transform of the distribution of the lifetime of the system $F(t)$ by

$$\hat{F}(s) = \int_0^{+\infty} e^{-st} dF(t),\ s \ge 0,$$

then

$$\begin{aligned} \hat{F}(s) &= E\{e^{-sX}\} = E\{e^{-s(X_1 + \cdots + X_n)}\} \\ &= \prod_{i=1}^{n} E\{e^{-sX_i}\} = \prod_{i=1}^{n} \int_0^{+\infty} e^{-st} dF_i(t) \\ &= \prod_{i=1}^{n} \frac{\lambda_i}{s + \lambda_i} = \sum_{i=1}^{n} c_i \frac{\lambda_i}{s + \lambda_i},\ s \ge 0, \end{aligned} \tag{2.2.4}$$

where

$$c_i = \prod_{\substack{k=1 \\ k \ne i}}^{n} \frac{\lambda_k}{\lambda_k - \lambda_i},\ i = 1, 2, \ldots, n. \tag{2.2.5}$$

By inversion of Laplace–Stieltjes transform on both sides of (2.2.4), we can obtain

$$F(t) = \sum_{i=1}^{n} c_i (1 - e^{-\lambda_i t}) = \sum_{i=1}^{n} c_i - \sum_{i=1}^{n} c_i e^{-\lambda_i t}.$$

Since $t \to +\infty$, we have $F(t) \to 1$, then $\sum_{i=1}^{n} c_i = 1$. The reliability and MTTF of the cold standby system are

$$\begin{cases} R(t) = \sum_{i=1}^{n}\left[\prod_{\substack{k=1\\k\neq i}}^{n}\frac{\lambda_k}{\lambda_k-\lambda_i}\right]\mathrm{e}^{-\lambda_i t} \\ \mathrm{MTTF} = \sum_{i=1}^{n}\frac{1}{\lambda_i}. \end{cases} \tag{2.2.6}$$

In particular, if the cold standby system consists of two components, we have

$$\begin{cases} R(t) = \frac{\lambda_2}{\lambda_2-\lambda_1}\mathrm{e}^{-\lambda_1 t} + \frac{\lambda_1}{\lambda_1-\lambda_2}\mathrm{e}^{-\lambda_2 t} \\ \mathrm{MTTF} = \frac{1}{\lambda_1}+\frac{1}{\lambda_2}. \end{cases} \tag{2.2.7}$$

2.2.2 *The Imperfect Conversion Switch Case: The Lifetime of the Conversion Switch Has 0–1 Distribution*

In practical problems, the conversion switch may fail, so the quality of the conversion switch is an important factor that affects the reliability of the cold standby system. The impact of the conversion switch on the cold standby system may have various types. Then we discuss the imperfect conversion switch cases.

Consider the cold standby system consisting of n components and a conversion switch. At the initial moment, one component is in operating and the remaining components are in cold standby. When the operating component fails, the conversion switch will convert to the component in cold standby immediately. We assume the conversion switch is imperfect and its lifetime has 0–1 distribution. That is, when the conversion switch is used, the probability of "the conversion switch is normal" is p, and the probability of "the conversion switch is fail" is $q = 1 - p$. The cold standby system fails in one of the following two situations:

(1) When the operating component fails, the conversion switch fails when it is used, and the cold standby system fails at this moment.

(2) Under the condition that the conversion switch is still normal after it has been used for $n - 1$ times, the cold standby system fails when all n components fail.

Denote that the lifetimes of the n components by $X_1, X_2, \ldots, X_n$, and suppose all components have independent and identical exponential distribution $1 - \mathrm{e}^{-\lambda t}$ and are independent of the lifetime of the conversion switch. To obtain the reliability of the cold standby system, we introduce a variable v:

$$v = \begin{cases} j, & \text{if the conversion switch fails in } j\text{th swiching}, j = 1, 2, \ldots, n-1 \\ n, & \text{if the conversion switch is still normal until no perfect component left.} \end{cases}$$

By the definition of v, it is easy to see that

$$\Pr\{v = j\} = p^{j-1}q,\ j = 1, 2, \ldots, n-1,$$
$$\Pr\{v = n\} = p^{n-1}.$$

Since

$$\sum_{j=1}^{n} \Pr\{\ v = j\} \ = 1,$$

it is easy to see that v is a random variable, we can arrive at

$$\begin{aligned} E\{v\} &= \sum_{j=1}^{n} j \Pr\{v = j\} \\ &= \sum_{j=1}^{n-1} jp^{j-1}q + np^{n-1} = \frac{1}{q}(1 - p^n). \end{aligned} \tag{2.2.8}$$

The lifetime of the cold standby system can be expressed as a random variable X

$$X \ = \ X_1 + X_2 + \cdots + X_v.$$

Since the lifetime of the conversion switch is independent of $X_1, X_2, \ldots, X_n$, then the reliability of the cold standby system is

$$\begin{aligned} R(t) &= \Pr\{X_1 + X_2 + \cdots + X_v > t\} \\ &= \sum_{j=1}^{n} \Pr\{X_1 + X_2 + \cdots + X_v > t | v = j\} \Pr\{v = j\} \\ &= \sum_{j=1}^{n-1} \Pr\{X_1 + X_2 + \cdots + X_j > t\} p^{j-1} q \\ &+ \Pr\{X_1 + X_2 + \cdots + X_n > t\} p^{n-1}. \end{aligned} \tag{2.2.9}$$

It follows from formula (2.2.3) that

$$\Pr\{X_1 + X_2 + \cdots + X_j > t\} = \sum_{i=0}^{j-1} \frac{(\lambda t)^i}{i!} \mathrm{e}^{-\lambda t}, j = 1, 2, \ldots, n. \tag{2.2.10}$$

By Eqs. (2.2.9) and (2.2.10), we can arrive at the reliability of the cold standby system

$$
\begin{aligned}
R(t) &= \sum_{j=1}^{n-1} p^{j-1} q \sum_{i=0}^{j-1} \frac{(\lambda t)^i}{i!} \mathrm{e}^{-\lambda t} + p^{n-1} \sum_{i=0}^{n-1} \frac{(\lambda t)^i}{i!} \mathrm{e}^{-\lambda t} \\
&= \sum_{j=0}^{n-2} p^j q \sum_{i=0}^{j} \frac{(\lambda t)^i}{i!} \mathrm{e}^{-\lambda t} + p^{n-1} \sum_{i=0}^{n-1} \frac{(\lambda t)^i}{i!} \mathrm{e}^{-\lambda t} \\
&= \sum_{i=0}^{n-2} \frac{(\lambda t)^i}{i!} \mathrm{e}^{-\lambda t} q \sum_{j=i}^{n-2} p^j + p^{n-1} \sum_{i=0}^{n-1} \frac{(\lambda t)^i}{i!} \mathrm{e}^{-\lambda t} \\
&= \sum_{i=0}^{n-1} \frac{(\lambda p t)^i}{i!} \mathrm{e}^{-\lambda t}.
\end{aligned} \tag{2.2.11}
$$

Since $X_1, X_2, \ldots, X_n$ and v are independent, by formula (2.2.8), the MTTF of the cold standby system is

$$
\begin{aligned}
\mathrm{MTTF} &= E\{X_1 + X_2 + \cdots + X_v\} \\
&= \sum_{j=1}^{n} E\{X_1 + X_2 + \cdots + X_v | v = j\} \Pr\{v = j\} \\
&= \sum_{j=1}^{n} j E\{X_1\} \Pr\{v = j\} \\
&= \frac{1}{\lambda} E\{v\} = \frac{1}{\lambda q}(1 - p^n).
\end{aligned} \tag{2.2.12}
$$

If the failure rates of components are different, the reliability and MTTF can be obtained similarly, but the expression is more complicated. We only give the following results for the two component case, if

$$
\Pr\{v = j\} = \begin{cases} q, & \text{if } j = 1 \\ p, & \text{if } j = 2, \end{cases}
$$

then we have

$$
\begin{aligned}
R(t) &= \Pr\left\{ \sum_{j=1}^{v} X_j > t \right\} \\
&= q \Pr\{X_1 > t\} + p \Pr\{X_1 + X_2 > t\} \\
&= q\mathrm{e}^{-\lambda_1 t} + p\left(\frac{\lambda_2}{\lambda_2 - \lambda_1} \mathrm{e}^{-\lambda_1 t} + \frac{\lambda_1}{\lambda_1 - \lambda_2} \mathrm{e}^{-\lambda_2 t} \right) \\
&= \mathrm{e}^{-\lambda_1 t} + \frac{p\lambda_1}{\lambda_1 - \lambda_2} (\mathrm{e}^{-\lambda_2 t} - \mathrm{e}^{-\lambda_1 t})
\end{aligned}
$$

and

$$\mathrm{MTTF} = \frac{1}{\lambda_1} + p\frac{1}{\lambda_2}.$$

If $p = 1$, that is just the perfect conversion switch case, all the results are consistent with the results in the perfect conversion switch case.

2.2.3 *The Imperfect Conversion Switch Case: The Lifetime of the Conversion Switch Has an Exponential Distribution*

Suppose that the lifetime of the conversion switch X_K follows an exponential distribution with parameter λ_K and is independent of the lifetimes of components. The remaining assumptions are the same as those in Sect. 2.2.2. In addition, the influence of conversion switch on the lifetime of the cold standby system may have two different forms:

(1) When the conversion switch fails, the cold standby system fails immediately. Obviously, the lifetime of the cold standby system is

$$X = \min[X_1 + X_2 + \cdots + X_n, X_K].$$

Therefore, the reliability and MTTF of the cold standby system are

$$\begin{aligned} R(t) &= \Pr\{\min[X_1 + X_2 + \cdots + X_n, X_K] > t\} \\ &= \Pr\{X_K > t\}\Pr\{X_1 + X_2 + \cdots + X_n > t\} \\ &= \mathrm{e}^{-\lambda_K t} \sum_{k=0}^{n-1} \frac{(\lambda t)^k}{k!} \mathrm{e}^{-\lambda t} \\ &= \mathrm{e}^{-(\lambda+\lambda_K)t} \sum_{k=0}^{n-1} \frac{(\lambda t)^k}{k!} \end{aligned} \tag{2.2.13}$$

and

$$\begin{aligned} \mathrm{MTTF} &= \int_0^{+\infty} R(t)\mathrm{d}t \\ &= \sum_{k=0}^{n-1} \frac{\lambda^k}{k!} \int_0^{+\infty} t^k \mathrm{e}^{-(\lambda+\lambda_K)t} \mathrm{d}t \end{aligned}$$

$$= \sum_{k=0}^{n-1} \frac{\lambda^k}{(\lambda + \lambda_K)^{k+1}} \cdot \frac{1}{k!} \int_0^{+\infty} x^k e^{-x} dx$$

$$= \frac{1}{\lambda + \lambda_K} \sum_{k=0}^{n-1} \left(\frac{\lambda}{\lambda + \lambda_K} \right)^k$$

$$= \frac{1}{\lambda_K} \left[1 - \left(\frac{\lambda}{\lambda + \lambda_K} \right)^n \right]. \tag{2.2.14}$$

respectively.

(2) When the conversion switch fails, the cold standby system does not fail immediately. That is, when the operating component fails and requires the conversion switch, the cold standby system fails due to the conversion switch failure.

For convenience, we only consider the two-component case. Suppose that the lifetimes of the two components (denoted by X_1 and X_2) and the lifetime of the conversion switch (denoted by X_K) follow exponential distributions with parameters λ_1, λ_2 and λ_K, respectively, and they are independent of each other. At the initial moment, component 1 starts into operation and component 2 is in cold standby.

When component 1 fails and requires the conversion switch, the cold standby system fails ($X_K < X_1$); if the conversion switch has failed at this time, then the lifetime of the cold standby system is X_1.

When component 1 fails and the conversion switch is normal ($X_K > X_1$), it is replaced by component 2 and enters the operating state. The cold standby system fails until component 2 fails. At this moment, the lifetime of the cold standby system is $X_1 + X_2$.

From the above description, we can see that the lifetime of the cold standby system X is

$$X = X_1 + X_2 \cdot I_{\{X_K > X_1\}}, \tag{2.2.15}$$

where $I_{\{X_K > X_1\}}$ is the indicator function of the random event $\{X_K > X_1\}$, i.e.,

$$I_{\{X_K > X_1\}} = \begin{cases} 1, & \text{if} X_K > X_1 \\ 0, & \text{if} X_K \le X_1 \end{cases}.$$

Thus

$$\begin{aligned} 1 - R(t) &= \Pr\{X \le t\} \\ &= \Pr\{X_1 \le t, X_K \le X_1\} + \Pr\{X_1 + X_2 \le t, X_K > X_1\} \\ &= \int_0^t \Pr\{X_K \le u\} d \Pr\{X_1 \le u\} + \int_0^t \Pr\{X_2 \le t - u, X_K > u\} d \Pr\{X_1 \le u\} \end{aligned}$$

$$= \int_0^t (1 - e^{-\lambda_K u})\lambda_1 e^{-\lambda_1 u} du + \int_0^t (1 - e^{-\lambda_2 (t-u)}) e^{-\lambda_K u} \lambda_1 e^{-\lambda_1 u} du$$

$$= 1 - e^{-\lambda_1 t} - \frac{\lambda_1}{\lambda_K + \lambda_1 - \lambda_2}\left[e^{-\lambda_2 t} - e^{-(\lambda_1 + \lambda_K)t}\right].$$

Then the reliability and MTTF of the cold standby system are

$$\begin{cases} R(t) = e^{-\lambda_1 t} + \dfrac{\lambda_1}{\lambda_K + \lambda_1 - \lambda_2}\left[e^{-\lambda_2 t} - e^{-(\lambda_1 + \lambda_K)t}\right] \\ \text{MTTF} = \dfrac{1}{\lambda_1} + \dfrac{\lambda_1}{\lambda_2(\lambda_1 + \lambda_K)}. \end{cases} \tag{2.2.16}$$

In particular, if $\lambda_K = 0$, Eqs. (2.2.13), (2.2.14) and (2.2.16) are consistent with the results in perfect conversion switch case in Sect. 2.2.1.

2.3 Warm Standby Systems

2.3.1 *The Perfect Conversion Switch Case*

The difference between the warm standby system and the cold standby system is that the components may fail during the standby period. In addition, the lifetime of component in standby period is different from that in operating period.

Consider the warm standby system composed of n identical components. The lifetime of component in operating and the lifetime of component in standby follow exponential distributions with parameters λ and μ, respectively. At the initial moment, one component is in operation and the other components are in warm standby, which means all components are subject to failure. When the operating component fails, it is replaced by the component in standby until all standby components fail and the warm standby system fails at this moment. We assume that

(1) The conversion switch is completely reliable and the transfer is instantaneous.
(2) The lifetimes of components in operating are irrelevant to how long they have been in standby, which follow the exponential distribution $1 - e^{-\lambda t}$, $t \geq 0$.
(3) The lifetimes in operating and the lifetimes in warm standby of components are independent of each other.

In order to obtain the reliability and MTTF of the warm standby system, we denote the failure time of the ith failed component by S_i, $i = 1, 2, \ldots, n$, and $S_0 = 0$. Obviously,

$$S_n = \sum_{i=1}^{n} (S_i - S_{i-1})$$

is the failure time of the warm standby system. In time interval (S_{i-1}, S_i), $i-1$ components have failed, and another $n-i+1$ components are normal in which one component is in operating and the other $n-i$ components are in warm standby. Since the exponential distribution is memoryless, $S_i - S_{i-1}$ follow the exponential distributions with parameters $\lambda+(n-i)\mu, i = 1, 2, \ldots, n$, which are independent of each other. Therefore, the warm standby system is equivalent to a cold standby system composed of n independent components. In the equivalent cold standby system, the lifetime of component i follows the exponential distribution with parameter $\lambda_i = \lambda + (n-i)\mu$, $i = 1, 2, \ldots, n$. If $\mu > 0$, by formula (2.2.6), we have

$$\begin{cases} R(t) = \Pr\{S_n > t\} \\ \quad = \sum\limits_{i=1}^{n} \left[\prod\limits_{\substack{k=1 \\ k\neq i}}^{n} \frac{\lambda+(n-k)\mu}{(i-k)\mu} \right] e^{-[\lambda+(n-i)\mu]t} \\ \quad = \sum\limits_{i=0}^{n-1} \left[\prod\limits_{\substack{k=0 \\ k\neq i}}^{n-1} \frac{\lambda+k\mu}{(k-i)\mu} \right] e^{-(\lambda+i\mu)t} \\ \text{MTTF} = \sum\limits_{i=1}^{n} \frac{1}{\lambda_i} = \sum\limits_{i=1}^{n} \frac{1}{\lambda+(n-i)\mu} = \sum\limits_{i=0}^{n-1} \frac{1}{\lambda+i\mu}. \end{cases} \tag{2.3.1}$$

If $\mu = 0$, the formula (2.2.3) can be used directly to obtain the reliability and MTTF of the warm standby system. If $\mu = \lambda$, the warm standby system changes to a parallel system composed of n identical components. In this case, (2.3.1) changes to (2.1.10).

If the distributions of the lifetimes of components are different, the reliability of the warm standby system is very complicated. We only discuss the two-component case. At the initial moment, component 1 is in operating and component 2 is in warm standby. The lifetimes of component 1 and component 2 are denoted by X_1 and X_2, respectively, and the lifetime of component 2 in warm standby is denoted by Y_2. We also assume that X_1, X_2 and Y_2 follow exponential distributions with parameters λ_1, λ_2 and μ, respectively. The lifetime of the warm standby system is

$$X = X_1 + X_2 \cdot I_{\{Y_2 > X_1\}}.$$

Therefore, the reliability and MTTF of the warm standby system are

$$\begin{cases} R(t) = \mathrm{e}^{-\lambda_1 t} + \frac{\lambda_1}{\lambda_1-\lambda_2+\mu}\left[\mathrm{e}^{-\lambda_2 t} - \mathrm{e}^{-(\lambda_1+\mu)t}\right] \\ \text{MTTF} = \frac{1}{\lambda_1} + \frac{1}{\lambda_2}\left(\frac{\lambda_1}{\lambda_1+\mu}\right). \end{cases} \tag{2.3.2}$$

2.3.2 The Imperfect Conversion Switch Case: The Lifetime of the Conversion Switch Has 0–1 Distribution

Consider the conversion switch is not completely reliable and the lifetime of the conversion switch has 0–1 distribution. That is, when we use the conversion switch, the probability of "the switch is normal" is p. For convenience, we only consider the two nonidentical component cases, and the remaining assumptions are the same as those in Sect. 2.3.1. Denote

$$X_K = \begin{cases} 1, & \text{the conversion switch is normal when the conversion switch is used} \\ 0, & \text{the conversion switch is failed when the conversion switch is used.} \end{cases}$$

Then the lifetime of the warm standby system can be expressed as

$$X = X_1 + X_2 \cdot I_{\{Y_2 > X_1\}} \cdot I_{\{X_K = 1\}}.$$

Therefore, the reliability of the warm standby system can be obtained by using the total probability formula and independence of random variables

$$\begin{aligned} R(t) &= \Pr\{X > t\} \\ &= \Pr\{X_1 > t, X_K = 0\} + \Pr\left\{X_1 + X_2 \cdot I_{\{Y_2 > X_1\}} > t, X_K = 1\right\} \\ &= q\Pr\{X_1 > t\} + p\Pr\left\{X_1 + X_2 \cdot I_{\{Y_2 > X_1\}} > t\right\} \\ &= \mathrm{e}^{-\lambda_1 t} + p\frac{\lambda_1}{\lambda_1 - \lambda_2 + \mu}\left[\mathrm{e}^{-\lambda_2 t} - \mathrm{e}^{-(\lambda_1 + \mu)t}\right], \end{aligned} \tag{2.3.3}$$

the last step of this equation is given by Eq. (2.3.2). The MTTF of the warm standby system is

$$\mathrm{MTTF} = \frac{1}{\lambda_1} + p\frac{\lambda_1}{\lambda_2(\lambda_1 + \mu)}.$$

2.3.3 The Imperfect Conversion Switch Case: The Lifetime of the Conversion Switch Has an Exponential Distribution

Consider the lifetime of the conversion switch X_K follows an exponential distribution with parameter λ_K and is independent of the lifetimes of components. The other assumptions are the same as those in Sect. 2.3.2. At this moment, the effect of the conversion switch on the warm standby system also has two different forms:

(1) When the conversion switch fails, the warm standby system fails immediately. In this case, the lifetime of the warm standby system is

$$X = \min\left\{X_1 + I_{\{Y_2 > X_1\}} \cdot X_2,\ X_K\right\}.$$

According to Eq. (2.3.2), the reliability and MTTF of the warm standby system can be obtained immediately, i.e.,

$$\begin{cases} R(t) = \Pr\{X_K > t\}\Pr\left\{X_1 + X_2 \cdot I_{\{Y_2 > X_1\}} > t\right\} \\ \quad = \mathrm{e}^{-\lambda_K t}\left\{\mathrm{e}^{-\lambda_1 t} + \dfrac{\lambda_1}{\lambda_1 - \lambda_2 + \mu}\left[\mathrm{e}^{-\lambda_2 t} - \mathrm{e}^{-(\lambda_1 + \mu)t}\right]\right\} \\ \mathrm{MTTF} = \dfrac{1}{\lambda_1 + \lambda_K} + \dfrac{\lambda_1}{(\lambda_2 + \lambda_K)(\lambda_1 + \mu + \lambda_K)}. \end{cases} \tag{2.3.4}$$

(2) When the conversion switch fails, the warm standby system does not fail immediately. As the operating component fails and requires the conversion switch, the system fails due to the conversion switch failure. Denote the lifetime of conversion switch by X_K, and the lifetime of the warm standby system is

$$X = X_1 + X_2 \cdot I_{\{Y_2 > X_1\}} \cdot I_{\{X_K > X_1\}}.$$

Then

$$\begin{aligned} 1 - R(t) &= \Pr\{\, X \le t\} \\ &= \Pr\{\, X \le t,\, Y_2 < X_1\} + \Pr\{\, X \le t,\, Y_2 > X_1, X_K < X_1\} \\ &\quad + \Pr\{X \le t,\, Y_2 > X_1, X_K > X_1\} \\ &= \Pr\{X_1 \le t,\, Y_2 < X_1\} + \Pr\{\, X_1 \le t,\, Y_2 > X_1, X_K < X_1\} \\ &\quad + \Pr\{X_1 + X_2 \le t,\, Y_2 > X_1, X_K > X_1\} \\ &= \int_0^t (1 - \mathrm{e}^{-\mu u})\lambda_1 \mathrm{e}^{-\lambda_1 u}\mathrm{d}u + \int_0^t \mathrm{e}^{-\mu u}(1 - \mathrm{e}^{-\lambda_K u})\lambda_1 \mathrm{e}^{-\lambda_1 u}\mathrm{d}u \\ &\quad + \int_0^t (1 - \mathrm{e}^{-\lambda_2 (t-u)})\mathrm{e}^{-\mu u}\mathrm{e}^{-\lambda_K u}\lambda_1 \mathrm{e}^{-\lambda_1 u}\mathrm{d}u \\ &= 1 - \mathrm{e}^{-\lambda_1 t} - \frac{\lambda_1}{\lambda_1 + \lambda_K + \mu - \lambda_2}\left[\mathrm{e}^{-\lambda_2 t} - \mathrm{e}^{-(\lambda_1 + \lambda_K + \mu)t}\right]. \end{aligned}$$

Therefore, the reliability and MTTF of the warm standby system are

$$\begin{cases} R(t) = \mathrm{e}^{-\lambda_1 t} + \frac{\lambda_1}{\lambda_1 + \lambda_K + \mu - \lambda_2}\left[\mathrm{e}^{-\lambda_2 t} - \mathrm{e}^{-(\lambda_1 + \lambda_K + \mu)t}\right] \\ \mathrm{MTTF} = \frac{1}{\lambda_1} + \frac{\lambda_1}{\lambda_2(\lambda_1 + \lambda_K + \mu)}. \end{cases} \tag{2.3.5}$$

2.4 Shock Models and the Bivariate Exponential Distribution

The lifetime distribution can be extended to multiple dimensions. For convenience, we only discuss the bivariate exponential distribution, which can be derived from the fatal shock model.

2.4.1 *Shock Models*

Assume that the system is composed of two components, which may fail after being subjected to external shocks. There are three independent Poisson processes P_1, P_2 and P_3 that control the occurrence of shocks, and their strengths are λ_1, λ_2 and λ_{12}, respectively. If the shock is caused by P_1, it only leads to the failure of component 1 with probability p_1. If the shock is caused by P_2, it only leads to the failure of component 2 with probability p_2. If the shock is caused by P_3, it has an impact on both components. It leads to the failure of only component 1 or component 2 with probabilities p_{01} or p_{10}, respectively; leads to the failure of both components with probability p_{00}; and does not lead to the failure of any component with probability p_{11}. Clearly, $p_{00} + p_{01} + p_{10} + p_{11} = 1$.

Let X and Y be the lifetimes of component 1 and component 2, respectively. Denote

$$\overline{F}(x, y) = \Pr\{X > x, Y > y\}, \quad x, y \geq 0. \tag{2.4.1}$$

Then, we use the properties of the Poisson process to derive the expression of $\overline{F}(x, y)$. Let $N_i(s, t)$ be the number of shocks caused by Poisson process P_i, $i = 1, 2, 3$ in $(s, t]$. If $0 \leq x \leq y$, we have

$$
\begin{aligned}
\overline{F}(x, y) = \Pr\{&N_1(0, x) = i, \\
&i \text{ shocks do not cause the failure of component 1}, \ i = 0, 1, \ldots; \\
&N_2(0, y) = j, \\
&j \text{ shocks do not cause the failure of component 2}, \ j = 0, 1, \ldots; \\
&N_3(0, x) = k, \\
&k \text{ shocks do not cause the failure of both component 1 and 2}, \ k = 0, 1, \ldots; \\
&N_3(x, y) = l, \\
&l \text{ shocks do not cause the failure of component 2}, \ l = 0, 1, \ldots\} \\
= &\sum_{i=0}^{\infty} \frac{(\lambda_1 x)^i}{i!} e^{-\lambda_1 x}(1 - p_1)^i \cdot \sum_{j=0}^{\infty} \frac{(\lambda_2 y)^j}{j!} e^{-\lambda_2 y}(1 - p_2)^j \\
&\cdot \sum_{k=0}^{\infty} \frac{(\lambda_{12} x)^k}{k!} e^{-\lambda_{12} x} p_{11}^k \cdot \sum_{l=0}^{\infty} \frac{[\lambda_{12}(y - x)]^l}{l!} e^{-\lambda_{12}(y-x)} (p_{01} + p_{11})^l \\
= &\exp\{-\lambda_1 p_1 x - \lambda_2 p_2 y - \lambda_{12}(1 - p_{11})x - \lambda_{12}(p_{00} + p_{10})(y - x)\} \\
= &\exp\{-(\lambda_1 p_1 + \lambda_{12} p_{01})x - [\lambda_2 p_2 + \lambda_{12}(p_{00} + p_{10})]y\}.
\end{aligned}
\tag{2.4.2}
$$

If $0 \le y \le x$, we have

$$\overline{F}(x, y) = \exp\{-[\lambda_1 p_1 + \lambda_{12}(p_{00} + p_{01})]x - (\lambda_2 p_2 + \lambda_{12} p_{10})y\}. \tag{2.4.3}$$

By Eqs. (2.4.2) and (2.4.3), we can obtain

$$\overline{F}(x, y) = \Pr\{X > x, Y > y\} = \exp\{-\delta_1 x - \delta_2 y - \delta_{12} \max(x, y)\}. \tag{2.4.4}$$

$$\delta_1 = \lambda_1 p_1 + \lambda_{12} p_{01}, \quad \delta_2 = \lambda_2 p_2 + \lambda_{12} p_{10}, \quad \delta_{12} = \lambda_{12} p_{00}. \tag{2.4.5}$$

In particular, if $p_1 = p_2 = p_{00} = 1$, the occurrence of each shock will cause the failure of corresponding component. Based on that, we can arrive at

$$\overline{F}(x, y) = \exp\{-\lambda_1 x - \lambda_2 y - \lambda_{12} \max(x, y)\}. \tag{2.4.6}$$

The case of formula (2.4.6) is called the fatal shock model, and the case of formula (2.4.4) is called the non-fatal shock model. Then, the bivariate exponential distribution is derived from the fatal shock model.

2.4.2 The Bivariate Exponential Distribution

Definition 2.4.1 If the random variable (X, Y) has joint survival probability

$$\overline{F}(x, y) = \exp\{-\lambda_1 x - \lambda_2 y - \lambda_{12} \max(x, y)\},\ \ x, y \geq 0, \tag{2.4.7}$$

then (X, Y) is called bivariate exponential distribution with parameters $\lambda_1,\ \lambda_2$ and λ_{12}, denoted by $(X, Y) \sim BVE(\lambda_1, \lambda_2, \lambda_{12})$.

The fatal shock model can give (X, Y) an intuitive meaning. Consider a system composed by two components which affected by three independent sources of shocks. The shock caused by P_1 only damages component 1, which occurs at random time U_1 and

$$\Pr\{\ U_1 > t\}\ = \mathrm{e}^{-\lambda_1 t}.$$

The shock caused by P_2 only damages component 2, which occurs at random time U_2 and

$$\Pr\{\ U_2 > t\}\ = \mathrm{e}^{-\lambda_2 t}.$$

The shock caused by P_3 damages both component 1 and component 2, which occurs at time U_{12} and

$$\Pr\{U_{12} > t\} = \mathrm{e}^{-\lambda_{12} t}.$$

Therefore, the lifetimes of component 1 and component 2 are

$$X = \min(U_1, U_{12}) \text{ and } Y = \min(U_2, U_{12}), \tag{2.4.8}$$

respectively. The joint survival probability of (X, Y) is $\overline{F}(x, y)$ in Eq. (2.4.7).

The marginal distributions of X and Y can be easily arrived from Definition 2.4.1; then,

$$\begin{cases} \overline{F}_1(x) = \Pr\{X > x\} = \mathrm{e}^{-(\lambda_1+\lambda_{12})x},\ x \geq 0 \\ \overline{F}_2(y) = \Pr\{Y > y\} = \mathrm{e}^{-(\lambda_2+\lambda_{12})y},\ y \geq 0, \end{cases} \tag{2.4.9}$$

It is easy to see that the marginal distributions of X and Y have exponential distributions. It follows from Eq. (2.4.7) that the joint distribution function of (X, Y) is.

$$F(x, y) = \Pr\{X \leq x, Y \leq y\} = 1 - \overline{F}_1(x) - \overline{F}_2(y) + \overline{F}(x, y). \tag{2.4.10}$$

Then we can arrive at

$$EX = \frac{1}{\lambda_1 + \lambda_{12}}, \quad \text{Var}X = \frac{1}{(\lambda_1 + \lambda_{12})^2},$$
$$EY = \frac{1}{\lambda_2 + \lambda_{12}}, \quad \text{Var}Y = \frac{1}{(\lambda_2 + \lambda_{12})^2},$$

$$E(XY) = \frac{1}{\lambda}\left(\frac{1}{\lambda_1 + \lambda_{12}} + \frac{1}{\lambda_2 + \lambda_{12}}\right),$$

where

$$\lambda = \lambda_1 + \lambda_2 + \lambda_{12}.$$

Therefore, the correlation coefficient of X and Y is

$$\rho = \text{Cov}(X, Y) = \frac{\lambda_{12}}{\lambda}.$$

The moment function of X and Y is

$$Ee^{-(sX+tY)} = \frac{(\lambda + s + t)(\lambda_1 + \lambda_{12})(\lambda_2 + \lambda_{12}) + \lambda_{12}st}{(\lambda + s + t)(\lambda_1 + \lambda_{12} + s)(\lambda_2 + \lambda_{12} + t)}.$$

For the exponential distribution, one of the most significant properties is "memoryless". That is, when the lifetime of a product follows an exponential distribution, if it is normal after it has been used for time t, the residual lifetime of the product is like a new one and follows the original exponential distribution. It can be expressed in mathematical form as follows.

Theorem 2.4.1 *If X is a nonnegative and nondegenerate random variable, $F(t)$ is the distribution function of X, then the sufficient and necessary conditions of*

$$F(t) = 1 - e^{-\lambda t}, \ \lambda > 0, \ t \geq 0$$

are

$$\Pr\{X > s + t | X > t\} = \Pr\{X > s\}, \ \forall s, t \geq 0.$$

Then, we will show that the bivariate exponential distribution also has this property.

Lemma 2.4.1 *The bivariate exponential distribution defined by formula (2.4.7) has the memoryless property; that is, for any $x, y, t \geq 0$, there is*

$$\Pr\{X > x + t, Y > y + t | X > t, Y > t\} = \Pr\{X > x, Y > y\} \qquad (2.4.11)$$

or

$$\overline{F}(x+t, y+t) = \overline{F}(x, y)\overline{F}(t, t). \tag{2.4.12}$$

We can prove that $\overline{F}(x, y)$ *has the form of Eq.* (2.4.7) *if it satisfies formula* (2.4.12) *and has exponential marginal distributions. For this reason, we introduce a lemma first.*

Lemma 2.4.2 *If Eq.* (2.4.12) *holds, then for any* $\theta > 0$,

$$\overline{F}(x, y) = \begin{cases} e^{-\theta y}\overline{F}_1(x-y), & \text{if } x \geq y \\ e^{-\theta x}\overline{F}_2(y-x), & \text{if } x \leq y, \end{cases} \tag{2.4.13}$$

where $F_1(x) = F(x, +\infty)$ *and* $F_2(y) = F(+\infty, y)$ *are the marginal distributions of* X *and* Y*, respectively.*

Proof In formula (2.4.12), let $x = y = s$, there is

$$\overline{F}(s+t, s+t) = \overline{F}(s, s)\overline{F}(t, t).$$

According to Theorem 2.4.1, we can arrive at $\overline{F}(s, s) = e^{-\theta s}$. For any $\theta > 0$, let $y = 0$ in Eq. (2.4.12); we have

$$\overline{F}(x+t, t) = \overline{F}(x, 0)\overline{F}(t, t) = \overline{F}_1(x)e^{-\theta t},$$

i.e.,

$$\overline{F}(x, y) = e^{-\theta y}\overline{F}_1(x-y), \quad x \geq y.$$

The other formula can be proved in the same way. The theorem is proved.

Theorem 2.4.2 *The bivariate exponential distribution is the only bivariate distribution that satisfies formula (2.4.12) and has exponential marginal distributions.*

Proof Necessity has been proved in Lemma 2.4.1. Sufficiency is proved below. Let the joint distribution function of the bivariate nonnegative random variable (X, Y) be $F(x, y)$, and let the marginal distributions of (X, Y) be $F_1(x)$ and $F_2(y)$, and satisfy Eq. (2.4.12). By assumption, we have

$$\overline{F}_1(x) = e^{-\delta_1 x}, \ \overline{F}_2(y) = e^{-\delta_2 y}, \ x, y \geq 0, \ \delta_1, \delta_2 > 0.$$

By Lemma 2.4.2, $\overline{F}(x, y)$ can be written as

$$\overline{F}(x, y) = \begin{cases} e^{-\theta y - \delta_1(x-y)}, & x \geq y \\ e^{-\theta x - \delta_2(y-x)}, & x \leq y, \end{cases} \tag{2.4.14}$$

where $\theta > 0$. Let $\lambda_1 = \theta - \delta_2$ and $\lambda_2 = \theta - \delta_1$. Since $\overline{F}(x, y)$ is monotonically decreased to x and y, respectively, by formula (2.4.14), we can arrive at $\lambda_1 \geq 0$ and $\lambda_2 \geq 0$. Let $\lambda_{12} = \delta_1 + \delta_2 - \theta$. Then, $\lambda_{12} \geq 0$ is proved below. Denote

$$G(x) = F(x, x).$$

Clearly,

$$\begin{aligned} G(x) &= F(x, x) \\ &= 1 - \overline{F}_1(x) - \overline{F}_2(x) + \overline{F}(x, x) \\ &= 1 - \mathrm{e}^{-\delta_1 x} - \mathrm{e}^{-\delta_2 x} + \mathrm{e}^{-\theta x}, \ x \geq 0, \end{aligned}$$

which is a distribution function with density function

$$g(x) = \delta_1 \mathrm{e}^{-\delta_1 x} + \delta_2 \mathrm{e}^{-\delta_2 x} - \theta \mathrm{e}^{-\theta x} \geq 0, \ x \geq 0.$$

Let $x \to 0^+$, and we have $\delta_1 + \delta_2 - \theta \geq 0$. Since $\lambda_1 = \theta - \delta_2$, $\lambda_2 = \theta - \delta_1$ and $\lambda_{12} = \delta_1 + \delta_2 - \theta$, then

$$\theta = \lambda_1 + \lambda_2 + \lambda_{12}, \ \delta_1 = \lambda_1 + \lambda_{12}, \ \delta_2 = \lambda_2 + \lambda_{12}.$$

By inserting the above formula into (2.4.14), we can arrive at Eq. (2.4.7). The theorem is proved.

Since Eq. (2.4.12) can be written as

$$\frac{\overline{F}(x + t, y + t)}{\overline{F}(x, y)} = \overline{F}(t, t), \tag{2.4.15}$$

therefore, the "memoryless" of the bivariate exponential distribution can be interpreted as "the survival probability of a series system composed of two components with ages x and y is the same as that of a new system." Therefore, the distribution of the lifetime of a series system composed by two old components whose lifetimes are exponential marginal distributions is irrelevant to the age of the components, if and only if the joint distribution of the lifetimes of the two components is a bivariate exponential distribution.

The bivariate exponential distribution has the following properties.

Theorem 2.4.3 *If* $(X, Y) \sim BVE(\lambda_1, \lambda_2, \lambda_{12})$,*then*

(1) $\Pr\{\ \min(X, Y) \leq t\} = 1 - e^{-\lambda t}, \ t \geq 0, \ \lambda = \lambda_1 + \lambda_2 + \lambda_{12}$.
(2) $\min(X, Y)$ is independent of the events $\{X < Y\}$, $\{X > Y\}$ and $\{X = Y\}$.
(3) $\min(X, Y)$ is independent of $|X - Y|$.

Proof

(1) By formula (2.4.8), $\min(X, Y) = \min(U_1, U_2, U_{12})$, the result is proved

(2) To prove $\min(X, Y)$ and $\{X < Y\}$ are independent, we should prove that for any $t \geq 0$,

$$\Pr\{\min(X, Y) > t | X < Y\} = \Pr\{\min(X, Y) > t\} = \mathrm{e}^{-\lambda t} \tag{2.4.16}$$

holds. On the other hand,

$$\begin{aligned}
&\Pr\{\min(X, Y) > t,\ X < Y\} \\
&= \Pr\{t < X < Y\} \\
&= \Pr\{t < U_1 < U_2,\ t < U_1 < U_{12}\} \\
&= \int_t^{+\infty} \Pr\{t < U_1 < U_2,\ t < U_1 < U_{12} | U_1 = u\} \mathrm{d}\Pr\{U_1 \leq u\} \\
&= \int_t^{+\infty} \Pr\{u < U_2,\ u < U_{12}\} \mathrm{d}\Pr\{U_1 \leq u\} \\
&= \int_t^{+\infty} \mathrm{e}^{-\lambda_2 u} \mathrm{e}^{-\lambda_{12} u} \lambda_1 \mathrm{e}^{-\lambda_1 u} \mathrm{d}u = \frac{\lambda_1}{\lambda} \mathrm{e}^{-\lambda t},
\end{aligned}$$

$$\begin{aligned}
\Pr\{X < Y\} &= \Pr\{U_1 < \min(U_2, U_{12})\} \\
&= \int_0^{+\infty} \Pr\{u < \min(U_2, U_{12})\} \mathrm{d}\Pr\{U_1 \leq u\} = \frac{\lambda_1}{\lambda}
\end{aligned}$$

Therefore, formula (2.4.16) holds. Other results can be proved by the similar method.

(3) We should prove that for any $t_1, t_2 \geq 0$,

$$\begin{aligned}
&\Pr\{\min(X, Y) > t_1,\ |X - Y| > t_2\} \\
&\quad = \Pr\{\min(X, Y) > t_1\} \Pr\{|X - Y| > t_2\}
\end{aligned} \tag{2.4.17}$$

holds. On the other hand,

$$\begin{aligned}
&\Pr\{\min(X, Y) > t_1,\ |X - Y| > t_2\} \\
&= \Pr\{\min(X, Y) > t_1, X - Y > t_2\} + \Pr\{\min(X, Y) > t_1, Y - X > t_2\} \\
&= \Pr\{Y > t_1, X > Y + t_2\} + \Pr\{X > t_1, Y > X + t_2\}
\end{aligned}$$

$$= \Pr\{U_2 > t_1, \min(U_1, U_{12}) > U_2 + t_2\} + \Pr\{U_1 > t_1, \min(U_2, U_{12}) > U_1 + t_2\}$$
$$= \mathrm{I} + \mathrm{II},$$

where

$$\begin{aligned}
\mathrm{I} &= \int_{t_1}^{+\infty} \Pr\{\min(U_1, U_{12}) > U_2 + t_2 | U_2 = u\} d\Pr\{U_2 \le u\} \\
&= \int_{t_1}^{+\infty} \Pr\{\min(U_1, U_{12}) > u + t_2\} \lambda_2 \mathrm{e}^{-\lambda_2 u} du \\
&= \frac{\lambda_2}{\lambda} \mathrm{e}^{-\lambda t_1} \mathrm{e}^{-(\lambda_1 + \lambda_{12}) t_2}
\end{aligned}$$

and

$$\mathrm{II} = \frac{\lambda_1}{\lambda} \mathrm{e}^{-\lambda t_1} \mathrm{e}^{-(\lambda_2 + \lambda_{12}) t_2}.$$

Then by Eq. (2.4.17), we have

$$\Pr\{\min(X, Y) > t_1, \ |X - Y| > t_2\} = \mathrm{e}^{-\lambda t_1} \left\{ \frac{\lambda_2}{\lambda} \mathrm{e}^{-(\lambda_1 + \lambda_{12}) t_2} + \frac{\lambda_1}{\lambda} \mathrm{e}^{-(\lambda_2 + \lambda_{12}) t_2} \right\}.$$

In the above equation, let t_1 and t_2 equal to 0, and we have

$$\Pr\{|X - Y| > t_2\} = \frac{\lambda_2}{\lambda} \mathrm{e}^{-(\lambda_1 + \lambda_{12}) t_2} + \frac{\lambda_1}{\lambda} \mathrm{e}^{-(\lambda_2 + \lambda_{12}) t_2}$$

and

$$\Pr\{\ \min(X, Y) > t_1\} \ = \mathrm{e}^{-\lambda t_1}.$$

Then, Eq. (2.4.17) holds. The theorem is proved.

We will discuss the decomposition of the bivariate exponential distribution $F(x, y)$. According to Lebesgue's decomposition theorem, $\overline{F}(x, y)$ can be decomposed into the absolutely continuous part F_a and the singular part F_s. For any $x, y \ge 0$, denote

$$\overline{F}(x, y) = \alpha \overline{F}_a(x, y) + (1 - \alpha) \overline{F}_s(x, y),$$

where $0 \le \alpha \le 1$. F_a is absolutely continuous, and

$$\frac{\partial^2}{\partial x \partial y} F_s(x, y) = 0$$

holds almost everywhere on the two-dimensional Lebesgue measure; the following theorem determines F_a and F_s, respectively.

Theorem 2.4.4 *If $F(x, y)$ is $BVE(\lambda_1, \lambda_2, \lambda_{12})$, then for any $x, y \geq 0$,*

$$\overline{F}(x, y) = \frac{\lambda_1 + \lambda_2}{\lambda} \overline{F}_a(x, y) + \frac{\lambda_{12}}{\lambda} \overline{F}_s(x, y), \tag{2.4.18}$$

where

$$\overline{F}_s(x, y) = \mathrm{e}^{-\lambda \max(x, y)} \tag{2.4.19}$$

is singular part, and the absolutely continuous part is

$$\overline{F}_a(x, y) = \frac{\lambda}{\lambda_1 + \lambda_2} \mathrm{e}^{-[\lambda_1 x + \lambda_2 y + \lambda_{12} \max(x, y)]} - \frac{\lambda_{12}}{\lambda_1 + \lambda_2} \mathrm{e}^{-\lambda \max(x, y)}. \tag{2.4.20}$$

Proof By the full probability formula, we have

$$\begin{aligned} \overline{F}(x, y) &= \Pr\{X > x,\ Y > y,\ U_{12} > \min(U_1, U_2)\} \\ &\quad + \Pr\{X > x,\ Y > y,\ U_{12} \leq \min(U_1, U_2)\} \\ &= \mathrm{I} + \mathrm{II}, \end{aligned} \tag{2.4.21}$$

where

$$\begin{aligned} \mathrm{II} &= \Pr\{U_{12} > x,\ U_{12} > y,\ U_{12} \leq \min(U_1, U_2)\} \\ &= \Pr\{U_{12} > \max(x, y),\ U_{12} \leq \min(U_1, U_2)\} \\ &= \int_{\max(x, y)}^{+\infty} \Pr\{u \leq \min(U_1, U_2)\} \mathrm{d}\Pr\{U_{12} \leq u\} \\ &= \frac{\lambda_{12}}{\lambda} \mathrm{e}^{-\lambda \max(x, y)} = \frac{\lambda_{12}}{\lambda} \overline{F}_s(x, y). \end{aligned}$$

By inserting this formula into (2.4.21), we get

$$\begin{aligned} \mathrm{I} &= \overline{F}(x, y) - \mathrm{II} \\ &= \mathrm{e}^{-\lambda_1 x - \lambda_2 y - \lambda_{12} \max(x, y)} - \frac{\lambda_{12}}{\lambda} \mathrm{e}^{-\lambda \max(x, y)} \end{aligned}$$

$$= \frac{\lambda_1 + \lambda_2}{\lambda} \left\{ \frac{\lambda}{\lambda_1 + \lambda_2} \mathrm{e}^{-\lambda_1 x - \lambda_2 y - \lambda_{12} \max(x,y)} - \frac{\lambda_{12}}{\lambda_1 + \lambda_2} \mathrm{e}^{-\lambda \max(x,y)} \right\}$$
$$= \frac{\lambda_1 + \lambda_2}{\lambda} \overline{F}_a(x, y).$$

Then, $\overline{F}_a(x, y)$ can be expressed as the form of the indefinite integral

$$\overline{F}_a(x, y) = \int_x^{+\infty} \int_y^{+\infty} f_a(u, v) \mathrm{d}v \mathrm{d}u,$$

where

$$f_a(x, y) = \begin{cases} \dfrac{\lambda \lambda_1 (\lambda_2 + \lambda_{12})}{\lambda_1 + \lambda_2} \mathrm{e}^{-\lambda_1 x - (\lambda_2 + \lambda_{12}) y}, & 0 \le x < y \\ \dfrac{\lambda \lambda_2 (\lambda_1 + \lambda_{12})}{\lambda_1 + \lambda_2} \mathrm{e}^{-(\lambda_1 + \lambda_{12}) x - \lambda_2 y}, & 0 \le y < x. \end{cases}$$

Then, function $\overline{F}_a(x, y)$ is absolutely continuous. Since

$$\overline{F}_s(x, y) = \begin{cases} \mathrm{e}^{-\lambda x}, & x > y \\ \mathrm{e}^{-\lambda y}, & x < y, \end{cases}$$

in addition to the two-dimensional Lebesgue zero measure set $\{ (x, y) : x = y \}$, we have

$$\frac{\partial^2}{\partial x \partial y} \overline{F}_s(x, y) = 0.$$

So $\overline{F}_s(x, y)$ is singular. The theorem is proved.

Chapter 3
Nonrepairable Systems with Fuzzy Lifetimes

In this chapter, the lifetime of a nonrepairable system is considered as a nonnegative fuzzy variable on the credibility space $(\Theta, \mathcal{P}(\Theta), \mathrm{Cr})$. Then, some basic mathematical models of nonrepairable systems are established; then, reliability and MTTF are given for nonrepairable systems.

Definition 3.1 Let X be the fuzzy lifetime of a nonrepairable system. The reliability of the nonrepairable system is defined by

$$R(t) = \mathrm{Cr}\{X > t\}.$$

Definition 3.2 Let X be the fuzzy lifetime of a nonrepairable system. The MTTF of the nonrepairable system is defined by

$$\mathrm{MTTF} = \int_0^{+\infty} R(t)\mathrm{d}t.$$

3.1 Series and Parallel Systems

3.1.1 The Series System

Consider a series system composed of n independent components. Assume that the lifetime of component i is a fuzzy variable X_i on the credibility space $(\Theta_i, \mathcal{P}(\Theta_i), \mathrm{Cr}_i)$, $i = 1, 2, \ldots, n$. Obviously, the lifetime of the series system is $X = \min\{X_1, X_2, \ldots, X_n\}$, which is a fuzzy variable on the product credibility space $(\Theta, \mathcal{P}(\Theta), \mathrm{Cr})$, where $\Theta = \Theta_1 \times \Theta_2 \times \cdots \times \Theta_n$ and $\mathrm{Cr} = \mathrm{Cr}_1 \wedge \mathrm{Cr}_2 \wedge \cdots \wedge \mathrm{Cr}_n$.

Y. Liu, *Reliability Theory Based on Uncertain Lifetimes*,
https://doi.org/10.1007/978-981-16-0995-4_3

Theorem 3.1.1 *Let X_i be the fuzzy lifetime of component i, $i = 1, 2, \ldots, n$. The reliability of the series system is*

$$R(t) = \min_{1 \le i \le n} \mathrm{Cr}\{X_i > t\}. \tag{3.1.1}$$

Proof It follows from Definition 3.1 and Definition 1.2.20 that

$$\begin{aligned} R(t) &= \mathrm{Cr}\{\min\{X_1, X_2, \ldots, X_n\} > t\} \\ &= \mathrm{Cr}\{X_1 > t, X_2 > t, \ldots, X_n > t\} \\ &= \mathrm{Cr}\left\{ \bigcap_{1 \le i \le n} \{X_i > t\} \right\} \\ &= \min_{1 \le i \le n} \mathrm{Cr}\{X_i > t\}. \end{aligned}$$

The theorem is proved.

Theorem 3.1.2 *Let X_i be the fuzzy lifetime of component i, $i = 1, 2, \ldots, n$. The MTTF of the series system is*

$$\mathrm{MTTF} = \min_{1 \le i \le n} E[X_i].$$

Proof It follows from Definition 3.2 and Theorem 3.1.1 that

$$\mathrm{MTTF} = \int_0^{+\infty} \min_{1 \le i \le n} \mathrm{Cr}\{X_i > t\} \mathrm{d}t. \tag{3.1.2}$$

It is clear that

$$\int_0^{+\infty} \min_{1 \le i \le n} \mathrm{Cr}\{ X_i > t\} \, \mathrm{d}t \le \int_0^{+\infty} \mathrm{Cr}\{X_i > t\} \mathrm{d}t.$$

$\mathrm{MTTF} = \min_{1 \le i \le n} E[X_i]$. Since i is any number in $\{ 1,2, \ldots, n\}$, then

$$\int_0^{+\infty} \min_{1 \le i \le n} \mathrm{Cr}\{X_i > t\} \mathrm{d}t \le \min_{1 \le i \le n} \int_0^{+\infty} \mathrm{Cr}\{X_i > t\} \mathrm{d}t. \tag{3.1.3}$$

On the other hand, for any $i \in \{1, 2, \ldots, n\}$,

$$\min_{1\le i\le n}\int_0^{+\infty} \mathrm{Cr}\{X_i > t\}\,\mathrm{d}t \le \int_0^{+\infty} \mathrm{Cr}\{X_i > t\}\mathrm{d}t$$

holds, which implies that

$$\min_{1\le i\le n}\int_0^{+\infty} \mathrm{Cr}\{X_i > t\}\,\mathrm{d}t \le \int_0^{+\infty} \mathrm{Cr}\{X_k > t\}\mathrm{d}t,$$

where $\mathrm{Cr}\{X_k > t\} = \min_{1\le i\le n} \mathrm{Cr}\{X_i > t\}$, i.e.,

$$\min_{1\le i\le n}\int_0^{+\infty} \mathrm{Cr}\{X_i > t\}\,\mathrm{d}t \le \int_0^{+\infty} \min_{1\le i\le n} \mathrm{Cr}\{X_i > t\}\mathrm{d}t. \tag{3.1.4}$$

It follows from Eqs. (3.1.3) and (3.1.4) that

$$\int_0^{+\infty} \min_{1\le i\le n} \mathrm{Cr}\{X_i > t\}\,\mathrm{d}t = \min_{1\le i\le n}\int_0^{+\infty} \mathrm{Cr}\{X_i > t\}\mathrm{d}t. \tag{3.1.5}$$

From Definition 1.2.22, we have

$$E[X_i] = \int_0^{+\infty} \mathrm{Cr}\{X_i > t\}\mathrm{d}t. \tag{3.1.6}$$

By Eqs. (3.1.2), (3.1.5) and (3.1.6),

$$\mathrm{MTTF} = \min_{1\le i\le n} E[X_i].$$

The theorem is proved.

Example 3.1.1 Suppose that X_i, $i = 1, 2, \ldots, n$ are independent and identically distributed triangular fuzzy variables (a, b, c). By Theorem 3.1.1 and Theorem 3.1.2, we have

$$R(t) = \begin{cases} 1, & \text{if } t \le a \\ \frac{2b-a-t}{2(b-a)}, & \text{if } a \le t \le b \\ \frac{c-t}{2(c-b)}, & \text{if } b \le t \le c \\ 0, & \text{if } t \ge c \end{cases}$$

and

$$\text{MTTF} = \min_{1 \le i \le n} E[X_i] = \frac{1}{4}(a + 2b + c).$$

Example 3.1.2 Suppose that X_i, $i = 1, 2, \ldots, n$ are independent and identically distributed trapezoidal fuzzy variables (a, b, c, d). By Theorem 3.1.1 and Theorem 3.1.2, we have

$$R(t) = \begin{cases} 1, & \text{if } t \le a \\ \frac{2b-a-t}{2(b-a)}, & \text{if } a \le t \le b \\ \frac{1}{2}, & \text{if } b \le t \le c \\ \frac{d-t}{2(d-c)}, & \text{if } c \le t \le d \\ 0, & \text{if } t \ge d \end{cases}$$

and

$$\text{MTTF} = \min_{1 \le i \le n} E[X_i] = \frac{1}{4}(a + b + c + d).$$

3.1.2 The Parallel System

Consider a parallel system composed of n independent components. Assume that the lifetime of component i is a fuzzy variable X_i on the credibility space $(\Theta_i, \mathcal{P}(\Theta_i), \text{Cr}_i)$, $i = 1, 2, \ldots, n$. Obviously, the lifetime of the parallel system is $X = \max\{X_1, X_2, \ldots, X_n\}$, which is a fuzzy variable on the product credibility space $(\Theta, \mathcal{P}(\Theta), \text{Cr})$, where $\Theta = \Theta_1 \times \Theta_2 \times \cdots \times \Theta_n$ and $\text{Cr} = \text{Cr}_1 \wedge \text{Cr}_2 \wedge \cdots \wedge \text{Cr}_n$.

Theorem 3.1.3 *Let X_i be the fuzzy lifetime of component i, $i = 1, 2, \ldots, n$. The reliability of the parallel system is*

$$R(t) = \max_{1 \le i \le n} \text{Cr}\{X_i > t\}. \tag{3.1.7}$$

Proof It follows from Definition 3.1 and Lemma 1.2.15 that

$$\begin{aligned} R(t) &= \text{Cr}\{\max\{X_1, X_2, \ldots, X_n\} > t\} \\ &= \text{Cr}\{X_1 > t, X_2 > t, \ldots, X_n > t\} \\ &= \text{Cr}\left\{\bigcup_{1 \le i \le n} \{X_i > t\}\right\} \\ &= \max_{1 \le i \le n} \text{Cr}\{X_i > t\}. \end{aligned}$$

The theorem is proved.

Theorem 3.1.4 *Let X_i be the fuzzy lifetime of component i, $i = 1, 2, \ldots, n$. The MTTF of the parallel system is*

$$\text{MTTF} = \max_{1 \le i \le n} E[X_i].$$

Proof It follows from Definition 3.2 and Theorem 3.1.3 that

$$\text{MTTF} = \int_0^{+\infty} \max_{1 \le i \le n} \text{Cr}\{X_i > t\}\text{d}t. \tag{3.1.8}$$

Note that

$$\int_0^{+\infty} \text{Cr}\{X_i > t\}\text{d}t \le \max_{1 \le i \le n} \int_0^{+\infty} \text{Cr}\{ X_i > t\} \text{ d}t.$$

Since i is any number in $\{ 1,2, \ldots, n\}$,

$$\int_0^{+\infty} \max_{1 \le i \le n} \text{Cr}\{ X_i > t\} \text{ d}t \le \max_{1 \le i \le n} \int_0^{+\infty} \text{Cr}\{X_i > t\}\text{d}t. \tag{3.1.9}$$

On the other hand, for any $i \in \{1, 2, \ldots, n\}$,

$$\int_0^{+\infty} \max_{1 \le i \le n} \text{Cr}\{X_i > t\}\text{d}t \ge \int_0^{+\infty} \text{Cr}\{X_i > t\}\text{d}t.$$

So

$$\int_0^{+\infty} \max_{1 \le i \le n} \text{Cr}\{X_i > t\}\text{d}t \ge \int_0^{+\infty} \text{Cr}\{X_k > t\}\text{d}t,$$

where $\int_0^{+\infty} \text{Cr}\{X_k > t\}\text{d}t = \max_{1 \le i \le n} \int_0^{+\infty} \text{Cr}\{X_i > t\}\text{d}t$. Then

$$\int_0^{+\infty} \max_{1 \le i \le n} \text{Cr}\{X_i > t\}\text{d}t \ge \max_{1 \le i \le n} \int_0^{+\infty} \text{Cr}\{X_i > t\}\text{d}t. \tag{3.1.10}$$

Form Eqs. (3.1.9) and (3.1.10), we have

$$\int_0^{+\infty} \max_{1\le i\le n} \mathrm{Cr}\{\ X_i > t\}\ \mathrm{d}t = \max_{1\le i\le n} \int_0^{+\infty} \mathrm{Cr}\{X_i > t\}\mathrm{d}t. \tag{3.1.11}$$

It follows form Definition 1.2.22 that

$$E[X_i] = \int_0^{+\infty} \mathrm{Cr}\{X_i > t\}\mathrm{d}t. \tag{3.1.12}$$

By Eqs. (3.1.8), (3.1.11) and (3.1.12), we have

$$\mathrm{MTTF} = \max_{1\le i\le n} E[X_i].$$

The theorem is proved.

3.1.3 The Series–Parallel System

Consider a series–parallel system which is a series system consisting of m subsystems, and each subsystem is composed of n parallel components. Suppose that X_{ij} is the lifetime of component j in ith subsystem, which is a fuzzy variable on the credibility space $(\Theta_{ij}, \mathcal{P}(\Theta_{ij}), \mathrm{Cr}_{ij})$, $i = 1, 2, \ldots, m,\ j = 1, 2, \ldots, n$. We also assume that the lifetimes of components are mutually independent. It is easy to see that the lifetime of the series–parallel system is $X = \min_{1\le i\le m}\left(\max_{1\le j\le n} X_{ij}\right)$, which is also a fuzzy variable on the product credibility space $(\Theta, \mathcal{P}(\Theta), \mathrm{Cr})$, where $\Theta = \Theta_{11} \times \Theta_{12} \times \cdots \times \Theta_{mn}$ and $\mathrm{Cr} = \mathrm{Cr}_{11} \wedge \mathrm{Cr}_{12} \wedge \cdots \wedge \mathrm{Cr}_{mn}$.

Theorem 3.1.5 *Let X_{ij} be the fuzzy lifetime of component j in ith subsystems, $i = 1, 2, \ldots, m,\ j = 1, 2, \ldots, n$.The reliability of the series–parallel system is*

$$R(t) = \min_{1\le i\le m} \max_{1\le j\le n} \mathrm{Cr}\{X_{ij} > t\}.$$

Proof It follows from Definition 3.1, Definition 1.2.20 and Lemma 1.2.15 that

$$\begin{aligned} R(t) &= \mathrm{Cr}\left\{\min_{1\le i\le m}\left(\max_{1\le j\le n} X_{ij}\right)\right\} \\ &= \mathrm{Cr}\left\{\bigcap_{1\le i\le m} \bigcup_{1\le j\le n} \{X_{ij} > t\}\right\} \end{aligned}$$

$$= \min_{1\le i\le m} \mathrm{Cr}\left\{\bigcup_{1\le j\le n}\{X_{ij} > t\}\right\}$$

$$= \min_{1\le i\le m}\max_{1\le j\le n} \mathrm{Cr}\{X_{ij} > t\}.$$

The theorem is proved.

Theorem 3.1.6 *Let X_{ij} be the fuzzy lifetime of component j in ith subsystems, $i = 1, 2, \ldots, m,\ j = 1, 2, \ldots, n$.The MTTF of the series–parallel system is*

$$\mathrm{MTTF} = \min_{1\le i\le m}\max_{1\le j\le n} E[X_{ij}].$$

Proof The result follows immediately from Theorem 3.1.2 and Theorem 3.1.4.

3.1.4 The Parallel–Series System

Consider a parallel-series system which is a parallel system consisting of m subsystems, and each subsystem is composed of n series components. Suppose that X_{ij} is the lifetime of component j in ith subsystem, which is a fuzzy variable on the credibility space $(\Theta_{ij}, \mathcal{P}(\Theta_{ij}), \mathrm{Cr}_{ij})$, $i = 1, 2, \ldots, m$, $j = 1, 2, \ldots, n$. We also assume that the lifetimes of components are mutually independent. It is easy to see that the lifetime of the parallel-series system is $X = \max_{1\le i\le m}\left(\min_{1\le j\le n} X_{ij}\right)$, which is also a fuzzy variable on the product credibility space $(\Theta, \mathcal{P}(\Theta), \mathrm{Cr})$, where $\Theta = \Theta_{11} \times \Theta_{12} \times \cdots \times \Theta_{mn}$ and $\mathrm{Cr} = \mathrm{Cr}_{11} \wedge \mathrm{Cr}_{12} \wedge \cdots \wedge \mathrm{Cr}_{mn}$.

Theorem 3.1.7 *Let X_{ij} be the fuzzy lifetime of component j in ith subsystems, $i = 1, 2, \ldots, m,\ j = 1, 2, \ldots, n$. The reliability of the parallel-series system is*

$$R(t) = \max_{1\le i\le m}\min_{1\le j\le n} \mathrm{Cr}\{X_{ij} > t\}.$$

Proof It follows from Definition 3.1, Definition 1.2.20 and Lemma 1.2.15 that

$$R(t) = \mathrm{Cr}\left\{\max_{1\le i\le m}\left(\min_{1\le j\le n} X_{ij}\right)\right\}$$

$$= \mathrm{Cr}\left\{\bigcup_{1\le i\le m}\bigcap_{1\le j\le n}\{X_{ij} > t\}\right\}$$

$$= \max_{1\le i\le m} \mathrm{Cr}\left\{\bigcap_{1\le j\le n}\{X_{ij} > t\}\right\}$$

$$= \max_{1 \le i \le m} \min_{1 \le j \le n} \text{Cr}\{X_{ij} > t\}.$$

The theorem is proved.

Theorem 3.1.8 *Let X_{ij} be the fuzzy lifetime of component j in ith subsystems, $i = 1, 2, \ldots, m,\ j = 1, 2, \ldots, n$. The MTTF of the parallel-series system is*

$$\text{MTTF} = \max_{1 \le i \le m} \min_{1 \le j \le n} E[X_{ij}].$$

Proof The result follows immediately from Theorem 3.1.2 and Theorem 3.1.4.

3.2 Cold Standby Systems

Consider a cold standby system consists of n independent components. At the initial moment, one component is in operation and the remaining $n - 1$ components are in standby. When the operating component fails, the standby component is replaced one by one until all components fail and the cold standby system fails. The cold standby system means that the components will not fail or deteriorate during the standby period.

3.2.1 The Perfect Conversion Switch Case

Suppose that the switching mechanism is completely reliable and the transfer is instantaneous. Let X_i be the lifetime of component i on the credibility space $(\Theta_i, \mathcal{P}(\Theta_i), \text{Cr}_i)$, $i = 1, 2, \ldots, n$. We also assume that X_i, $i = 1, 2, \ldots, n$ are independent fuzzy variables. The lifetime of the cold standby system can be expressed by a sum of the lifetimes of the n components, i.e., $X = X_1 + X_2 + \cdots + X_n$, which is a fuzzy variable on the product credibility space $(\Theta, \mathcal{P}(\Theta), \text{Cr})$, where $\Theta = \Theta_1 \times \Theta_2 \times \cdots \times \Theta_n$ and $\text{Cr} = \text{Cr}_1 \wedge \text{Cr}_2 \wedge \cdots \wedge \text{Cr}_n$.

Theorem 3.2.1 *Let X_i be the fuzzy lifetime of component i, $i = 1, 2, \ldots, n$. The reliability of the cold standby system is*

$$R(t) = \text{Cr}\{X_1 + X_2 + \cdots + X_n > t\}.$$

Theorem 3.2.2 *Let X_i be the fuzzy lifetime of component i, $i = 1, 2, \ldots, n$. The MTTF of the cold standby system is*

$$\text{MTTF} = \sum_{i=1}^{n} E[X_i].$$

Proof It follows from Lemma 1.2.26 that

$$\text{MTTF} = E[X_1 + X_2 + \cdots + X_n] = \sum_{i=1}^{n} E[X_i].$$

The theorem is proved.

Example 3.2.1 Let X_i be triangular fuzzy variables (a_i, b_i, c_i), $i = 1, 2, \ldots, n$. Then, the lifetime of the cold standby system is also a triangular fuzzy variable $\left(\sum_{i=1}^{n} a_i, \sum_{i=1}^{n} b_i, \sum_{i=1}^{n} c_i\right)$. It follows from Theorem 3.2.1 and Theorem 3.2.2 that

$$R(t) = \begin{cases} 1, & \text{if } t \le \sum_{i=1}^{n} a_i \\ \frac{\sum_{i=1}^{n}(2b_i - a_i) - t}{\sum_{i=1}^{n}(b_i - a_i)}, & \text{if } \sum_{i=1}^{n} a_i \le t \le \sum_{i=1}^{n} b_i \\ \frac{\sum_{i=1}^{n} c_i - t}{2\sum_{i=1}^{n}(c_i - b_i)}, & \text{if } \sum_{i=1}^{n} b_i \le t \le \sum_{i=1}^{n} c_i \\ 0, & \text{if } t \ge \sum_{i=1}^{n} c_i \end{cases}$$

and

$$\text{MTTF} = \sum_{i=1}^{n} E[X_i] = \frac{1}{4}\sum_{i=1}^{n}(a_i + 2b_i + c_i).$$

3.2.1.1 The Imperfect Conversion Switch Case

In this case, we assume the conversion switch may deteriorate. The lifetime of the conversion switch is considered to be a continuous positive fuzzy variable X_K on the credibility space $(\Theta_K, \mathcal{P}(\Theta_K), \text{Cr}_K)$. The cold standby system fails immediately when there is no operating component left or the conversion switch fails. Then the lifetime of the cold standby system is $X = \min\{X_1 + X_2 + \cdots + X_n,\ X_K\}$, which is a fuzzy variable on the product credibility space $(\Theta, \mathcal{P}(\Theta), \text{Cr})$, where $\Theta = \Theta_1 \times \Theta_2 \times \cdots \times \Theta_n \times \Theta_K$ and $\text{Cr} = \text{Cr}_1 \wedge \text{Cr}_2 \wedge \cdots \wedge \text{Cr}_n \wedge \text{Cr}_K$.

Theorem 3.2.3 *Let X_i be the fuzzy lifetime of component i, $i = 1, 2, \ldots, n$ and X_K be the fuzzy lifetime of the conversion switch. The reliability of the cold standby system is*

$$R(t) = \min\{\text{Cr}\{X_1 + X_2 + \cdots + X_n > t\},\ \ \text{Cr}\{X_K > t\}\}.$$

Theorem 3.2.4 *Let X_i be the fuzzy lifetime of component i, $i = 1, 2, \ldots, n$ and X_K be the fuzzy lifetime of the conversion switch. The MTTF of the cold standby system is*

$$\text{MTTF} = \min\{E[X_1 + X_2 + \cdots + X_n],\ E[X_K]\}.$$

Chapter 4
Nonrepairable Systems with Random Fuzzy Lifetimes

In real life, there are more cases that randomness and fuzziness coexist in a nonrepairable system. In this chapter, the lifetimes of components are considered as independent random fuzzy variables, and the basic mathematical models of nonrepairable systems are established, including series systems, parallel systems, series–parallel systems, parallel–series systems, cold standby systems and warm standby systems. Then, reliability indices such as reliability and MTTF are given. In addition, a random fuzzy shock model and a random fuzzy fatal shock model are proposed, respectively. Furthermore, the bivariate random fuzzy exponential distribution is derived from the random fuzzy fatal shock model. Finally, some properties of the bivariate random fuzzy exponential distribution are proposed.

To evaluate the reliability of random fuzzy nonrepairable systems, the reliability and MTTF are redefined firstly.

Definition 4.1 Let X be the random fuzzy lifetime of a nonrepairable system on the credibility space $(\Theta, \mathcal{P}(\Theta), \mathrm{Cr})$. Then, the reliability of the nonrepairable system is defined by

$$R(t) = \mathrm{Ch}\{X > t\}.$$

Definition 4.2 Let X be the random fuzzy lifetime of a nonrepairable system on the credibility space $(\Theta, \mathcal{P}(\Theta), \mathrm{Cr})$. Then, the MTTF of the nonrepairable system is defined by

$$\mathrm{MTTF} = \int_0^{+\infty} \mathrm{Cr}\{\theta \in \Theta | E[X(\theta)] \geq r\}\mathrm{d}r.$$

Y. Liu, *Reliability Theory Based on Uncertain Lifetimes*,
https://doi.org/10.1007/978-981-16-0995-4_4

4.1 Series and Parallel Systems

4.1.1 The Series System

Consider a series system composed of n independent components. Assume that the lifetime of component i is a random fuzzy variable X_i on the credibility space $(\Theta_i, \mathcal{P}(\Theta_i), \mathrm{Cr}_i)$, $i = 1, 2, \ldots, n$. Obviously, the lifetime of the series system is $X = \min\{X_1, X_2, \ldots, X_n\}$, which is a random fuzzy variable on the product credibility space $(\Theta, \mathcal{P}(\Theta), \mathrm{Cr})$, where $\Theta = \Theta_1 \times \Theta_2 \times \cdots \times \Theta_n$ and $\mathrm{Cr} = \mathrm{Cr}_1 \wedge \mathrm{Cr}_2 \wedge \cdots \wedge \mathrm{Cr}_n$.

Theorem 4.1.1 *Let $X_i, i = 1, 2, \ldots, n$ be random fuzzy variables. Assume that the α-pessimistic values and the α-optimistic values of $E[X_i(\theta_i)], \theta_i \in \Theta_i, i = 1, 2, \ldots, n$ are continuous almost everywhere with respect to $\alpha, \alpha \in (0, 1]$. The reliability of the random fuzzy series system is*

$$R(t) = E\left[\prod_{i=1}^{n} \Pr\{\omega \in \Omega | X_i(\theta_i)(\omega) > t\}\right]. \tag{4.1.1}$$

Proof It follows from Definition 4.1, Definition 1.3.7 and Theorem 1.2.23 that

$$\begin{aligned} R(t) &= \mathrm{Ch}\{X < t\} \\ &= \int_0^1 \mathrm{Cr}\{\theta \in \Theta | \Pr\{X(\theta) < t\} \geq p\} \mathrm{d}p \\ &= \frac{1}{2} \int_0^1 \left(\Pr{}_{\alpha}^{L}\{\omega \in \Omega | X(\theta)(\omega) < t\} + \Pr{}_{\alpha}^{U}\{\omega \in \Omega | X(\theta)(\omega) < t\}\right) \mathrm{d}\alpha. \end{aligned} \tag{4.1.2}$$

Let $A_i = \{\theta_i \in \Theta_i | \mu\{\theta_i\} \geq \alpha\}, i = 1, 2, \ldots, n$. Since the α-pessimistic values and the α-optimistic values of fuzzy variables $E[X_i(\theta_i)]$, $\theta_i \in \Theta_i, i = 1, 2, \ldots, n$ are continuous almost everywhere with respect to $\alpha, \alpha \in (0, 1]$, then there at least exist points $\theta_i^{'}, \theta_i^{''} \in A_i, i = 1, 2, \ldots, n$ such that

$$E[X_i(\theta_i^{'})] = E[X_i(\theta_i)]_{\alpha}^{L} \text{ and } E[X_i(\theta_i^{''})] = E[X_i(\theta_i)]_{\alpha}^{U}.$$

For any $\theta_{i,\alpha} \in A_i$, we have

$$E[X_i(\theta_i^{'})] \leq E[X_i(\theta_{i,\alpha})] \leq E[X_i(\theta_i^{''})], i = 1, 2, \ldots, n. \tag{4.1.3}$$

Hence, by Theorem 1.1.21, we have

$$X_i(\theta_i^{'}) \preceq_d X_i(\theta_{i,\alpha}) \preceq_d X_i(\theta_i^{''}), \forall \theta_{i,\alpha} \in A_i, i = 1, 2, \ldots, n. \tag{4.1.4}$$

It follows from Definition 1.1.31 that

$$\begin{aligned}\Pr\left\{\omega \in \Omega | X_i(\theta_i^{'})(\omega) > t\right\} &\leq \Pr\{\omega \in \Omega | X_i(\theta_{i,\alpha})(\omega) > t\} \\ &\leq \Pr\left\{\omega \in \Omega | X_i(\theta_i^{''})(\omega) > t\right\}.\end{aligned} \tag{4.1.5}$$

It is easy to see that

$$\begin{aligned}\prod_{i=1}^{n} \Pr\left\{\omega \in \Omega | X_i(\theta_i^{'})(\omega) > t\right\} &\leq \prod_{i=1}^{n} \Pr\{\omega \in \Omega | X_i(\theta_{i,\alpha})(\omega) > t\} \\ &\leq \prod_{i=1}^{n} \Pr\left\{\omega \in \Omega | X_i(\theta_i^{''})(\omega) > t\right\}.\end{aligned} \tag{4.1.6}$$

That is,

$$\begin{aligned}&\Pr\left\{\omega \in \Omega | \min\left\{X_1(\theta_1^{'})(\omega), \ldots, X_n(\theta_n^{'})(\omega)\right\} > t\right\} \\ &\leq \Pr\{\omega \in \Omega | \min\{X_1(\theta_{1,\alpha})(\omega), \ldots, X_n(\theta_{n,\alpha})(\omega)\} > t\} \\ &\leq \Pr\left\{\omega \in \Omega | \min\left\{X_1(\theta_1^{''})(\omega), \ldots, X_n(\theta_n^{''})(\omega)\right\} > t\right\}.\end{aligned} \tag{4.1.7}$$

Since $\theta_{i,\alpha}$ are arbitrary points in $A_i, i = 1, 2, \ldots, n$, by Eq. (4.1.7), we have

$$\begin{aligned}&\Pr{}_{\alpha}^{L}\{\omega \in \Omega | X(\theta)(\omega) > t\} \\ &\quad = \Pr\left\{\omega \in \Omega | \min\left\{X_1(\theta_1^{'})(\omega), \ldots, X_n(\theta_n^{'})(\omega)\right\} > t\right\} \\ &\quad = \prod_{i=1}^{n} \Pr\left\{\omega \in \Omega | X_i(\theta_i^{'})(\omega) > t\right\}\end{aligned} \tag{4.1.8}$$

and

$$\begin{aligned}&\Pr{}_{\alpha}^{U}\{\omega \in \Omega | X(\theta)(\omega) < t\} \\ &\quad = \Pr\left\{\omega \in \Omega | \min\left\{X_1(\theta_1^{''})(\omega), \ldots, X_n(\theta_n^{''})(\omega)\right\} < t\right\} \\ &\quad = \prod_{i=1}^{n} \Pr\left\{\omega \in \Omega | X_i(\theta_i^{''})(\omega) < t\right\}.\end{aligned} \tag{4.1.9}$$

On the other hand, by Eq. (4.1.6), we have

$$\Pr{}_{\alpha}^{L}\{\omega \in \Omega | X_i(\theta_i)(\omega) > t\} = \Pr\left\{\omega \in \Omega | X_i(\theta_i^{'})(\omega) > t\right\} \tag{4.1.10}$$

and

$$\Pr{}_{\alpha}^{U}\{\omega\in\Omega|X_i(\theta_i)(\omega)>t\}=\Pr\left\{\omega\in\Omega|X_i(\theta_i^{''})(\omega)>t\right\} \tag{4.1.11}$$

for $i=1,2,\ldots,n$. By Eqs. (4.1.8)–(4.1.11), we have

$$\Pr{}_{\alpha}^{L}\{\omega\in\Omega|X(\theta)(\omega)>t\}=\prod_{i=1}^{n}\Pr{}_{\alpha}^{L}\{\omega\in\Omega|X_i(\theta_i)(\omega)>t\} \tag{4.1.12}$$

and

$$\Pr{}_{\alpha}^{U}\{\omega\in\Omega|X(\theta)(\omega)>t\}=\prod_{i=1}^{n}\Pr{}_{\alpha}^{U}\{\omega\in\Omega|X_i(\theta_i)(\omega)>t\}. \tag{4.1.13}$$

By Eqs. (4.1.2), (4.1.12) and (4.1.13), we have

$$\begin{aligned}
R(t)&=\frac{1}{2}\int_0^1\left(\Pr{}_{\alpha}^{L}\{\omega\in\Omega|X(\theta)(\omega)>t\}+\Pr{}_{\alpha}^{U}\{\omega\in\Omega|X(\theta)(\omega)>t\}\right)\mathrm{d}\alpha\\
&=\frac{1}{2}\int_0^1\left(\prod_{i=1}^{n}\Pr{}_{\alpha}^{L}\{\omega\in\Omega|X_i(\theta_i)(\omega)>t\}+\prod_{i=1}^{n}\Pr{}_{\alpha}^{U}\{\omega\in\Omega|X_i(\theta_i)(\omega)>t\}\mathrm{d}\alpha\right)\\
&=\frac{1}{2}\int_0^1\left(\left[\prod_{i=1}^{n}\Pr\{\omega\in\Omega|X_i(\theta_i)(\omega)>t\}\right]_{\alpha}^{L}+\left[\prod_{i=1}^{n}\Pr\{\omega\in\Omega|X_i(\theta_i)(\omega)>t\}\right]_{\alpha}^{U}\right)\mathrm{d}\alpha\\
&=E\left[\prod_{i=1}^{n}\Pr\{\omega\in\Omega|X_i(\theta_i)(\omega)>t\}\right].
\end{aligned} \tag{4.1.14}$$

The proof is complete.

Remark 4.1.1 If $X_i, i=1,2,\ldots,n$ degenerate to random variables, the result in Theorem 4.1.1 degenerates to the form

$$\begin{aligned}
R(t)&=\Pr\{\omega\in\Omega|\min\{X_1(\omega),X_2(\omega),\ldots,X_n(\omega)\}>t\}\\
&=\prod_{i=1}^{n}\Pr\{\omega\in\Omega|X_i(\omega)>t\},
\end{aligned} \tag{4.1.15}$$

which is consistent with the result in stochastic case.

Remark 4.1.2 If $X_i, i=1,2,\ldots,n$ degenerate to fuzzy variables, the result in Theorem 4.1.1 degenerates to the form

$$\begin{aligned}R(t) &= \mathrm{Cr}\{\theta \in \Theta \mid \min\{X_1(\theta_1), X_2(\theta_2), \ldots, X_n(\theta_n)\} > t\} \\ &= \min_{1\le i\le n} \mathrm{Cr}\{\theta \in \Theta \mid X_i(\theta_i) > t\}. \end{aligned} \tag{4.1.16}$$

which is consistent with the result in fuzzy case.

Theorem 4.1.2 *Let $X_i, i = 1, 2, \ldots, n$ be random fuzzy variables. Assume that the α-pessimistic values and the α-optimistic values of $E[X_i(\theta_i)]$, $\theta_i \in \Theta_i, i = 1, 2, \ldots, n$ are continuous almost everywhere with respect to $\alpha, \alpha \in (0, 1]$. The MTTF of the random fuzzy series system is*

$$\begin{aligned}\mathrm{MTTF} = \frac{1}{2}\int_0^1\int_0^{+\infty}\Bigg\{&\prod_{i=1}^n \mathrm{Pr}_\alpha^L\{\omega \in \Omega \mid X_i(\theta_i)(\omega) \ge t\} \\ &+ \prod_{i=1}^n \mathrm{Pr}_\alpha^U\{\omega \in \Omega \mid X_i(\theta_i)(\omega) \ge t\}\Bigg\}\mathrm{d}t\mathrm{d}\alpha. \end{aligned} \tag{4.1.17}$$

Proof It follows from Definition 4.2 and Theorem 1.2.23 that

$$\begin{aligned}\mathrm{MTTF} &= \int_0^{+\infty} \mathrm{Cr}\{\theta \in \Theta \mid E[X(\theta)] \ge r\}\mathrm{d}r \\ &= \frac{1}{2}\int_0^1 \left(E[X(\theta)]_\alpha^L + E[X(\theta)]_\alpha^U\right)\mathrm{d}\alpha. \end{aligned} \tag{4.1.18}$$

By Eq. (4.1.7), we can arrive at

$$\begin{aligned}&\min\left\{X_1(\theta_1^{'})(\omega), \ldots, X_n(\theta_n^{'})(\omega)\right\} \\ &\le_d \min\left\{X_1(\theta_{1,\alpha})(\omega), \ldots, X_n(\theta_{n,\alpha})(\omega)\right\} \\ &\le_d \min\left\{X_1(\theta_1^{''})(\omega), \ldots, X_n(\theta_n^{''})(\omega)\right\}. \end{aligned} \tag{4.1.19}$$

It follows from Theorem 1.1.21 that

$$\begin{aligned}&E\left[\min\left\{X_1(\theta_1^{'})(\omega), \ldots, X_n(\theta_n^{'})(\omega)\right\}\right] \\ &\le E\left[\min\left\{X_1(\theta_{1,\alpha})(\omega), \ldots, X_n(\theta_{n,\alpha})(\omega)\right\}\right] \\ &\le E\left[\min\left\{X_1(\theta_1^{''})(\omega), \ldots, X_n(\theta_n^{''})(\omega)\right\}\right]. \end{aligned} \tag{4.1.20}$$

Since $\theta_{i,\alpha}$ are arbitrary points in $A_i, i = 1, 2, \ldots, n$, we have

$$
\begin{aligned}
E[X(\theta)]_\alpha^L &= E\left[\min\left\{X_1(\theta_1^{'})(\omega), \ldots, X_n(\theta_n^{'})(\omega)\right\}\right] \\
&= \int_0^{+\infty} \prod_{i=1}^n \Pr\left\{\omega \in \Omega | X_i(\theta_i^{'})(\omega) \geq t \, \mathrm{d}t\right\}
\end{aligned} \tag{4.1.21}
$$

and

$$
\begin{aligned}
E[X(\theta)]_\alpha^U &= E\left[\min\left\{X_1(\theta_1^{''})(\omega), \ldots, X_n(\theta_n^{''})(\omega)\right\}\right] \\
&= \int_0^{+\infty} \prod_{i=1}^n \Pr\left\{\omega \in \Omega | X_i(\theta_i^{''})(\omega) \geq t\right\} \mathrm{d}t.
\end{aligned} \tag{4.1.22}
$$

By Eqs. (4.1.10), (4.1.11), (4.1.18), (4.1.21) and (4.1.22), we have

$$
\begin{aligned}
\text{MTTF} &= \frac{1}{2} \int_0^1 \left(E[X(\theta)]_\alpha^L + E[X(\theta)]_\alpha^U\right) \mathrm{d}\alpha \\
&= \frac{1}{2} \int_0^1 \left\{ \int_0^{+\infty} \prod_{i=1}^n \Pr\left\{\omega \in \Omega | X_i(\theta_i^{'})(\omega) \geq t\right\} \mathrm{d}t \right. \\
&\quad \left. + \int_0^{+\infty} \prod_{i=1}^n \Pr\left\{\omega \in \Omega | X_i(\theta_i^{''})(\omega) \geq t\}\mathrm{d}t\right\} \right\} \mathrm{d}\alpha \\
&= \frac{1}{2} \int_0^1 \int_0^{+\infty} \left\{ \prod_{i=1}^n \Pr{}_\alpha^L \{\omega \in \Omega | X_i(\theta_i)(\omega) \geq t\} \right. \\
&\quad \left. + \prod_{i=1}^n \Pr{}_\alpha^U \{\omega \in \Omega | X_i(\theta_i)(\omega) \geq t\} \right\} \mathrm{d}t \mathrm{d}\alpha.
\end{aligned} \tag{4.1.23}
$$

The proof is complete.

Remark 4.1.3 If $X_i, i = 1, 2, \ldots, n$ degenerate to random variables, the result in Theorem 4.1.2 degenerates to the form.

$$
\text{MTTF} = \int_0^{+\infty} \prod_{i=1}^n \Pr\{\omega \in \Omega | X_i(\omega) \geq t\} \mathrm{d}t, \tag{4.1.24}
$$

which is consistent with the result in stochastic case.

Remark 4.1.4 If X_i, $i = 1, 2, \ldots, n$ degenerate to fuzzy variables, the result in Theorem 4.1.2 degenerates to the form

$$\begin{aligned} \text{MTTF} &= \int_0^{+\infty} \text{Cr}\{\theta \in \Theta | \min\{X_1(\theta_1), X_2(\theta_2), \ldots, X_n(\theta_n)\} \geq t\} \text{d}t \\ &= \min_{1 \leq i \leq n} E[X_i], \end{aligned} \tag{4.1.25}$$

which is consistent with the result in fuzzy case.

Example 4.1.1 If the random fuzzy variables $X_i \sim \exp(\lambda_i)$, where λ_i are fuzzy variables on $(\Theta_i, \mathcal{P}(\Theta_i), \text{Cr}_i)$, $i = 1, 2, \ldots, n$. then we can arrive at

$$\text{Pr}_\alpha^L\{\omega \in \Omega | X(\theta)(\omega) > t\} = \prod_{i=1}^n \exp(-\lambda_{i,\alpha}^U t) = \exp\left(-\sum_{i=1}^n \lambda_{i,\alpha}^U t\right)$$

and

$$\text{Pr}_\alpha^U\{\omega \in \Omega | X(\theta)(\omega) > t\} = \prod_{i=1}^n \exp(-\lambda_{i,\alpha}^L t) = \exp\left(-\sum_{i=1}^n \lambda_{i,\alpha}^L t\right).$$

By Theorem 4.1.1 and Theorem 4.1.2, we have

$$\begin{aligned} R(t) &= \frac{1}{2} \int_0^1 \left(\text{Pr}_\alpha^L\{\omega \in \Omega | X(\theta)(\omega) > t\} + \text{Pr}_\alpha^U\{\omega \in \Omega | X(\theta)(\omega) > t\}\right) \text{d}\alpha \\ &= \frac{1}{2} \int_0^1 \left(\exp\left(-\sum_{i=1}^n \lambda_{i,\alpha}^U t\right) + \exp\left(-\sum_{i=1}^n \lambda_{i,\alpha}^L t\right)\right) \text{d}\alpha \\ &= E\left[\exp\left(-\sum_{i=1}^n t\lambda_i\right)\right] \end{aligned}$$

and

$$\text{MTTF} = \int_0^{+\infty} E\left[\exp\left(-\sum_{i=1}^n t\lambda_i\right)\right] \text{d}t$$

in which "E" is the expected value operator of fuzzy variables.

4.1.2 The Parallel System

Consider a parallel system composed of n independent components. Assume that the lifetime of component i is a random fuzzy variable X_i on the credibility space $(\Theta_i, \mathcal{P}(\Theta_i), \mathrm{Cr}_i), i = 1, 2, \ldots, n$. Obviously, the lifetime of the parallel system is $X = \max\{X_1, X_2, \ldots, X_n\}$, which is a random fuzzy variable on the product credibility space $(\Theta, \mathcal{P}(\Theta), \mathrm{Cr})$, where $\Theta = \Theta_1 \times \Theta_2 \times \cdots \times \Theta_n$ and $\mathrm{Cr} = \mathrm{Cr}_1 \wedge \mathrm{Cr}_2 \wedge \cdots \wedge \mathrm{Cr}_n$.

Theorem 4.1.3 *Let $X_i, i = 1, 2, \ldots, n$ be random fuzzy variables. Assume that the α-pessimistic values and the α-optimistic values of $E[X_i(\theta_i)], \theta_i \in \Theta_i, i = 1, 2, \ldots, n$ are continuous almost everywhere with respect to $\alpha, \alpha \in (0, 1]$. The reliability of the random fuzzy parallel system is*

$$R(t) = 1 - E\left[\prod_{i=1}^{n}(1 - \Pr\{\omega \in \Omega | X_i(\theta_i)(\omega) > t\})\right]. \tag{4.1.26}$$

Proof It follows from Definition 4.1, Definition 1.3.7 and Theorem 1.2.23 that

$$\begin{aligned} R(t) &= \mathrm{Ch}\{X > t\} \\ &= \int_0^1 \mathrm{Cr}\{\theta \in \Theta | \Pr\{X(\theta) > t\} \geq p\}\mathrm{d}p \\ &= \frac{1}{2}\int_0^1 \left(\Pr{}_{\alpha}^{L}\{\omega \in \Omega | X(\theta)(\omega) > t\} + \Pr{}_{\alpha}^{U}\{\omega \in \Omega | X(\theta)(\omega) > t\}\right)\mathrm{d}\alpha. \end{aligned} \tag{4.1.27}$$

Let $A_i = \{\theta_i \in \Theta_i | \mu\{\theta_i\} \geq \alpha\}, i = 1, 2, \ldots, n$. Since the α-pessimistic values and the α-optimistic values of fuzzy variables $E[X_i(\theta_i)], \theta_i \in \Theta_i, i = 1, 2, \ldots, n$ are continuous almost everywhere with respect to $\alpha, \alpha \in (0, 1]$, then there at least exist points $\theta_i^{'}, \theta_i^{''} \in A_i, i = 1, 2, \ldots, n$ such that

$$E[X_i(\theta_i^{'})] = E[X_i(\theta_i)]_{\alpha}^{L} \text{ and } E[X_i(\theta_i^{''})] = E[X_i(\theta_i)]_{\alpha}^{U}.$$

For any $\theta_{i,\alpha} \in A_i$, we have

$$E[X_i(\theta_i^{'})] \leq E[X_i(\theta_{i,\alpha})] \leq E[X_i(\theta_i^{''})], \ i = 1, 2, \ldots, n. \tag{4.1.28}$$

Hence, by Theorem 1.1.21, we have

$$X_i(\theta_i^{'}) \leq_d X_i(\theta_{i,\alpha}) \leq_d X_i(\theta_i^{''}), \ \forall \theta_{i,\alpha} \in A_i, \ i = 1, 2, \ldots, n. \tag{4.1.29}$$

It follows from Definition 1.1.31 that

$$\Pr\left\{\omega \in \Omega | X_i(\theta_i^{'})(\omega) > t\right\} \le \Pr\{\omega \in \Omega | X_i(\theta_{i,\alpha})(\omega) > t\}$$
$$\le \Pr\left\{\omega \in \Omega | X_i(\theta_i^{''})(\omega) > t\right\}, \ i = 1, 2, \ldots, n. \tag{4.1.30}$$

It is easy to see that

$$\begin{aligned} & 1 - \prod_{i=1}^{n}\left[1 - \Pr\left\{\omega \in \Omega | X_i(\theta_i^{'})(\omega) > t\right\}\right] \\ & \le 1 - \prod_{i=1}^{n}[1 - \Pr\{\omega \in \Omega | X_i(\theta_{i,\alpha})(\omega) > t\}] \\ & \le 1 - \prod_{i=1}^{n}\left[1 - \Pr\left\{\omega \in \Omega | X_i(\theta_i^{''})(\omega) > t\right\}\right]. \end{aligned} \tag{4.1.31}$$

That is,

$$\begin{aligned} & \Pr\left\{\omega \in \Omega | \max\left\{X_1(\theta_1^{'})(\omega), \ldots, X_n(\theta_n^{'})(\omega)\right\} > t\right\} \\ & \le \Pr\{\omega \in \Omega | \max\{X_1(\theta_{1,\alpha})(\omega), \ldots, X_n(\theta_{n,\alpha})(\omega)\} > t\} \\ & \le \Pr\left\{\omega \in \Omega | \max\left\{X_1(\theta_1^{''})(\omega), \ldots, X_n(\theta_n^{''})(\omega)\right\} > t\right\}. \end{aligned} \tag{4.1.32}$$

Since $\theta_{i,\alpha}$ are arbitrary points in $A_i, \ i = 1, 2, \ldots, n$, by Eq. (4.1.32), we have

$$\begin{aligned} & \Pr{}_{\alpha}^{L}\{\omega \in \Omega | X(\theta)(\omega) > t\} \\ & = Pr\left\{\omega \in \Omega | \max\left\{X_1(\theta_1^{'})(\omega), \ldots, X_n(\theta_n^{'})(\omega)\right\} > t\right\} \\ & = 1 - \prod_{i=1}^{n}\left[1 - \Pr\left\{\omega \in \Omega | X_i(\theta_i^{'})(\omega) > t\right\}\right] \end{aligned} \tag{4.1.33}$$

and

$$\begin{aligned} & \Pr{}_{\alpha}^{U}\{\omega \in \Omega | X(\theta)(\omega) > t\} \\ & = \Pr\left\{\omega \in \Omega | \max\left\{X_1(\theta_1^{''})(\omega), \ldots, X_n(\theta_n^{''})(\omega)\right\} > t\right\} \\ & = 1 - \prod_{i=1}^{n}\left[1 - \Pr\left\{\omega \in \Omega | X_i(\theta_i^{''})(\omega) > t\right\}\right]. \end{aligned} \tag{4.1.34}$$

On the other hand, by Eq. (4.1.30), we have

$$\Pr{}_{\alpha}^{L}\{\omega \in \Omega | X_i(\theta_i)(\omega) > t\} = \Pr\left\{\omega \in \Omega | X_i(\theta_i^{'})(\omega) > t\right\} \tag{4.1.35}$$

and

$$\Pr{}_{\alpha}^{U}\{\omega \in \Omega | X_i(\theta_i)(\omega) > t\} = \Pr\left\{\omega \in \Omega | X_i(\theta_i^{''})(\omega) > t\right\} \tag{4.1.36}$$

for $i = 1, 2, \ldots, n$. By Eqs. (4.1.33)–(4.1.36), we have

$$\begin{aligned}&\Pr{}_{\alpha}^{L}\{\omega \in \Omega | X(\theta)(\omega) > t\}\\&= 1 - \prod_{i=1}^{n}\left[1 - \Pr{}_{\alpha}^{L}\{\omega \in \Omega | X_i(\theta_i)(\omega) > t\}\right]\end{aligned} \tag{4.1.37}$$

and

$$\begin{aligned}&\Pr{}_{\alpha}^{U}\{\omega \in \Omega | X(\theta)(\omega) > t\}\\&= 1 - \prod_{i=1}^{n}\left[1 - \Pr{}_{\alpha}^{U}\{\omega \in \Omega | X_i(\theta_i)(\omega) > t\}\right].\end{aligned} \tag{4.1.38}$$

By Eqs. (4.1.27), (4.1.37) and (4.1.38), we have

$$\begin{aligned}R(t) &= \frac{1}{2}\int_0^1 \left(\Pr{}_{\alpha}^{L}\{\omega \in \Omega | X(\theta)(\omega) > t\} + \Pr{}_{\alpha}^{U}\{\omega \in \Omega | X(\theta)(\omega) > t\}\right)\mathrm{d}\alpha\\
&= \frac{1}{2}\int_0^1 \Bigg(1 - \prod_{i=1}^{n}\left[1 - \Pr{}_{\alpha}^{L}\{\omega \in \Omega | X_i(\theta_i)(\omega) > t\}\right]\\
&\quad + 1 - \prod_{i=1}^{n}\left[1 - \Pr{}_{\alpha}^{U}\{\omega \in \Omega | X_i(\theta_i)(\omega) > t\}\right]\Bigg)\mathrm{d}\alpha\\
&= 1 - \frac{1}{2}\int_0^1 \Bigg(\left\{\prod_{i=1}^{n}[1 - \Pr\{\omega \in \Omega | X_i(\theta_i)(\omega) > t\}]\right\}_{\alpha}^{U}\\
&\quad + \left\{\prod_{i=1}^{n}[1 - \Pr\{\omega \in \Omega | X_i(\theta_i)(\omega) > t\}]\right\}_{\alpha}^{L}\Bigg)\mathrm{d}\alpha\\
&= 1 - E\left[\prod_{i=1}^{n}(1 - \Pr\{\omega \in \Omega | X_i(\theta_i)(\omega) > t\})\right].\end{aligned} \tag{4.1.39}$$

The proof is completed.

Remark 4.1.5 If $X_i,\ i = 1, 2, \ldots, n$ degenerate to random variables, the result in Theorem 4.1.3 degenerates to the form

$$\begin{aligned}R(t) &= \Pr\{\omega \in \Omega | \max\{X_1(\omega), X_2(\omega), \ldots, X_n(\omega)\} > t\} \\ &= 1 - \prod_{i=1}^{n} [1 - \Pr\{\omega \in \Omega | X_i(\omega) > t\}],\end{aligned} \tag{4.1.40}$$

which is consistent with the result in stochastic case.

Remark 4.1.6 If $X_i,\ i = 1, 2, \ldots, n$ degenerate to fuzzy variables, the result in Theorem 4.1.3 degenerates to the form

$$\begin{aligned}R(t) &= \mathrm{Cr}\{\theta \in \Theta | \max\{X_1(\theta_1), X_2(\theta_2), \ldots, X_n(\theta_n)\} > t\} \\ &= \max_{1 \le i \le n} \mathrm{Cr}\{\theta \in \Theta | X_i(\theta_i) > t\},\end{aligned} \tag{4.1.41}$$

which is consistent with the result in fuzzy case.

Theorem 4.1.4 *Let $X_i,\ i = 1, 2, \ldots, n$ be random fuzzy variables. Assume that the α-pessimistic values and the α-optimistic values of $E[X_i(\theta_i)],\ \theta_i \in \Theta_i,\ i = 1, 2, \ldots, n$ are continuous almost everywhere with respect to $\alpha,\ \alpha \in (0, 1]$. The MTTF of the random fuzzy parallel system is*

$$\begin{aligned}\mathrm{MTTF} = \frac{1}{2} \int_0^1 \int_0^{+\infty} \Bigg\{ 2 &- \prod_{i=1}^{n} [1 - \Pr_\alpha^L\{\omega \in \Omega | X_i(\theta_i)(\omega) \ge t\}] \\ &- \prod_{i=1}^{n} [1 - \Pr_\alpha^U\{\omega \in \Omega | X_i(\theta_i)(\omega) \ge t\}] \Bigg\} \mathrm{d}t\mathrm{d}\alpha.\end{aligned} \tag{4.1.42}$$

Proof It follows from Definition 4.2 and Theorem 1.2.23 that

$$\begin{aligned}\mathrm{MTTF} &= \int_0^{+\infty} \mathrm{Cr}\{\theta \in \Theta | E[X(\theta)] \ge r\} \mathrm{d}r \\ &= \frac{1}{2} \int_0^1 \left(E[X(\theta)]_\alpha^L + E[X(\theta)]_\alpha^U \right) \mathrm{d}\alpha.\end{aligned} \tag{4.1.43}$$

By Eq. (4.1.32), we can arrive at

$$\max\left\{ X_1(\theta_1')(\omega), \ldots, X_n(\theta_n')(\omega) \right\}$$

$$\begin{aligned}&\leq_d \max\{X_1(\theta_{1,\alpha})(\omega),\ldots,X_n(\theta_{n,\alpha})(\omega)\}\\&\leq_d \max\left\{X_1(\theta_1^{''})(\omega),\ldots,X_n(\theta_n^{''})(\omega)\right\}.\end{aligned}\tag{4.1.44}$$

It follows form Theorem 1.1.21 that

$$\begin{aligned}&E\left[\max\left\{X_1(\theta_1^{'})(\omega),\ldots,X_n(\theta_n^{'})(\omega)\right\}\right]\\&\leq E\left[\max\{X_1(\theta_{1,\alpha})(\omega),\ldots,X_n(\theta_{n,\alpha})(\omega)\}\right]\\&\leq E\left[\max\left\{X_1(\theta_1^{''})(\omega),\ldots,X_n(\theta_n^{''})(\omega)\right\}\right].\end{aligned}\tag{4.1.45}$$

Since $\theta_{i,\alpha}$ are arbitrary points in $A_i,\ i=1,2,\ldots,n$, we have

$$\begin{aligned}E[X(\theta)]_\alpha^L&=E\left[\max\left\{X_1(\theta_1^{'})(\omega),\ldots,X_n(\theta_n^{'})(\omega)\right\}\right]\\&=\int_0^{+\infty}\left\{1-\prod_{i=1}^n\left[1-\Pr\left\{\omega\in\Omega|X_i(\theta_i^{'})(\omega)\geq t\right\}\right]\right\}\mathrm{d}t\end{aligned}\tag{4.1.46}$$

and

$$\begin{aligned}E[X(\theta)]_\alpha^U&=E\left[\max\left\{X_1(\theta_1^{''})(\omega),\ldots,X_n(\theta_n^{''})(\omega)\right\}\right]\\&=\int_0^{+\infty}\left\{1-\prod_{i=1}^n\left[1-\Pr\left\{\omega\in\Omega|X_i(\theta_i^{''})(\omega)\geq t\right\}\right]\right\}\mathrm{d}t.\end{aligned}\tag{4.1.47}$$

By Eqs. (4.1.35), (4.1.36), (4.1.43), (4.1.46) and (4.1.47), we have

$$\begin{aligned}\mathrm{MTTF}&=\frac{1}{2}\int_0^1\left(E[X(\theta)]_\alpha^L+E[X(\theta)]_\alpha^U\right)\mathrm{d}\alpha\\&=\frac{1}{2}\int_0^1\left\{\int_0^{+\infty}\left\{1-\prod_{i=1}^n\left[1-\Pr\left\{\omega\in\Omega|X_i(\theta_i^{'})(\omega)\geq t\right\}\right]\right\}\mathrm{d}t\right.\\&\quad\left.+\int_0^{+\infty}\left\{1-\prod_{i=1}^n\left[1-\Pr\left\{\omega\in\Omega|X_i(\theta_i^{''})(\omega)\geq t\right\}\right]\right\}Et\right\}\mathrm{d}\alpha\\&=\frac{1}{2}\int_0^1\left\{\int_0^{+\infty}\left\{1-\prod_{i=1}^n\left[1-\mathrm{Pr}_\alpha^L\{\omega\in\Omega|X_i(\theta_i)(\omega)\geq t\}\right]\right\}\mathrm{d}t\right.\end{aligned}$$

$$+ \int_0^{+\infty} \left\{ 1 - \prod_{i=1}^{n} \left[1 - \mathrm{Pr}_\alpha^U \{ \omega \in \Omega | X_i(\theta_i)(\omega) \geq t \} \right] \right\} \mathrm{d}t \Bigg\} \mathrm{d}\alpha$$

$$= \frac{1}{2} \int_0^1 \int_0^{+\infty} \left\{ 2 - \prod_{i=1}^{n} \left[1 - \mathrm{Pr}_\alpha^L \{ \omega \in \Omega | X_i(\theta_i)(\omega) \geq t \} \right] \right.$$

$$\left. - \prod_{i=1}^{n} \left[1 - \mathrm{Pr}_\alpha^U \{ \omega \in \Omega | X_i(\theta_i)(\omega) \geq t \} \right] \right\} \mathrm{d}t \mathrm{d}\alpha. \quad (4.1.48)$$

The proof is completed.

Remark 4.1.7 If $X_i, \ i = 1, 2, \ldots, n$ degenerate to random variables, the result in Theorem 4.1.4 degenerates to the form

$$\mathrm{MTTF} = \int_0^{+\infty} \left\{ 1 - \prod_{i=1}^{n} \left[1 - \mathrm{Pr} \{ \omega \in \Omega | X_i(\theta_i)(\omega) \geq t \} \right] \right\} \mathrm{d}t, \quad (4.1.49)$$

which is consistent with the result in stochastic case.

Remark 4.1.8 If $X_i, \ i = 1, 2, \ldots, n$ degenerate to fuzzy variables, the result in Theorem 4.1.4 degenerates to the form

$$\mathrm{MTTF} = \int_0^{+\infty} \mathrm{Cr}\{\theta \in \Theta | \max\{X_1(\theta_1), X_2(\theta_2), \ldots, X_n(\theta_n)\} \geq t\} \mathrm{d}t$$

$$= \max_{1 \leq i \leq n} E[X_i], \quad (4.1.50)$$

which is consistent with the result in fuzzy case.

Example 4.1.2 If the random fuzzy variables $X_i \sim \exp(\lambda_i)$, where λ_i are fuzzy variables on $(\Theta_i, \mathcal{P}(\Theta_i), \mathrm{Cr}_i)$, $i = 1, 2, \ldots, n$. Then we can arrive at

$$\mathrm{Pr}_\alpha^L \{ \omega \in \Omega | X(\theta)(\omega) > t \} = 1 - \prod_{i=1}^{n} \left[1 - \exp(-\lambda_{i,\alpha}^U t) \right]$$

and

$$\mathrm{Pr}_\alpha^U \{ \omega \in \Omega | X(\theta)(\omega) > t \} = 1 - \prod_{i=1}^{n} \left[1 - \exp(-\lambda_{i,\alpha}^L t) \right].$$

By Theorem 4.1.3 and Theorem 4.1.4, we have

$$\begin{aligned}R(t) &= \frac{1}{2}\int_0^1 \left(\Pr_\alpha^L\{\omega\in\Omega|X(\theta)(\omega)>t\} + \Pr_\alpha^U\{\omega\in\Omega|X(\theta)(\omega)>t\}\right)\mathrm{d}\alpha \\ &= \frac{1}{2}\int_0^1 \left(1-\prod_{i=1}^n[1-\exp(-\lambda_{i,\alpha}^U t)] + 1 - \prod_{i=1}^n[1-\exp(-\lambda_{i,\alpha}^L t)]\right)\mathrm{d}\alpha \\ &= 1 - E\left[\prod_{i=1}^n (1-\exp(-\lambda_i t))\right]\end{aligned}$$

and

$$\mathrm{MTTF} = \int_0^{+\infty}\left\{1 - E\left[\prod_{i=1}^n (1-\exp(-\lambda_i t))\right]\right\}\mathrm{d}t,$$

in which "E" is the expected value operator of fuzzy variables.

4.1.3 The Series–Parallel System

Consider a series–parallel system which is a series system consisting of m subsystems, and each subsystem is composed of n parallel components. Suppose that X_{ij} is the lifetime of component j in i th subsystem, which is a random fuzzy variable on the credibility space $(\Theta_{ij}, \mathcal{P}(\Theta_{ij}), \mathrm{Cr}_{ij})$, $i = 1, 2, \ldots, m,\ j = 1, 2, \ldots, n$. We also assume that the lifetimes of components are mutually independent. It is easy to see that the lifetime of the series–parallel system is $X = \min_{1\le i\le m}\left(\max_{1\le j\le n} X_{ij}\right)$, which is a random fuzzy variable on the product credibility space $(\Theta, \mathcal{P}(\Theta), \mathrm{Cr})$, where $\Theta = \Theta_{11}\times\Theta_{12}\times\cdots\times\Theta_{mn}$ and $\mathrm{Cr} = \mathrm{Cr}_{11}\wedge\mathrm{Cr}_{12}\wedge\cdots\wedge\mathrm{Cr}_{mn}$.

Theorem 4.1.5 *Let X_{ij}, $i = 1, 2, \ldots, m$, $j = 1, 2, \ldots, n$ be random fuzzy variables. Assume that the α-pessimistic values and the α-optimistic values of $E[X_{ij}(\theta_{ij})]$, $i = 1, 2, \ldots, m$, $j = 1, 2, \ldots, n$ are continuous almost everywhere with respect to α, $\alpha\in(0, 1]$. The reliability of the series–parallel system is*

$$R(t) = E\left[\prod_{i=1}^m\left\{1-\prod_{j=1}^n[1-\Pr\{\omega\in\Omega|X_{ij}(\theta_{ij})(\omega)>t\}]\right\}\right].$$

Remark 4.1.9 If X_{ij}, $i = 1, 2, \ldots, m$, $j = 1, 2, \ldots, n$ degenerate to random variables, the result in Theorem 4.1.5 degenerates to the form

$$R(t) = \prod_{i=1}^{m}\left\{1 - \prod_{j=1}^{n}[1 - \Pr\{\omega \in \Omega | X_{ij}(\omega) > t\}]\right\},$$

which is consistent with the result in stochastic case.

Remark 4.1.10 If $X_{ij},\ i = 1, 2, \ldots, m,\ j = 1, 2, \ldots, n$ degenerate to fuzzy variables, the result in Theorem 4.1.5 degenerates to the form

$$R(t) = \mathrm{Cr}\left\{\min_{1 \le i \le m}\left(\max_{1 \le j \le n} X_{ij}\right) > t\right\},$$

which is consistent with the result in fuzzy case.

Theorem 4.1.6 *Let $X_{ij},\ i = 1, 2, \ldots, m,\ j = 1, 2, \ldots, n$ be random fuzzy variables. Assume that the α-pessimistic values and the α-optimistic values of $E[X_{ij}(\theta_{ij})],\ i = 1, 2, \ldots, m,\ j = 1, 2, \ldots, n$ are continuous almost everywhere with respect to α, $\alpha \in (0, 1]$. The MTTF of the random fuzzy series–parallel system is*

$$\begin{aligned}\mathrm{MTTF} = \frac{1}{2}\int_0^1\int_0^{+\infty}&\left\{\prod_{i=1}^{m}\left\{1 - \prod_{j=1}^{n}[1 - \Pr{}_{\alpha}^{L}\{\omega \in \Omega | X_{ij}(\theta_{ij})(\omega) \ge t\}]\right\}\right.\\ &\left.+\prod_{i=1}^{m}\left\{1 - \prod_{j=1}^{n}[1 - \Pr{}_{\alpha}^{U}\{\omega \in \Omega | X_{ij}(\theta_{ij})(\omega) \ge t\}]\right\}\right\}\mathrm{d}t\mathrm{d}\alpha.\end{aligned}$$

Remark 4.1.11 If $X_{ij},\ i = 1, 2, \ldots, m,\ j = 1, 2, \ldots, n$ degenerate to random variables, the result in Theorem 4.1.6 degenerates to the form.

$$\mathrm{MTTF} = \int_0^{+\infty}\prod_{i=1}^{m}\left\{1 - \prod_{j=1}^{n}[1 - \Pr\{\omega \in \Omega | X_{ij}(\theta_{ij})(\omega) \ge t\}]\right\}\mathrm{d}t,$$

which is consistent with the result in stochastic case.

Remark 4.1.12 If $X_{ij},\ i = 1, 2, \ldots, m,\ j = 1, 2, \ldots, n$ degenerate to fuzzy variables, the result in Theorem 4.1.6 degenerates to the form

$$\begin{aligned}\mathrm{MTTF} &= \int_0^{+\infty}\mathrm{Cr}\left\{\theta \in \Theta | \min_{1 \le i \le m}\left(\max_{1 \le j \le n} X_{ij}(\theta_{ij})\right) \ge t\right\}\mathrm{d}t\\ &= \min_{1 \le i \le m}\max_{1 \le i \le n} E[X_i],\end{aligned}$$

which is consistent with the result in fuzzy case.

4.1.4 The Parallel–Series System

Consider a parallel–series system which is a parallel system consisting of m subsystems, and each subsystem is composed of n series components. Suppose that X_{ij} is the lifetime of component j in i th subsystem, which is a random fuzzy variable on the credibility space $(\Theta_{ij}, \mathcal{P}(\Theta_{ij}), \mathrm{Cr}_{ij})$, $i = 1, 2, \ldots, m$, $j = 1, 2, \ldots, n$. We also assume that the lifetimes of components are mutually independent. It is easy to see that the lifetime of the parallel–series system is $X = \max\limits_{1 \le i \le m} \left(\min\limits_{1 \le j \le n} X_{ij} \right)$, which is a random fuzzy variable on the product credibility space $(\Theta, \mathcal{P}(\Theta), \mathrm{Cr})$, where $\Theta = \Theta_{11} \times \Theta_{12} \times \cdots \times \Theta_{mn}$ and $\mathrm{Cr} = \mathrm{Cr}_{11} \wedge \mathrm{Cr}_{12} \wedge \cdots \wedge \mathrm{Cr}_{mn}$.

Theorem 4.1.7 *Let X_{ij}, $i = 1, 2, \ldots, m$, $j = 1, 2, \ldots, n$ be random fuzzy variables. Assume that the α-pessimistic values and the α-optimistic values of $E[X_{ij}(\theta_{ij})]$, $i = 1, 2, \ldots, m$, $j = 1, 2, \ldots, n$ are continuous almost everywhere with respect to α, $\alpha \in (0, 1]$. The reliability of the parallel–series system is*

$$R(t) = 1 - E\left[\prod_{i=1}^{m}\left\{1 - \prod_{j=1}^{n} \Pr\{\omega \in \Omega | X_{ij}(\theta_{ij})(\omega) > t\}\right\}\right].$$

Remark 4.1.13 If X_{ij}, $i = 1, 2, \ldots, m$, $j = 1, 2, \ldots, n$ degenerate to random variables, the result in Theorem 4.1.7 degenerates to the form

$$R(t) = 1 - \prod_{i=1}^{m}\left\{1 - \prod_{j=1}^{n} \Pr\{\omega \in \Omega | X_{ij}(\omega) > t\}\right\},$$

which is consistent with the result in stochastic case.

Remark 4.1.14 If X_{ij}, $i = 1, 2, \ldots, m$, $j = 1, 2, \ldots, n$ degenerate to fuzzy variables, the result in Theorem 4.1.7 degenerates to the form

$$R(t) = \mathrm{Cr}\left\{\max_{1 \le i \le m}\left(\min_{1 \le j \le n} X_{ij}\right) > t\right\},$$

which is consistent with the result in fuzzy case.

Theorem 4.1.8 *Let X_{ij}, $i = 1, 2, \ldots, m$, $j = 1, 2, \ldots, n$ be random fuzzy variables. Assume that the α-pessimistic values and the α-optimistic values of $E[X_{ij}(\theta_{ij})]$, $i = 1, 2, \ldots, m$, $j = 1, 2, \ldots, n$ are continuous almost everywhere with respect to α, $\alpha \in (0, 1]$. The MTTF of the random fuzzy parallel–series system is*

$$\text{MTTF} = \frac{1}{2}\int_0^1\int_0^{+\infty}\left\{2-\prod_{i=1}^{m}\left[1-\prod_{j=1}^{n}\text{Pr}_{\alpha}^{L}\{\omega\in\Omega|X_{ij}(\theta_{ij})(\omega)\geq t\}\right]\right.$$
$$\left.-\prod_{i=1}^{m}\left[1-\prod_{j=1}^{n}\text{Pr}_{\alpha}^{U}\{\omega\in\Omega|X_{ij}(\theta_{ij})(\omega)\geq t\}\right]\right\}\mathrm{d}t\mathrm{d}\alpha.$$

Remark 4.1.15 If X_{ij}, $i = 1, 2, \ldots, m$, $j = 1, 2, \ldots, n$ degenerate to random variables, the result in Theorem 4.1.8 degenerates to the form

$$\text{MTTF} = \int_0^{+\infty}\left\{1-\prod_{i=1}^{m}\left[1-\prod_{j=1}^{n}Pr\{\omega\in\Omega|X_{ij}(\omega)\geq t\}\right]\right\}\mathrm{d}t,$$

which is consistent with the result in stochastic case.

Remark 4.1.16 If X_{ij}, $i = 1, 2, \ldots, m$, $j = 1, 2, \ldots, n$ degenerate to fuzzy variables, the result in Theorem 4.1.8 degenerates to the form

$$\begin{aligned}\text{MTTF} &= \int_0^{+\infty}\text{Cr}\left\{\theta\in\Theta|\max_{1\leq i\leq m}\left(\min_{1\leq j\leq n}X_{ij}\right)\geq t\right\}\mathrm{d}t\\ &= \max_{1\leq i\leq m}\min_{1\leq i\leq n}E[X_i],\end{aligned}$$

which is consistent with the result in fuzzy case.

Example 4.1.3 Consider a lighting lamp system composed by lamp L_1, lamp L_2 and lamp L_3, and see Fig. 4.1. The lifetimes of lamp L_1, lamp L_2 and lamp L_3 are denoted by X_1, X_2 and X_3, respectively, and the lifetime of the lighting lamp system is denoted by X. We also assume X_1, X_2 and X_3 are independent random fuzzy variables and $X_i(\lambda_i) \sim \text{EXP}(\lambda_i)$, $i = 1, 2, 3$, where $\lambda_1 = (1, 2, 3)$, $\lambda_2 = (0, 1, 2)$ and $\lambda_3 = (0, 1, 2)$. We can arrive at

$$\begin{cases}\lambda_{1,\alpha}^{L} = 1+\alpha\\ \lambda_{1,\alpha}^{U} = 3-\alpha\end{cases},\quad\begin{cases}\lambda_{2,\alpha}^{L} = \alpha\\ \lambda_{2,\alpha}^{U} = 2-\alpha\end{cases},\quad\begin{cases}\lambda_{3,\alpha}^{L} = \alpha\\ \lambda_{3,\alpha}^{U} = 2-\alpha\end{cases}.$$

Then, we have

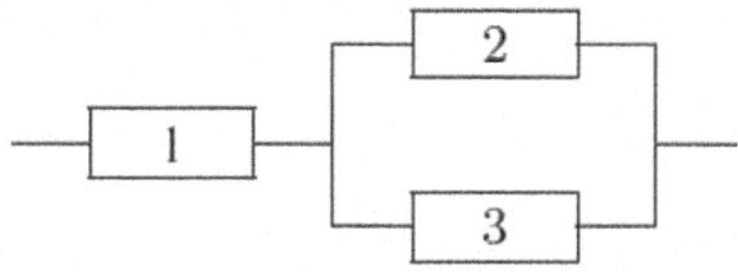

Fig. 4.1 Lighting lamp system

$$\begin{aligned}
&\Pr{}_{\alpha}^{L}\{X > t\} \\
&= \Pr{}_{\alpha}^{L}\{X_1 > t\} \cdot \left\{1 - \left(1 - \Pr{}_{\alpha}^{L}\{X_2 > t\}\right)\left(1 - \Pr{}_{\alpha}^{L}\{X_3 > t\}\right)\right\} \\
&= e^{-(3-\alpha)t}\left\{1 - \left(1 - e^{-(2-\alpha)t}\right)\left(1 - e^{-(2-\alpha)t}\right)\right\} \\
&= 2e^{-(5-2\alpha)t} - e^{-(7-3\alpha)t}
\end{aligned}$$

and

$$\begin{aligned}
&\Pr{}_{\alpha}^{U}\{X > t\} \\
&= \Pr{}_{\alpha}^{U}\{X_1 > t\} \cdot \left\{1 - \left(1 - \Pr{}_{\alpha}^{U}\{X_2 > t\}\right)\left(1 - \Pr{}_{\alpha}^{U}\{X_3 > t\}\right)\right\} \\
&= e^{-(1+\alpha)t}\left\{1 - \left(1 - e^{-\alpha t}\right)\left(1 - e^{-\alpha t}\right)\right\} \\
&= 2e^{-(1+2\alpha)t} - e^{-(1+3\alpha)t}.
\end{aligned}$$

Then, the reliability of the lighting lamp system is

$$\begin{aligned}
R(t) &= \frac{1}{2}\int_0^1 \left(\Pr{}_{\alpha}^{L}\{X > t\} + \Pr{}_{\alpha}^{U}\{X > t\}\right)\mathrm{d}\alpha \\
&= \frac{1}{2}\int_0^1 \left[2e^{-(5-2\alpha)t} - e^{-(7-3\alpha)t} + 2e^{-(1+2\alpha)t} - e^{-(1+3\alpha)t}\right]\mathrm{d}\alpha \\
&= \frac{1}{6t}\left(2e^{-t} - 3^{-5t} + e^{-7t}\right).
\end{aligned}$$

The MTTF of the lighting lamp system is

$$\text{MTTF} = \int_0^{+\infty} \frac{1}{6t}\left(2e^{-t} - 3e^{-5t} + e^{-7t}\right)\mathrm{d}t \approx 0.4804.$$

Example 4.1.4 A more elaborate example is a stereo hi-fi system with the following components: (1) FM tuner; (2) record changer; (3) amplifier; (4) speaker A; and (5) speaker B. The hi-fi system is functioning if we can obtain music (monaural or stereo) through FM or records. The system structure is illustrated in Fig. 4.2. Let X_i be the random fuzzy lifetime of component i, $i = 1, 2, \ldots, 5$. We also assume $X_i(\lambda_i) \sim \exp(\lambda_i)$, $i = 1, 2, \ldots, 5$, where $\lambda_1 = (1, 2, 3)$, $\lambda_2 =$

Fig. 4.2 Hi-fi system

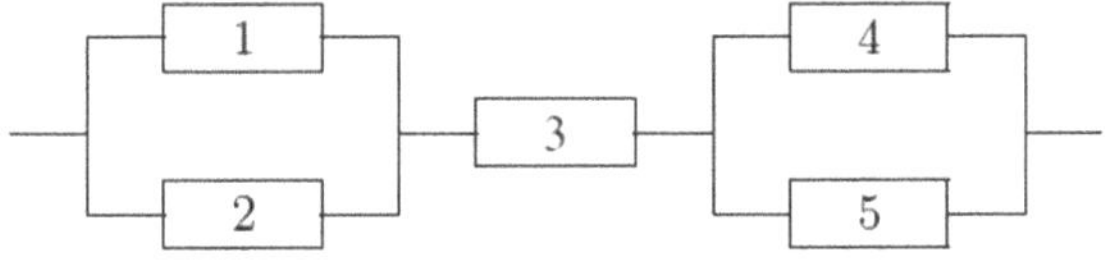

(1.5, 2.5, 3.5), $\lambda_3 = (0, 1, 2)$, $\lambda_4 = (1, 2, 3)$ and $\lambda_5 = (0.5, 1.5, 2.5)$. The lifetime of the hi-fi system is denoted by X.

We first compute

$$\begin{cases} \lambda_{1,\alpha}^L = 1+\alpha \\ \lambda_{1,\alpha}^U = 3-\alpha \end{cases}, \begin{cases} \lambda_{2,\alpha}^L = 1.5+\alpha \\ \lambda_{2,\alpha}^U = 3.5-\alpha \end{cases}, \begin{cases} \lambda_{3,\alpha}^L = \alpha \\ \lambda_{3,\alpha}^U = 2-\alpha \end{cases}, \begin{cases} \lambda_{4,\alpha}^L = 1+\alpha \\ \lambda_{4,\alpha}^U = 3-\alpha \end{cases}, \begin{cases} \lambda_{5,\alpha}^L = 0.5+\alpha \\ \lambda_{5,\alpha}^U = 2.5-\alpha \end{cases}.$$

Then, we can arrive at

$$\begin{aligned} &\Pr{}_\alpha^L\{X < t\} \\ &= \left\{1 - \left(1 - \Pr{}_\alpha^L\{X_1 < t\}\right)\left(1 - \Pr{}_\alpha^L\{X_2 < t\}\right)\right\} \cdot \Pr{}_\alpha^L\{X_3 < t\} \\ &\quad \cdot \left\{1 - \left(1 - \Pr{}_\alpha^L\{X_4 < t\}\right)\left(1 - \Pr{}_\alpha^L\{X_5 < t\}\right)\right\} \\ &= \left\{1 - \left(1 - e^{-(3-\alpha)t}\right)\left(1 - e^{-(3.5-\alpha)t}\right)\right\} e^{-(2-\alpha)t} \cdot \left\{1 - \left(1 - e^{-(3-\alpha)t}\right)\left(1 - e^{-(2.5-\alpha)t}\right)\right\} \\ &= e^{-(2-\alpha)t}\left[e^{-(3.5-\alpha)t} + e^{-(3-\alpha)t} - e^{-(6.5-2\alpha)t}\right] \cdot \left[e^{-(3-\alpha)t} + e^{-(2.5-\alpha)t} - e^{-(3.5-2\alpha)t}\right] \end{aligned}$$

and

$$\begin{aligned} &\Pr{}_\alpha^U\{X < t\} \\ &= \left\{1 - \left(1 - \Pr{}_\alpha^U\{X_1 < t\}\right)\left(1 - \Pr{}_\alpha^L\{X_2 < t\}\right)\right\} \cdot \Pr{}_\alpha^L\{X_3 < t\} \\ &\quad \cdot \left\{1 - \left(1 - \Pr{}_\alpha^L\{X_4 < t\}\right)\left(1 - \Pr{}_\alpha^L\{X_5 < t\}\right)\right\} \\ &= \left\{1 - \left(1 - e^{-(1+\alpha)t}\right)\left(1 - e^{-(1.5+\alpha)t}\right)\right\} e^{-\alpha t} \cdot \left\{1 - \left(1 - e^{-(1+\alpha)t}\right)\left(1 - e^{-(0.5+\alpha)t}\right)\right\} \\ &= e^{-\alpha t}\left[e^{-(1+\alpha)t} + e^{-(1.5+\alpha)t} - e^{-(2.5+2\alpha)t}\right] \cdot \left[e^{-(1+\alpha)t} + e^{-(0.5+\alpha)t} - e^{-(1.5+2\alpha)t}\right]. \end{aligned}$$

The reliability of the hi-fi system is

$$\begin{aligned} R(t) = \frac{1}{2}\int_0^1 &\left\{e^{-(2-\alpha)t}\left[e^{-(3.5-\alpha)t} + e^{-(3-\alpha)t} - e^{-(6.5-2\alpha)t}\right] \times \left[e^{-(3-\alpha)t} + e^{-(2.5-\alpha)t} - e^{-(5.5-2\alpha)t}\right]\right. \\ &\left. + e^{-\alpha t}\left[e^{-(1+\alpha)t} + e^{-(1.5+\alpha)t} - e^{-(2.5+2\alpha)t}\right] \times \left[e^{-(1+\alpha)t} + e^{-(0.5+\alpha)t} - e^{-(1.5+2\alpha)t}\right]\right\} d\alpha \\ = \frac{1}{6t}&\left(2e^{-2t} + e^{-1.5t} + e^{-2.5t} - e^{-8.5t} - 2e^{-8t} - e^{-7.5t}\right) + \frac{1}{10t}\left(e^{-4t} - e^{-14t}\right) \\ + \frac{1}{8t}&\left(2e^{-10t} - 2e^{-6t} + e^{-10.5t} + e^{-11.5t} - e^{-2.5t} + 2e^{-7t} - 2e^{-3t} - e^{-3.5t}\right). \end{aligned}$$

The MTTF of the hi-fi system is

$$\mathrm{MTTF} = \int_0^{+\infty} \left[\frac{1}{6t}\left(2e^{-2t} + e^{-1.5t} + e^{-2.5t} - e^{-8.5t} - 2e^{-8t} - e^{-7.5t}\right) + \frac{1}{10t}\left(e^{-4t} - e^{-14t}\right)\right.$$

$$+\frac{1}{8t}\left(2e^{-10t}-2e^{-6t}+e^{-10.5t}+e^{-11.5t}-e^{-2.5t}+2e^{-7t}-2e^{-3t}-e^{-3.5t}\right)\Big]\mathrm{d}t$$
$$\approx 0.3921.$$

4.2 Cold Standby Systems

Consider a cold standby system composed of n independent components. At the initial moment, one component is in operating and the remaining $n-1$ components are in standby. When the operating component fails, the standby component is replaced one by one until all components fail and the cold system fails. Suppose that the failed components are nonrepairable and the components in standby do not deteriorate.

4.2.1 *The Perfect Conversion Switch Case*

Let X_i be the lifetime of component i on the credibility space $(\Theta_i, \mathcal{P}(\Theta_i), \mathrm{Cr}_i)$, $i = 1, 2, \ldots, n$. Suppose that the conversion switch is completely reliable and the conversion is instantaneous. The lifetime of the cold standby system can be expressed as the sum of the lifetimes of the n components, that is, $X = X_1 + X_2 + \cdots + X_n$, which is a random fuzzy variable on the product credibility space $(\Theta, \mathcal{P}(\Theta), \mathrm{Cr})$, where $\Theta = \Theta_1 \times \Theta_2 \times \cdots \times \Theta_n$ and $\mathrm{Cr} = \mathrm{Cr}_1 \wedge \mathrm{Cr}_2 \wedge \cdots \wedge \mathrm{Cr}_n$.

Theorem 4.2.1 *Let X_i, $i = 1, 2, \ldots, n$ be random fuzzy variables. Assume that the α-pessimistic values and the α-optimistic values of $E[X_i(\theta_i)]$, $\theta_i \in \Theta_i$, $i = 1, 2, \ldots, n$ are continuous almost everywhere with respect to α, $\alpha \in (0, 1]$. The reliability of random fuzzy cold standby system is*

$$R(t) = E[\Pr\{\omega \in \Omega | X_1(\theta_1)(\omega) + \cdots + X_n(\theta_n)(\omega) > t\}]. \tag{4.2.1}$$

Proof By Definition 4.1, Definition 1.3.7 and Proposition 1.2.23, we have

$$\begin{aligned}
R(t) &= \mathrm{Ch}\{X \geq t\} \\
&= \int_0^1 \mathrm{Cr}\{\theta \in \Theta | \Pr\{X(\theta) > t\} \geq p\}\mathrm{d}p \\
&= \frac{1}{2}\int_0^1 \left(\Pr{}_{\alpha}^{L}\{\omega \in \Omega | X(\theta)(\omega) > t\} + \Pr{}_{\alpha}^{U}\{\omega \in \Omega | X(\theta)(\omega) > t\}\right)\mathrm{d}\alpha
\end{aligned}$$

$$= \frac{1}{2}\int_0^1 \left(\Pr{}_\alpha^L\{\omega \in \Omega | X_1(\theta_1)(\omega) + \cdots + X_n(\theta_n)(\omega) > t\}\right.$$
$$\left. + \Pr{}_\alpha^U\{\omega \in \Omega | X_1(\theta_1)(\omega) + \cdots + X_n(\theta_n)(\omega) > t\}\right) \mathrm{d}\alpha$$
$$= E[\Pr\{\omega \in \Omega | X_1(\theta_1)(\omega) + \cdots + X_n(\theta_n)(\omega) > t\}]. \quad (4.2.2)$$

The theorem is proved.

Remark 4.2.1 If X_i, $i = 1, 2, \ldots, n$ degenerate to random variables, the result in Theorem 4.2.1 degenerates to the form

$$R(t) = \Pr\{\omega \in \Omega | X_1(\omega) + X_2(\omega) + \cdots + X_n(\omega) > t\}, \quad (4.2.3)$$

which is consistent with the result in stochastic case.

Remark 4.2.2 If X_i, $i = 1, 2, \ldots, n$ degenerate to fuzzy variables, the result in Theorem 4.2.1 degenerates to the form

$$R(t) = \mathrm{Cr}\{\theta \in \Theta | X_1(\theta_1) + X_2(\theta_2) + \cdots + X_n(\theta_n) > t\}, \quad (4.2.4)$$

which is consistent with the result in fuzzy case.

Theorem 4.2.2 *Let X_i, $i = 1, 2, \ldots, n$ be random fuzzy variables. Assume that the α-pessimistic values and the α-optimistic values of $E[X_i(\theta_i)]$, $\theta_i \in \Theta_i$, $i = 1, 2, \ldots, n$ are continuous almost everywhere with respect to α, $\alpha \in (0, 1]$. The MTTF of random fuzzy cold standby system is*

$$\mathrm{MTTF} = \frac{1}{2}\int_0^1 \int_0^{+\infty} \left\{\Pr{}_\alpha^L\left\{\omega \in \Omega | X_1(\theta_1)(\omega) + \cdots + X_n(\theta_n)(\omega) \geq t\right\}\right.$$
$$\left. + \Pr{}_\alpha^U\left\{\omega \in \Omega | X_1(\theta_1)(\omega) + \cdots + X_n(\theta_n)(\omega) \geq t\right\}\right\} \mathrm{d}t\,\mathrm{d}\alpha. \quad (4.2.5)$$

Proof It follows from Definition 4.2 and Proposition 1.2.23 that

$$\mathrm{MTTF} = \int_0^{+\infty} \mathrm{Cr}\{\theta \in \Theta | E[X(\theta)] \geq r\} \mathrm{d}r$$
$$= \frac{1}{2}\int_0^1 \left(E[X(\theta)]_\alpha^L + E[X(\theta)]_\alpha^U\right) \mathrm{d}\alpha. \quad (4.2.6)$$

Let $A_i = \{\theta_i \in \Theta_i | \mu\{\theta_i\} \geq \alpha\}$, $i = 1, 2, \ldots, n$. Since the α-pessimistic values and the α-optimistic values of fuzzy variables $E[X_i(\theta_i)]$, $\theta_i \in \Theta_i$, $i = 1, 2, \ldots, n$

are continuous almost everywhere with respect to α, $\alpha \in (0, 1]$, then there at least exist points $\theta_i^{'}, \theta_i^{''} \in A_i$ such that

$$E[X_i(\theta_i^{'})] = E[X_i(\theta_i)]_\alpha^L \text{ and } E[X_i(\theta_i^{''})] = E[X_i(\theta_i)]_\alpha^U.$$

For any $\theta_{i,\alpha} \in A_i$, $i = 1, 2, \ldots, n$, we have

$$E[X_i(\theta_i^{'})] \le E[X_i(\theta_{i,\alpha})] \le E[X_i(\theta_i^{''})]. \tag{4.2.7}$$

It follows from Definition 1.1.21 that

$$X_i(\theta_i^{'}) \le_d X_i(\theta_{i,\alpha}) \le_d X_i(\theta_i^{''}),\ \forall \theta_{i,\alpha} \in A_i,\ i = 1, 2, \ldots, n. \tag{4.2.8}$$

Hence, we have

$$\begin{aligned} X_1(\theta_1^{'}) + \cdots + X_n(\theta_n^{'}) &\le_d X_1(\theta_{1,\alpha}) + \cdots + X_n(\theta_{n,\alpha}) \\ &\le_d X_1(\theta_1^{''}) + \cdots + X_n(\theta_n^{''}). \end{aligned} \tag{4.2.9}$$

It follows from Definition 1.1.21 that

$$\begin{aligned} E[X_1(\theta_1^{'}) + \cdots + X_n(\theta_n^{'})] &\le E[X_1(\theta_{1,\alpha}) + \cdots + X_n(\theta_{n,\alpha})] \\ &\le E[X_1(\theta_1^{''}) + \cdots + X_n(\theta_n^{''})]. \end{aligned} \tag{4.2.10}$$

Since $\theta_{i,\alpha}$ are arbitrary points in A_i, $i = 1, 2, \ldots, n$, we have

$$\begin{aligned} E[X(\theta)]_\alpha^L &= E[X_1(\theta_1^{'}), \ldots, X_n(\theta_n^{'})] \\ &= \int_0^{+\infty} \Pr\left\{\omega \in \Omega | X_1(\theta_1^{'})(\omega) + \cdots + X_n(\theta_n^{'})(\omega) \ge t\right\} \mathrm{d}t \end{aligned} \tag{4.2.11}$$

and

$$\begin{aligned} E[X(\theta)]_\alpha^U &= E[X_1(\theta_1^{''}), \ldots, X_n(\theta_n^{''})] \\ &= \int_0^{+\infty} \Pr\left\{\omega \in \Omega | X_1(\theta_1^{''})(\omega) + \cdots + X_n(\theta_n^{''})(\omega) \ge t\right\} \mathrm{d}t. \end{aligned} \tag{4.2.12}$$

It follows from Eqs. (4.2.6), (4.2.11) and (4.2.12) that

$$\mathrm{MTTF} = \frac{1}{2} \int_0^1 \left(E[X(\theta)]_\alpha^L + E[X(\theta)]_\alpha^U\right) \mathrm{d}\alpha$$

$$= \frac{1}{2}\int_0^1 \left\{ \int_0^{+\infty} \Pr\left\{\omega \in \Omega | X_1(\theta_1^{'})(\omega) + \cdots + X_n(\theta_n^{'})(\omega) \geq t\right\} dt \right.$$

$$\left. + \int_0^{+\infty} \Pr\left\{\omega \in \Omega | X_1(\theta_1^{''})(\omega) + \cdots + X_n(\theta_n^{''})(\omega) \geq t\right\} dt \right\} d\alpha$$

$$= \frac{1}{2}\int_0^1 \int_0^{+\infty} \left\{ \Pr\left\{\omega \in \Omega | X_1(\theta_1^{'})(\omega) + \cdots + X_n(\theta_n^{'})(\omega) \geq t\right\} \right.$$

$$\left. + \Pr\left\{\omega \in \Omega | X_1(\theta_1^{''})(\omega) + \cdots + X_n(\theta_n^{''})(\omega) \geq t\right\}\right\} dt d\alpha. \tag{4.2.13}$$

On the other hand, by Eq. (4.2.9) and Definition 1.1.31, we have

$$\begin{aligned} &\Pr\left\{\omega \in \Omega | X_1(\theta_1^{'}) + \cdots + X_n(\theta_n^{'}) \geq t\right\} \\ &\leq \Pr\{\omega \in \Omega | X_1(\theta_{1,\alpha}) + \cdots + X_n(\theta_{n,\alpha}) \geq t\} \\ &\leq \Pr\left\{\omega \in \Omega | X_1(\theta_1^{''}) + \cdots + X_n(\theta_n^{''}) \geq t\right\}. \end{aligned} \tag{4.2.14}$$

Since $\theta_{i,\alpha}$ are arbitrary points in $A_i,\ i = 1, 2, \ldots, n$, we have

$$\begin{aligned} &\Pr{}_\alpha^L\{\omega \in \Omega | X_1(\theta_1)(\omega) + \cdots + X_n(\theta_n)(\omega) \geq t\} \\ &= \Pr\left\{\omega \in \Omega | X_1(\theta_1^{'}) + \cdots + X_n(\theta_n^{'}) \geq t\right\} \end{aligned} \tag{4.2.15}$$

and

$$\begin{aligned} &\Pr{}_\alpha^U\{\omega \in \Omega | X_1(\theta_1)(\omega) + \cdots + X_n(\theta_n)(\omega) \geq t\} \\ &= \Pr\left\{\omega \in \Omega | X_1(\theta_1^{''}) + \cdots + X_n(\theta_n^{''}) \geq t\right\}. \end{aligned} \tag{4.2.16}$$

It follows from Eqs. (4.2.13), (4.2.15) and (4.2.16) that

$$\begin{aligned} \text{MTTF} = &\frac{1}{2}\int_0^1 \int_0^{+\infty} \{\Pr{}_\alpha^L\{\omega \in \Omega | X_1(\theta_1)(\omega) + \cdots + X_n(\theta_n)(\omega) \geq t\} \\ &+ \Pr{}_\alpha^U\{\omega \in \Omega | X_1(\theta_1)(\omega) + \cdots + X_n(\theta_n)(\omega) \geq t\}\} dt d\alpha. \end{aligned} \tag{4.2.17}$$

The theorem is proved.

Remark 4.2.3 If $X_i,\ i = 1, 2, \ldots, n$ degenerate to random variables, the result in Theorem 4.2.2 degenerates to the form

$$\text{MTTF} = \int_0^{+\infty} \Pr\{\omega \in \Omega | X_1(\omega) + \cdots + X_n(\omega) \geq t\} \mathrm{d}t = \sum_{i=1}^{n} E[X_i], \quad (4.2.18)$$

which is consistent with the result in stochastic case.

Remark 4.2.4 If X_i, $i = 1, 2, \ldots, n$ degenerate to fuzzy variables, the result in Theorem 4.2.2 degenerates to the form

$$\text{MTTF} = \int_0^{+\infty} \text{Cr}\{\theta \in \Theta | X_1(\theta_1) + \cdots + X_n(\theta_n) \geq t\} \mathrm{d}t = \sum_{i=1}^{n} E[X_i], \quad (4.2.19)$$

which is consistent with the result in fuzzy case.

4.2.2 The Imperfect Conversion Switch Case: The Lifetime of the Conversion Switch has 0–1 Distribution

Let X_i be the lifetime of component i on the credibility space $(\Theta_i, \mathcal{P}(\Theta_i), \text{Cr}_i)$, $i = 1, 2, \ldots, n$. Suppose the failed components are nonrepairable and the conversion switch may fail. When the conversion switch is used, the probability of "the conversion switch is still operating" is p, and then the probability of "the conversion switch is failed" is $1 - p$. Let

$$v = \begin{cases} j, & \text{if the conversion switch fails in } j\text{th swiching}, \; j = 1, 2, \ldots, n-1 \\ n, & \text{if the conversion switch is still operating until no perfect component left.} \end{cases}$$

From the definition of v, it is easy to see that

$$\Pr\{v = j\} = p^{j-1}q, \; j = 1, 2, \ldots, n-1,$$
$$\Pr\{v = n\} = p^{n-1}.$$

Obviously, the lifetime of the cold standby system can be expressed as $X = X_1 + X_2 + \cdots + X_v$, which is a random fuzzy variable on the product credibility space $(\Theta, \mathcal{P}(\Theta), \text{Cr})$, where $\Theta = \Theta_1 \times \Theta_2 \times \cdots \times \Theta_n$ and $\text{Cr} = \text{Cr}_1 \wedge \text{Cr}_2 \wedge \cdots \wedge \text{Cr}_n$.

Theorem 4.2.3 *Let X_i, $i = 1, 2, \ldots, n$ be random fuzzy variables. Assume that the α-pessimistic values and the α-optimistic values of $E[X_i(\theta_i)]$, $\theta_i \in \Theta_i$, $i =$*

$1, 2, \ldots, n$ *are continuous almost everywhere with respect to* α, $\alpha \in (0, 1]$. *The reliability of the random fuzzy cold standby system is*

$$R(t) = \sum_{j=1}^{n-1} p^{j-1} q E\big[\Pr\big\{\omega \in \Omega | X_1(\theta_1)(\omega) + \cdots + X_j(\theta_j)(\omega) > t\big\}\big]$$
$$+ p^{n-1} E[\Pr\{\omega \in \Omega | X_1(\theta_1)(\omega) + \cdots + X_n(\theta_n)(\omega) > t\}].$$

Proof Let $A_i = \{\theta_i \in \Theta_i | \mu\{\theta_i\} \geq \alpha\}$, $i = 1, 2, \ldots, n$. Since the α-pessimistic values and the α-optimistic values of fuzzy variables $E[X_i(\theta_i)]$, $\theta_i \in \Theta_i$, $i = 1, 2, \ldots, n$ are continuous almost everywhere with respect to α, $\alpha \in (0, 1]$, then there at least exist points $\theta_i^{'}, \theta_i^{''} \in A_i$, $i = 1, 2, \ldots, n$ such that

$$E[X_i(\theta_i^{'})] = E[X_i(\theta_i)]_\alpha^L \text{ and } E[X_i(\theta_i^{''})] = E[X_i(\theta_i)]_\alpha^U.$$

For any $\theta_{i,\alpha} \in A_i$, we have

$$E[X_i(\theta_i^{'})] \leq E[X_i(\theta_{i,\alpha})] \leq E[X_i(\theta_i^{''})], \; i = 1, 2, \ldots, n.$$

It follows from Definition 1.1.21 that

$$X_i(\theta_i^{'}) \leq_d X_i(\theta_{i,\alpha}) \leq_d X_i(\theta_i^{''}), \; \forall \theta_{i,\alpha} \in A_i, \; i = 1, 2, \ldots, n.$$

Hence, for $j = 1, 2, \ldots, n$, we have

$$X_1(\theta_1^{'}) + \cdots + X_j(\theta_j^{'}) \leq_d X_1(\theta_{1,\alpha}) + \cdots + X_j(\theta_{j,\alpha})$$
$$\leq_d X_1(\theta_1^{''}) + \cdots + X_j(\theta_j^{''}). \quad (4.2.20)$$

It follows from Definition 1.1.21 that

$$\Pr\Big\{\omega \in \Omega | X_1(\theta_1^{'})(\omega) + \cdots + X_j(\theta_j^{'})(\omega) > t\Big\}$$
$$\leq \Pr\big\{\omega \in \Omega | X_1(\theta_{1,\alpha})(\omega) + \cdots + X_j(\theta_{j,\alpha})(\omega) > t\big\}$$
$$\leq \Pr\Big\{\omega \in \Omega | X_1(\theta_1^{''})(\omega) + \cdots + X_j(\theta_j^{''})(\omega) > t\Big\} \quad (4.2.21)$$

for any $j = 1, 2, \ldots, n$.

We can construct three cold standby systems in which the conversion switches are "0–1" mode. When the conversion switch in each system is used, the probability of "the conversion switch is operating" is p and the probability of "the conversion switch is failed" is $q = 1 - p$.

(1) Cold standby system *A*: Let $X_i(\theta_i^{'})$ be the lifetime of component i, $i = 1, 2, \ldots, n$.
(2) Cold standby system *B*: Let $X_i(\theta_i^{''})$ be the lifetime of component i, $i = 1, 2, \ldots, n$.

(3) Cold standby system C: Let $X_i(\theta_{i,\alpha})$ be the lifetime of component i, $i = 1, 2, \ldots, n$.

Let X_A, X_B and X_C be the lifetimes of cold standby systems A, B and C, respectively. It is easy to see that the cold standby systems A, B and C are all stochastic cold standby systems. By the result in classical reliability theory, we have

$$\begin{aligned}
&\Pr\{\omega \in \Omega | X_A(\omega) > t\} \\
&= \sum_{j=1}^{n-1} \Pr\left\{\omega \in \Omega | X_1(\theta_1^{'})(\omega) + \cdots + X_j(\theta_j^{'})(\omega) > t\right\} p^{j-1} q \\
&+ \Pr\left\{\omega \in \Omega | X_1(\theta_1^{'})(\omega) + \cdots + X_n(\theta_n^{'})(\omega) > t\right\} p^{n-1},
\end{aligned} \tag{4.2.22}$$

$$\begin{aligned}
&\Pr\{\omega \in \Omega | X_B(\omega) > t\} \\
&= \sum_{j=1}^{n-1} \Pr\left\{\omega \in \Omega | X_1(\theta_1^{''})(\omega) + \cdots + X_j(\theta_j^{''})(\omega) > t\right\} p^{j-1} q \\
&+ \Pr\left\{\omega \in \Omega | X_1(\theta_1^{''})(\omega) + \cdots + X_n(\theta_n^{''})(\omega) > t\right\} p^{n-1}
\end{aligned} \tag{4.2.23}$$

and

$$\begin{aligned}
&\Pr\{\omega \in \Omega | X_C(\omega) > t\} \\
&= \sum_{j=1}^{n-1} \Pr\left\{\omega \in \Omega | X_1(\theta_{1,\alpha})(\omega) + \cdots + X_j(\theta_{j,\alpha})(\omega) > t\right\} p^{j-1} q \\
&+ \Pr\left\{\omega \in \Omega | X_1(\theta_{1,\alpha})(\omega) + \cdots + X_n(\theta_{n,\alpha})(\omega) > t\right\} p^{n-1}.
\end{aligned} \tag{4.2.24}$$

By Eqs. (4.2.21)–(4.2.24), we have

$$\begin{aligned}
\Pr\{\omega \in \Omega | X_A(\omega) > t\} &\le \Pr\{\omega \in \Omega | X_C(\omega) > t\} \\
&\le \Pr\{\omega \in \Omega | X_B(\omega) > t\}.
\end{aligned} \tag{4.2.25}$$

Since $\theta_{i,\alpha}$ are arbitrary points in A_i, $i = 1, 2, \ldots, n$, we can arrive at

$$\begin{aligned}
&\Pr{}_{\alpha}^{L}\{\omega \in \Omega | X(\theta)(\omega) > t\} \\
&= \sum_{j=1}^{n-1} \Pr\left\{\omega \in \Omega | X_1(\theta_1^{'})(\omega) + \cdots + X_j(\theta_j^{'})(\omega) > t\right\} p^{j-1} q \\
&+ \Pr\left\{\omega \in \Omega | X_1(\theta_1^{'})(\omega) + \cdots + X_n(\theta_n^{'})(\omega) > t\right\} p^{n-1}
\end{aligned} \tag{4.2.26}$$

and

$$\begin{aligned}&\Pr{}^U_\alpha\{\omega \in \Omega | X(\theta)(\omega) > t\}\\&= \sum_{j=1}^{n-1} \Pr\left\{\omega \in \Omega | X_1(\theta_1^{''})(\omega) + \cdots + X_j(\theta_j^{''})(\omega) > t\right\} p^{j-1} q\\&\quad + \Pr\left\{\omega \in \Omega | X_1(\theta_1^{''})(\omega) + \cdots + X_n(\theta_n^{''})(\omega) > t\right\} p^{n-1}.\end{aligned} \tag{4.2.27}$$

On the other hand, from Eq. (4.2.21), we can arrive at

$$\begin{aligned}&\Pr{}^L_\alpha\left\{\omega \in \Omega | X_1(\theta_1)(\omega) + \cdots + X_j(\theta_j)(\omega) > t\right\}\\&= \Pr\left\{\omega \in \Omega | X_1(\theta_1^{'})(\omega) + \cdots + X_j(\theta_j^{'})(\omega) > t\right\}\end{aligned} \tag{4.2.28}$$

and

$$\begin{aligned}&\Pr{}^U_\alpha\left\{\omega \in \Omega | X_1(\theta_1)(\omega) + \cdots + X_j(\theta_j)(\omega) > t\right\}\\&= \Pr\left\{\omega \in \Omega | X_1(\theta_1^{''})(\omega) + \cdots + X_j(\theta_j^{''})(\omega) > t\right\}.\end{aligned} \tag{4.2.29}$$

for each $j = 1, 2, \ldots, n$. It follows from Eqs. (4.2.26)–(4.2.29) that

$$\begin{aligned}&\Pr{}^L_\alpha\{\omega \in \Omega | X(\theta)(\omega) > t\}\\&= \sum_{j=1}^{n-1} \Pr{}^L_\alpha\left\{\omega \in \Omega | X_1(\theta_1)(\omega) + \cdots + X_j(\theta_j)(\omega) > t\right\} p^{j-1} q\\&\quad + \Pr{}^L_\alpha\left\{\omega \in \Omega | X_1(\theta_1)(\omega) + \cdots + X_n(\theta_n)(\omega) > t\right\} p^{n-1}\end{aligned} \tag{4.2.30}$$

and

$$\begin{aligned}&\Pr{}^U_\alpha\{\omega \in \Omega | X(\theta)(\omega) > t\}\\&= \sum_{j=1}^{n-1} \Pr{}^U_\alpha\left\{\omega \in \Omega | X_1(\theta_1)(\omega) + \cdots + X_j(\theta_j)(\omega) > t\right\} p^{j-1} q\\&\quad + \Pr{}^U_\alpha\left\{\omega \in \Omega | X_1(\theta_1)(\omega) + \cdots + X_n(\theta_n)(\omega) > t\right\} p^{n-1}.\end{aligned} \tag{4.2.31}$$

By Definition 4.1, Definition 1.3.7 and Theorem 1.2.23, we have

$$R(t) = \int_0^1 \mathrm{Cr}\{\theta \in \Theta | \Pr\{X(\theta) > t\} \ge p\} \mathrm{d}p$$

$$= \frac{1}{2}\int_0^1 \left(\Pr{}_\alpha^L\{\omega \in \Omega | X(\theta)(\omega) > t\} + \Pr{}_\alpha^U\{\omega \in \Omega | X(\theta)(\omega) > t\}\right) d\alpha. \tag{4.2.32}$$

It follows from Eqs. (4.2.30)–(4-.2.32) that

$$\begin{aligned} R(t) = \frac{1}{2}\int_0^1 \Bigg(&\sum_{j=1}^{n-1} \Pr{}_\alpha^L\big\{\omega \in \Omega | X_1(\theta_1)(\omega) + \cdots + X_j(\theta_j)(\omega) > t\big\} p^{j-1} q \\ &+ \Pr{}_\alpha^L\{\omega \in \Omega | X_1(\theta_1)(\omega) + \cdots + X_{\mathrm{n}}(\theta_n)(\omega) > t\} p^{n-1} \\ &+ \sum_{j=1}^{n-1} \Pr{}_\alpha^U\big\{\omega \in \Omega | X_1(\theta_1)(\omega) + \cdots + X_j(\theta_j)(\omega) > t\big\} p^{j-1} q \\ &+ \Pr{}_\alpha^U\{\omega \in \Omega | X_1(\theta_1)(\omega) + \cdots + X_{\mathrm{n}}(\theta_n)(\omega) > t\} p^{n-1}\Bigg) d\alpha \\ = \sum_{j=1}^{n-1} \Bigg[&\frac{1}{2} p^{j-1} q \int_0^1 \Big(\Pr{}_\alpha^L\big\{\omega \in \Omega | X_1(\theta_1)(\omega) + \cdots + X_j(\theta_j)(\omega) > t\big\} \\ &+ \Pr{}_\alpha^U\big\{\omega \in \Omega | X_1(\theta_1)(\omega) + \cdots + X_j(\theta_j)(\omega) > t\big\}\Big) d\alpha\Bigg] \\ &+ \frac{1}{2} p^{n-1} \int_0^1 \Big(\Pr{}_\alpha^L\{\omega \in \Omega | X_1(\theta_1)(\omega) + \cdots + X_n(\theta_n)(\omega) > t\} \\ &+ \Pr{}_\alpha^U\{\omega \in \Omega | X_1(\theta_1)(\omega) + \cdots + X_n(\theta_n)(\omega) > t\}\Big) d\alpha \\ = \sum_{j=1}^{n-1} &p^{j-1} q E\big[\Pr\big\{\omega \in \Omega | X_1(\theta_1)(\omega) + \cdots + X_j(\theta_j)(\omega) > t\big\}\big] \\ &+ p^{n-1} E[\Pr\{\omega \in \Omega | X_1(\theta_1)(\omega) + \cdots + X_n(\theta_n)(\omega) > t\}]. \end{aligned}$$

The theorem is proved.

Remark 4.2.5 If X_i, $i = 1, 2, \ldots, n$ degenerate to random variables, the result in Theorem 4.2.3 degenerates to the form

$$\begin{aligned} R(t) = &\sum_{j=1}^{n-1} p^{j-1} q \Pr\big\{\omega \in \Omega | X_1(\theta_1)(\omega) + \cdots + X_j(\theta_j)(\omega) > t\big\} \\ &+ p^{n-1} \Pr\{\omega \in \Omega | X_1(\theta_1)(\omega) + \cdots + X_n(\theta_n)(\omega) > t\}, \end{aligned}$$

which is consistent with the result in stochastic case.

Theorem 4.2.4 *Let X_i, $i = 1, 2, \ldots, n$ be random fuzzy variables. Assume that the α-pessimistic values and the α-optimistic values of $E[X_i(\theta_i)]$, $\theta_i \in \Theta_i$, $i =$*

$1, 2, \ldots, n$ *are continuous almost everywhere with respect to* α, $\alpha \in (0, 1]$. *Then, MTTF of the cold standby system* is

$$\text{MTTF} = \sum_{j=1}^{n-1} p^{j-1} q E[X_1 + \cdots + X_j] + p^{n-1} E[X_1 + \cdots + X_n].$$

Proof It follows from Definition 4.2 and Theorem 1.2.23 that

$$\begin{aligned}\text{MTTF} &= \int_0^{+\infty} \text{Cr}\{\theta \in \Theta | E[X(\theta)] \geq r\} \mathrm{d}r \\ &= \frac{1}{2} \int_0^1 \left(E[X(\theta)]_\alpha^L + E[X(\theta)]_\alpha^U\right) \mathrm{d}\alpha. \end{aligned} \tag{4.2.33}$$

In the proof of Theorem 4.2.3, we have constructed three cold standby systems. Denote the MTTF of cold standby systems A, B and C by MTTF_A, MTTF_B and MTTF_C, respectively. It follows from the result in classical reliability theory that

$$\begin{aligned}\text{MTTF}_A &= \sum_{j=1}^{n-1} E[X_1(\theta_1^{'}) + \cdots + X_j(\theta_j^{'})] p^{j-1} q \\ &\quad + E[X_1(\theta_1^{'}) + \cdots + X_n(\theta_n^{'})] p^{n-1}, \end{aligned} \tag{4.2.34}$$

$$\begin{aligned}\text{MTTF}_B &= \sum_{j=1}^{n-1} E[X_1(\theta_1^{''}) + \cdots + X_j(\theta_j^{''})] p^{j-1} q \\ &\quad + E[X_1(\theta_1^{''}) + \cdots + X_n(\theta_n^{''})] p^{n-1} \end{aligned} \tag{4.2.35}$$

and

$$\begin{aligned}\text{MTTF}_C &= \sum_{j=1}^{n-1} E[X_1(\theta_{1,\alpha}) + \cdots + X_j(\theta_{j,\alpha})] p^{j-1} q \\ &\quad + E[X_1(\theta_{1,\alpha}) + \cdots + X_n(\theta_{n,\alpha})] p^{n-1}. \end{aligned} \tag{4.2.36}$$

On the other hand, by Eq. (4.2.20), we have

$$\begin{aligned} E[X_1(\theta_1^{'}) + \cdots + X_j(\theta_j^{'})] &\leq E[X_1(\theta_{1,\alpha}) + \cdots + X_j(\theta_{j,\alpha})] \\ &\leq E[X_1(\theta_1^{''}) + \cdots + X_j(\theta_j^{''})] \end{aligned} \tag{4.2.37}$$

for any $j = 1, 2, \ldots, n$. By Eqs. (4.2.34)–(4.2.37), we can arrive at

$$\mathrm{MTTF}_A \le \mathrm{MTTF}_C \le \mathrm{MTTF}_B.$$

Since $\theta_{i,\alpha}$ are arbitrary points in $A_i, i = 1, 2, \ldots, n$, then

$$\begin{aligned} E[X(\theta)]_\alpha^L &= \sum_{j=1}^{n-1} E[X_1(\theta_1^{'}) + \cdots + X_j(\theta_j^{'})]p^{j-1}q \\ &\quad + E[X_1(\theta_1^{'}) + \cdots + X_n(\theta_n^{'})]p^{n-1} \\ &= \sum_{j=1}^{n-1} E[X_1(\theta_1) + \cdots + X_j(\theta_j)]_\alpha^L p^{j-1}q \\ &\quad + E[X_1(\theta_1) + \cdots + X_n(\theta_n)]_\alpha^L p^{n-1} \end{aligned} \tag{4.2.38}$$

and

$$\begin{aligned} E[X(\theta)]_\alpha^U &= \sum_{j=1}^{n-1} E[X_1(\theta_1^{''}) + \cdots + X_j(\theta_j^{''})]p^{j-1}q \\ &\quad + E[X_1(\theta_1^{''}) + \cdots + X_n(\theta_n^{''})]p^{n-1} \\ &= \sum_{j=1}^{n-1} E[X_1(\theta_1) + \cdots + X_j(\theta_j)]_\alpha^U p^{j-1}q \\ &\quad + E[X_1(\theta_1) + \cdots + X_n(\theta_n)]_\alpha^U p^{n-1}. \end{aligned} \tag{4.2.39}$$

It follows from Eqs. (4.2.33), (4.2.38) and (4.2.39) that

$$\begin{aligned} \mathrm{MTTF} &= \frac{1}{2}\int_0^1 \left\{ \sum_{j=1}^{n-1} E[X_1(\theta_1) + \cdots + X_j(\theta_j)]_\alpha^L p^{j-1}q + E[X_1(\theta_1) + \cdots + X_n(\theta_n)]_\alpha^L p^{n-1} \right. \\ &\quad \left. + \sum_{j=1}^{n-1} E[X_1(\theta_1) + \cdots + X_j(\theta_j)]_\alpha^U p^{j-1}q + E[X_1(\theta_1) + \cdots + X_n(\theta_n)]_\alpha^U p^{n-1} \right\} \mathrm{d}\alpha \\ &= \sum_{j=1}^{n-1} \left\{ \frac{1}{2} p^{j-1} q \int_0^1 \left(E[X_1(\theta_1) + \cdots + X_j(\theta_j)]_\alpha^L + E[X_1(\theta_1) + \cdots + X_j(\theta_j)]_\alpha^U \right) \mathrm{d}\alpha \right\} \\ &\quad + \frac{1}{2} p^{n-1} \int_0^1 \left(E[X_1(\theta_1) + \cdots + X_n(\theta_n)]_\alpha^L + E[X_1(\theta_1) + \cdots + X_n(\theta_n)]_\alpha^U \right) \mathrm{d}\alpha \\ &= \sum_{j=1}^{n-1} p^{j-1} q E[X_1 + \cdots + X_j] + p^{n-1} E[X_1 + \cdots + X_n]. \end{aligned}$$

The theorem is proved.

Remark 4.2.6 If $X_i,\ i=1,2,\ldots,n$ degenerate to random variables, the result in Theorem 4.2.4 degenerates to the form

$$\text{MTTF}=\sum_{j=1}^{n-1}p^{j-1}qE[X_1+\cdots+X_j]+p^{n-1}E[X_1+\cdots+X_n],$$

which is consistent with the result in stochastic case.

Example 4.2.3 Suppose the cold standby system composed by two components. Let X_1 and X_2 be the lifetimes of component 1 and component 2, respectively. We assume $X_1\sim\exp(\lambda),\ X_2\sim\exp(\lambda)$, where $\lambda=(1,2,3)$. If $p=0.8$, by Theorem 4.2.3 and Theorem 4.2.4, we have

$$\begin{aligned}R(t)&=qE[\Pr\{X_1>t\}]+pE[\Pr\{X_1+X_2>t\}]\\&=\frac{1}{10t}(9+4t)e^{-t}-\frac{1}{10t}(9+12t)e^{-3t}\end{aligned}$$

and

$$\text{MTTF}=qE[X_1]+pE[X_1+X_2]=0.9\ln 3.$$

4.2.3 *The Imperfect Conversion Switch Case: The Lifetime of the Conversion Switch is Continuous*

In this case, we assume the conversion switch can deteriorate and the lifetime of the conversion switch is a continuous random fuzzy variable X_K, which is defined on the credibility space $(\Theta_K,\mathcal{P}(\Theta_K),\text{Cr}_K)$. The cold standby system may fail immediately when the conversion switch fails or there is no component left. Obviously, the lifetime of the random fuzzy cold standby system is $X=\min\{X_1+X_2+\cdots+X_n,X_K\}$, which is a random fuzzy variable on the product credibility space $(\Theta,\mathcal{P}(\Theta),\text{Cr})$, where $\Theta=\Theta_1\times\Theta_2\times\cdots\times\Theta_n\times\Theta_K$ and $\text{Cr}=\text{Cr}_1\wedge\text{Cr}_2\wedge\cdots\wedge\text{Cr}_n\wedge\text{Cr}_K$.

Theorem 4.2.5 *Let X_K and $X_i,\ i=1,2,\ldots,n$ be positive random fuzzy variables. Assume that the α-pessimistic values and the α-optimistic values of $E[X_K(\theta_K)]$, $\theta_K\in\Theta_K$ and $E[X_i(\theta_i)]$, $\theta_i\in\Theta_i,\ i=1,2,\ldots,n$ are continuous almost everywhere with respect to α, $\alpha\in(0,1]$. Then, the reliability of the random fuzzy cold standby system is*

$$R(t)=E[\Pr\{\omega\in\Omega|X_K(\theta_K)(\omega)>t\}\Pr\{\omega\in\Omega|X_1(\theta_1)(\omega)+\cdots+X_n(\theta_n)(\omega)>t\}].$$

Proof Let $A_K = \{\theta_K \in \Theta_K | \mu\{\theta_K\} \geq \alpha\}$ and $A_i = \{\theta_i \in \Theta_i | \mu\{\theta_i\} \geq \alpha\}$, $i = 1, 2, \ldots, n$. Since the α-pessimistic values and the α-optimistic values of fuzzy variables $E[X_K(\theta_K)]$, $\theta_K \in \Theta_K$ and $E[X_i(\theta_i)]$, $\theta_i \in \Theta_i$, $i = 1, 2, \ldots, n$ are continuous almost everywhere for any α, $\alpha \in (0, 1]$, there at least exist points $\theta_K^{'}, \theta_K^{''} \in A_K$ and $\theta_i^{'}, \theta_i^{''} \in A_i$, $i = 1, 2, \ldots, n$ such that

$$E[X_K(\theta_K^{'})] = E[X_K(\theta_K)]_{\alpha}^{L},$$
$$E[X_K(\theta_K^{''})] = E[X_K(\theta_K)]_{\alpha}^{U},$$
$$E[X_i(\theta_i^{'})] = E[X_i(\theta_i)]_{\alpha}^{L},$$
$$E[X_i(\theta_i^{''})] = E[X_i(\theta_i)]_{\alpha}^{U}.$$

For any $\theta_{K,\alpha} \in A_K$ and $\theta_{i,\alpha} \in A_i$, we have

$$E[X_K(\theta_K^{'})] \leq E[X_K(\theta_{K,\alpha})] \leq E[X_K(\theta_K^{''})]$$

and

$$E[X_i(\theta_i^{'})] \leq E[X_i(\theta_{i,\alpha})] \leq E[X_i(\theta_i^{''})],\ i = 1, 2, \ldots, n.$$

Hence, by Definition 1.1.21, it is easy to see that

$$X_K(\theta_K^{'}) \leq_d X_K(\theta_{K,\alpha}) \leq_d X_K(\theta_K^{''}),\ \forall \theta_{K,\alpha} \in A_K \tag{4.2.40}$$

and

$$X_i(\theta_i^{'}) \leq_d X_i(\theta_{i,\alpha}) \leq_d X_i(\theta_i^{''}),\ \forall \theta_{i,\alpha} \in A_i,\ i = 1, 2, \ldots, n. \tag{4.2.41}$$

It follows from Eq. (4.2.41) that

$$\begin{aligned} X_1(\theta_1^{'}) + \cdots + X_n(\theta_n^{'}) &\leq_d X_1(\theta_{1,\alpha}) + \cdots + X_n(\theta_{n,\alpha}) \\ &\leq_d X_1(\theta_1^{''}) + \cdots + X_n(\theta_n^{''}),\ \forall \theta_{i,\alpha} \in A_i,\ i = 1, 2, \ldots, n. \end{aligned} \tag{4.2.42}$$

From Definition 1.1.3, Eqs. (4.2.40) and (4.2.42), we have

$$\begin{aligned} \Pr\left\{\omega \in \Omega | X_K(\theta_K^{'})(\omega) > t\right\} &\leq \Pr\{\omega \in \Omega | X_K(\theta_{K,\alpha})(\omega) > t\} \\ &\leq \Pr\left\{\omega \in \Omega | X_K(\theta_K^{''})(\omega) > t\right\} \end{aligned} \tag{4.2.43}$$

and

$$\begin{aligned}
&\Pr\left\{\omega \in \Omega | X_1(\theta_1^{'})(\omega) + \cdots + X_n(\theta_n^{'})(\omega) > t\right\} \\
&\le \Pr\{\omega \in \Omega | X_1(\theta_{1,\alpha})(\omega) + \cdots + X_n(\theta_{n,\alpha})(\omega) > t\} \\
&\le \Pr\left\{\omega \in \Omega | X_1(\theta_1^{''})(\omega) + \cdots + X_n(\theta_n^{''})(\omega) > t\right\}.
\end{aligned} \tag{4.2.44}$$

By Eqs. (4.2.43) and (4.2.44), we have

$$\begin{aligned}
&\Pr\left\{\omega \in \Omega | \min\left\{X_1(\theta_1^{'})(\omega) + \cdots + X_n(\theta_n^{'})(\omega), X_K(\theta_K^{'})(\omega)\right\} > t\right\} \\
&\le \Pr\{\omega \in \Omega | \min\{X_1(\theta_{1,\alpha})(\omega) + \cdots + X_n(\theta_{n,\alpha})(\omega), X_K(\theta_{K,\alpha})(\omega)\} > t\} \\
&\le \Pr\left\{\omega \in \Omega | \min\left\{X_1(\theta_1^{''})(\omega) + \cdots + X_n(\theta_n^{''})(\omega), X_K(\theta_K^{''})(\omega)\right\} > t\right\}.
\end{aligned} \tag{4.2.45}$$

Since $\theta_{i,\alpha}$ are arbitrary points in A_i, $i = 1, 2, \ldots, n$, it follows from Eq. (4.2.25) that

$$\begin{aligned}
&\Pr{}_{\alpha}^{L}\{\omega \in \Omega | X(\theta)(\omega) > t\} \\
&= \Pr\left\{\omega \in \Omega | \min\left\{X_1(\theta_1^{'})(\omega) + \cdots + X_n(\theta_n^{'})(\omega), X_K(\theta_K^{'})(\omega) > t\right\}\right\} \\
&= \Pr\left\{\omega \in \Omega | X_K(\theta_K^{'})(\omega) > t\right\} \Pr\left\{\omega \in \Omega | X_1(\theta_1^{'})(\omega) + \cdots + X_n(\theta_n^{'})(\omega) > t\right\}
\end{aligned} \tag{4.2.46}$$

and

$$\begin{aligned}
&\Pr{}_{\alpha}^{U}\{\omega \in \Omega | X(\theta)(\omega) > t\} \\
&= \Pr\left\{\omega \in \Omega | \min\left\{X_1(\theta_1^{''})(\omega) + \cdots + X_n(\theta_n^{''})(\omega), X_K(\theta_K^{''})(\omega)\right\} > t\right\} \\
&= \Pr\left\{\omega \in \Omega | X_K(\theta_K^{''})(\omega) > t\right\} \Pr\left\{\omega \in \Omega | X_1(\theta_1^{''})(\omega) + \cdots + X_n(\theta_n^{''})(\omega) > t\right\}.
\end{aligned} \tag{4.2.47}$$

On the other hand, by Eqs. (4.2.43) and (4.2.44), we can arrive at

$$\Pr{}_{\alpha}^{L}\{\omega \in \Omega | X_K(\theta_K)(\omega) > t\} = \Pr\left\{\omega \in \Omega | X_K(\theta_K^{'})(\omega) > t\right\}, \tag{4.2.48}$$

$$\Pr{}_{\alpha}^{U}\{\omega \in \Omega | X_K(\theta_K)(\omega) > t\} = \Pr\left\{\omega \in \Omega | X_K(\theta_K^{''})(\omega) > t\right\}, \tag{4.2.49}$$

$$\begin{aligned}
&\Pr{}_{\alpha}^{L}\{\omega \in \Omega | X_1(\theta_1)(\omega) + \cdots + X_n(\theta_n)(\omega) > t\} \\
&= \Pr\left\{\omega \in \Omega | X_1(\theta_1^{'})(\omega) + \cdots + X_n(\theta_n^{'})(\omega) > t\right\}
\end{aligned} \tag{4.2.50}$$

and

$$\begin{aligned}&\Pr{}_{\alpha}^{U}\{\omega\in\Omega|X_1(\theta_1)(\omega)+\cdots+X_n(\theta_n)(\omega)>t\}\\&=\Pr\Big\{\omega\in\Omega|X_1(\theta_1^{''})(\omega)+\cdots+X_n(\theta_n^{''})(\omega)>t\Big\}.\end{aligned}\tag{4.2.51}$$

By Eqs. (4.2.46)–(4.2.51), it is easy to see that

$$\begin{aligned}&\Pr{}_{\alpha}^{L}\{\omega\in\Omega|X(\theta)(\omega)>t\}\\&=\Pr{}_{\alpha}^{L}\{\omega\in\Omega|X_K(\theta_K)(\omega)>t\}\Pr{}_{\alpha}^{L}\{\omega\in\Omega|X_1(\theta_1)(\omega)+\cdots+X_n(\theta_n)(\omega)>t\}\end{aligned}\tag{4.2.52}$$

and

$$\begin{aligned}&\Pr{}_{\alpha}^{U}\{\omega\in\Omega|X(\theta)(\omega)>t\}\\&=\Pr{}_{\alpha}^{U}\{\omega\in\Omega|X_K(\theta_K)(\omega)>t\}\Pr{}_{\alpha}^{U}\{\omega\in\Omega|X_1(\theta_1)(\omega)+\cdots+X_n(\theta_n)(\omega)>t\}.\end{aligned}\tag{4.2.53}$$

By Definition 4.1, Definition 1.3.7, Theorem 1.2.23, Eqs. (4.2.52) and (4.2.53), we have

$$\begin{aligned}R(t)&=\int_0^1\mathrm{Cr}\{\theta\in\Theta|\Pr\{X(\theta)>t\}\geq p\}\mathrm{d}p\\&=\frac{1}{2}\int_0^1\Big(\Pr{}_{\alpha}^{L}\{\omega\in\Omega|X(\theta)(\omega)>t\}+\Pr{}_{\alpha}^{U}\{\omega\in\Omega|X(\theta)(\omega)>t\}\Big)\mathrm{d}\alpha\\&=\frac{1}{2}\int_0^1\Big(\Pr{}_{\alpha}^{L}\{\omega\in\Omega|X_K(\theta_K)(\omega)>t\}\Pr{}_{\alpha}^{L}\{\omega\in\Omega|X_1(\theta_1)(\omega)+\cdots+X_n(\theta_n)(\omega)>t\}\\&\quad+\Pr{}_{\alpha}^{U}\{\omega\in\Omega|X_K(\theta_K)(\omega)>t\}\Pr{}_{\alpha}^{U}\{\omega\in\Omega|X_1(\theta_1)(\omega)+\cdots+X_n(\theta_n)(\omega)>t\}\Big)\mathrm{d}\alpha\\&=\frac{1}{2}\int_0^1\Big\{(\Pr\{\omega\in\Omega|X_K(\theta_K)(\omega)>t\}\Pr\{\omega\in\Omega|X_1(\theta_1)(\omega)+\cdots+X_n(\theta_n)(\omega)>t\})_{\alpha}^{L}\\&\quad+(\Pr\{\omega\in\Omega|X_K(\theta_K)(\omega)>t\}\Pr\{\omega\in\Omega|X_1(\theta_1)(\omega)+\cdots+X_n(\theta_n)(\omega)>t\})_{\alpha}^{U}\Big\}\mathrm{d}\alpha\\&=E[\Pr\{\omega\in\Omega|X_K(\theta_K)(\omega)>t\}\Pr\{\omega\in\Omega|X_1(\theta_1)(\omega)+\cdots+X_n(\theta_n)(\omega)>t\}].\end{aligned}$$

The theorem is proved.

Remark 4.2.7 If X_K and X_i, $i=1,2,\ldots,n$ degenerate to random variables, the result in Theorem 4.2.5 degenerates to the form

$$R(t)=\Pr\{\omega\in\Omega|X_K(\omega)\geq t\}\Pr\{\omega\in\Omega|X_1(\omega)+\cdots+X_n(\omega)>t\},$$

which is consistent with the result in stochastic case.

Theorem 4.2.6 *Let X_K and X_i, $i = 1, 2, \ldots, n$ be positive random fuzzy variables. Assume that the α-pessimistic values and the α-optimistic values of $E[X_K(\theta_K)]$, $\theta_K \in \Theta_K$ and $E[X_i(\theta_i)]$, $\theta_i \in \Theta_i$, $i = 1, 2, \ldots, n$ are continuous almost everywhere with respect to α, $\alpha \in (0, 1]$. Then, MTTF of the random fuzzy cold standby system is*

$$\begin{aligned}
&\text{MTTF}\\
&= \frac{1}{2}\int_0^1\int_0^{+\infty} \Big(\Pr{}_\alpha^L\{\omega \in \Omega | X_K(\theta_K)(\omega) \ge t\} \Pr{}_\alpha^L\{\omega \in \Omega | X_1(\theta_1)(\omega) + \cdots + X_n(\theta_n)(\omega) \ge t\}\\
&\quad + \Pr{}_\alpha^U\{\omega \in \Omega | X_K(\theta_K)(\omega) \ge t\} \Pr{}_\alpha^U\{\omega \in \Omega | X_1(\theta_1)(\omega) + \cdots + X_n(\theta_n)(\omega) \ge t\}\Big)\mathrm{d}t\mathrm{d}\alpha.
\end{aligned}$$

Proof It follows from Definition 4.2 and Theorem 1.2.23 that

$$\begin{aligned}
\text{MTTF} &= \int_0^{+\infty} \text{Cr}\{\theta \in \Theta | E[X(\theta)] \ge r\}\mathrm{d}r\\
&= \frac{1}{2}\int_0^1 \left(E[X(\theta)]_\alpha^L + E[X(\theta)]_\alpha^U\right)\mathrm{d}\alpha. \qquad (4.2.54)
\end{aligned}$$

By Eq. (4.2.25), it is easy to see that

$$\begin{aligned}
&\min\left\{X_1(\theta_1^{'}) + \cdots + X_n(\theta_n^{'}),\ X_K(\theta_K^{'})\right\}\\
&\le_d \min\left\{X_1(\theta_{1,\alpha}) + \cdots + X_n(\theta_{n,\alpha}),\ X_K(\theta_{K,\alpha})\right\}\\
&\le_d \min\left\{X_1(\theta_1^{''}) + \cdots + X_n(\theta_n^{''}),\ X_K(\theta_K^{''})\right\}.
\end{aligned}$$

It follows from Theorem 1.2.21 that

$$\begin{aligned}
&E\left[\min\left\{X_1(\theta_1^{'}) + \cdots + X_n(\theta_n^{'}),\ X_K(\theta_K^{'})\right\}\right]\\
&\le E\left[\min\{X_1(\theta_{1,\alpha}) + \cdots + X_n(\theta_{n,\alpha}),\ X_K(\theta_{K,\alpha})\}\right]\\
&\le E\left[\min\left\{X_1(\theta_1^{''}) + \cdots + X_n(\theta_n^{''}),\ X_K(\theta_K^{''})\right\}\right].
\end{aligned}$$

Since $\theta_{i,\alpha}$ are arbitrary points in the A_i, $i = 1, 2, \ldots, n$, then

$$\begin{aligned}
&E[X(\theta)]_\alpha^L\\
&= E\left[\min\left\{X_1(\theta_1^{'}) + \cdots + X_n(\theta_n^{'}),\ X_K(\theta_K^{'})\right\}\right]
\end{aligned}$$

$$= \int_0^{+\infty} \Pr\left\{\omega \in \Omega | X_K(\theta_K^{'})(\omega) \geq t\right\} \Pr\left\{\omega \in \Omega | X_1(\theta_1^{'})(\omega) + \cdots + X_n(\theta_n^{'})(\omega) \geq t\right\} \mathrm{d}t \tag{4.2.55}$$

and

$$\begin{aligned} & E[X(\theta)]_\alpha^U \\ & = E\left[\min\left\{X_1(\theta_1^{''}) + \cdots + X_n(\theta_n^{''}),\ X_K(\theta_K^{''})\right\}\right] \\ & = \int_0^{+\infty} \Pr\left\{\omega \in \Omega | X_K(\theta_K^{''})(\omega) \geq t\right\} \Pr\left\{\omega \in \Omega | X_1(\theta_1^{''})(\omega) + \cdots + X_n(\theta_n^{''})(\omega) \geq t\right\} \mathrm{d}t. \end{aligned} \tag{4.2.56}$$

By Eqs. (4.2.48)–(4.2.51) and (4.2.54)–(4.2.56), we have

$$\begin{aligned} & \text{MTTF} \\ & = \frac{1}{2} \int_0^1 \left\{ \int_0^{+\infty} \Pr\left\{\omega \in \Omega | X_K(\theta_K^{'})(\omega) \geq t\right\} \Pr\left\{\omega \in \Omega | X_1(\theta_1^{'})(\omega) + \cdots + X_n(\theta_n^{'})(\omega) \geq t\right\} \mathrm{d}t \right. \\ & \left. + \int_0^{+\infty} \Pr\left\{\omega \in \Omega | X_K(\theta_K^{''})(\omega) \geq t\right\} \Pr\left\{\omega \in \Omega | X_1(\theta_1^{''})(\omega) + \cdots + X_n(\theta_n^{''})(\omega) \geq t\right\} \mathrm{d}t \right\} \mathrm{d}\alpha \\ & = \frac{1}{2} \int_0^1 \int_0^{+\infty} \left(\Pr\left\{\omega \in \Omega | X_K(\theta_K^{'})(\omega) \geq t\right\} \Pr\left\{\omega \in \Omega | X_1(\theta_1^{'})(\omega) + \cdots + X_n(\theta_n^{'})(\omega) \geq t\right\} \mathrm{d}t \right. \\ & \left. + \Pr\left\{\omega \in \Omega | X_K(\theta_K^{'})(\omega) \geq t\right\} \Pr\left\{\omega \in \Omega | X_1(\theta_1^{''})(\omega) + \cdots + X_n(\theta_n^{''})(\omega) \geq t\right\} \right) \mathrm{d}t \mathrm{d}\alpha \\ & = \frac{1}{2} \int_0^1 \int_0^{+\infty} \left(\Pr{}_\alpha^L\{\omega \in \Omega | X_K(\theta_K)(\omega) \geq t\} \Pr{}_\alpha^L\{\omega \in \Omega | X_1(\theta_1)(\omega) + \cdots + X_n(\theta_n)(\omega) \geq t\} \right. \\ & \left. + \Pr{}_\alpha^U\{\omega \in \Omega | X_K(\theta_K)(\omega) \geq t\} \Pr{}_\alpha^U\{\omega \in \Omega | X_1(\theta_1)(\omega) + \cdots + X_n(\theta_n)(\omega) \geq t\} \right) \mathrm{d}t \mathrm{d}\alpha. \end{aligned}$$

The theorem is proved.

Remark 4.2.8 If X_K and $X_i,\ i = 1, 2, \ldots, n$ degenerate to random variables, the result in Theorem 4.2.6 degenerates to the form

$$\text{MTTF} = \int_0^{+\infty} \Pr\{\omega \in \Omega | X_K(\omega) \geq t\} \Pr\{\omega \in \Omega | X_1(\omega) + \cdots + X_n(\omega) \geq t\} \mathrm{d}t,$$

which is consistent with the result in stochastic case.

Example 4.2.4 Suppose the cold standby system composed by two components. Let X_1, X_2 and X_K be the lifetimes of component 1, component 2 and conversion switch, respectively. We assume $X_1 \sim \exp(\lambda)$, $X_2 \sim \exp(\lambda)$ and $X_K \sim \exp(\lambda_K)$, where $\lambda = (1, 2, 3)$ and $\lambda_K = (0, 1, 2)$. We also assume X_1, X_2 and X_K are independent. By Theorem 4.2.5 and Theorem 4.2.6, we have

$$R(t) = \frac{1}{8t}(3 + 2t) - \frac{3}{8t}(1 + 2t)e^{-5t}$$

and

$$\text{MTTF} = \frac{1}{5} + \frac{3}{4}\ln 5.$$

4.3 Warm Standby Systems

The difference between the warm standby system and the cold standby system is that the standby components in standby may fail during the standby period. The lifetime of each component during the warm standby period is longer than that during the operation period. Then, the reliability mathematical models of three random fuzzy warm standby systems are proposed.

4.3.1 The Perfect Conversion Switch and Identical Component Case

Consider a warm standby system composed by n identical components. The lifetimes of components in operating have a random fuzzy exponential distribution with parameter λ defined on the credibility space $(\Theta_1, \mathcal{P}(\Theta_1), \text{Cr}_1)$, and the lifetimes of components in warm standby have a random fuzzy exponential distribution with parameter μ defined on the credibility space $(\Theta_2, \mathcal{P}(\Theta_2), \text{Cr}_2)$. We assume one component is operating at $t = 0$, and the other components are in warm standby state. When the operating component fails, another component in standby is replaced. System failure occurs when no operating components are available. We also assume the conversion switch is completely reliable and the conversion is instantaneous. The lifetimes of components in operating and in warm standby are independent.

Let S_i be the failure moment of i th operating component, $i = 1, 2, \ldots, n$ and $S_0 = 0$. Then,

$$S_n = \sum_{i=1}^{n} (S_i - S_{i-1})$$

is the failure moment of the warm standby system. Since the random fuzzy exponential distribution has the memoryless property, $S_i - S_{i-1}$ have random fuzzy exponential distributions with parameters $\lambda + (n-i)\mu$, $i = 1, 2, \ldots, n$. For convenience, we discuss the system reliability on the product credibility space $(\Theta, \mathcal{P}(\Theta), \mathrm{Cr})$, where $\Theta = \Theta_1 \times \Theta_2$ and $\mathrm{Cr} = \mathrm{Cr}_1 \wedge \mathrm{Cr}_2$.

Theorem 4.3.1 *The reliability of the random fuzzy warm standby system is*

$$R(t) = \frac{1}{2}\sum_{i=0}^{n-1}\int_0^1 \left(\left[\prod_{k=0,k\neq i}^{n-1} \frac{\lambda_\alpha^L + k\mu_\alpha^L}{(k-i)\mu_\alpha^L} e^{-(\lambda_\alpha^L + i\mu_\alpha^L)t}\right] + \left[\prod_{k=0,k\neq i}^{n-1} \frac{\lambda_\alpha^U + k\mu_\alpha^U}{(k-i)\mu_\alpha^U} e^{-(\lambda_\alpha^U + i\mu_\alpha^U)t}\right]\right) \mathrm{d}\alpha.$$

Proof It follows from Definition 4.1, Definition 1.3.7 and Theorem 1.2.23 that

$$\begin{aligned} R(t) &= \mathrm{Ch}\{S_n > t\} \\ &= \int_0^1 \mathrm{Cr}\{\theta \in \Theta \mid \Pr\{S_n(\theta) > t\} \geq p\} \mathrm{d}p \\ &= \frac{1}{2}\int_0^1 \left(\Pr{}_\alpha^L\{\omega \in \Omega \mid S_n(\theta)(\omega) > t\} + \Pr{}_\alpha^U\{\omega \in \Omega \mid S_n(\theta)(\omega) > t\}\right) \mathrm{d}\alpha. \end{aligned} \tag{4.3.1}$$

Let $A_\alpha = \{\theta_1 \in \Theta_1 | \mu\{\theta_1\} \geq \alpha\}$ and $B_\alpha = \{\theta_2 \in \Theta_2 | \mu\{\theta_2\} \geq \alpha\}$. For $\forall \theta_1 \in A_\alpha$ and $\forall \theta_2 \in B_\alpha$, we can arrive at $\lambda_\alpha^L \leq \lambda(\theta_1) \leq \lambda_\alpha^U$ and $\mu_\alpha^L \leq \mu(\theta_2) \leq \mu_\alpha^U$. We can construct three warm standby systems:

(1) Warm standby system 1: The lifetimes of components in operating have an exponential distribution with parameter λ_α^L, and the lifetimes of components in warm standby have an exponential distribution with parameter μ_α^L.
(2) Warm standby system 2: The lifetimes of components in operating have an exponential distribution with parameter $\lambda(\theta_1)$, and the lifetimes of components in warm standby have an exponential distribution with parameter $\mu(\theta_2)$.
(3) Warm standby system 3: The lifetimes of components in operating have an exponential distribution with parameter λ_α^U, and the lifetimes of components in warm standby have an exponential distribution with parameter μ_α^U.

It is easy to see that system (1) and system (3) are two standard stochastic warm standby systems. For any fixed $\theta_1 \in A_\alpha$ and $\theta_2 \in B_\alpha$, system (2) is also a stochastic warm standby system. Let S_n^1, S_n^2 and S_n^3 be the failure moments of warm standby system (1), system (2) and system (3), respectively. It is easy to see that

$$\Pr\{S_n^3 > t\} \le \Pr\{S_n^2 > t\} \le \Pr\{S_n^1 > t\}.$$

Since θ_1 and θ_2 are arbitrary points in A_α and B_α, we have

$$\Pr{}_\alpha^L\{\omega \in \Omega | S_n(\theta)(\omega) > t\} = \Pr\{S_n^3 > t\} \tag{4.3.2}$$

and

$$\Pr{}_\alpha^U\{\omega \in \Omega | S_n(\theta)(\omega) > t\} = \Pr\{S_n^1 > t\}. \tag{4.3.3}$$

From the result in classical reliability theory, we can arrive at

$$\Pr\{S_n^1 > t\} = \sum_{i=0}^{n-1}\left[\prod_{k=0,k\ne i}^{n-1}\frac{\lambda_\alpha^L + k\mu_\alpha^L}{(k-i)\mu_\alpha^L}\right]e^{-(\lambda_\alpha^L + i\mu_\alpha^L)t} \tag{4.3.4}$$

and

$$\Pr\{S_n^3 > t\} = \sum_{i=0}^{n-1}\left[\prod_{k=0,k\ne i}^{n-1}\frac{\lambda_\alpha^U + k\mu_\alpha^U}{(k-i)\mu_\alpha^U}\right]e^{-(\lambda_\alpha^U + i\mu_\alpha^U)t}. \tag{4.3.5}$$

Then from Eqs. (4.3.1) to (4.3.5), we have

$$\begin{aligned}
R(t) &= \frac{1}{2}\int_0^1 \left(\Pr{}_\alpha^L\{\omega \in \Omega | S_n(\theta)(\omega) > t\} + \Pr{}_\alpha^U\{\omega \in \Omega | S_n(\theta)(\omega) > t\}\right)\mathrm{d}\alpha \\
&= \frac{1}{2}\int_0^1 \left(\Pr\{S_n^1 > t\} + \Pr\{S_n^3 > t\}\right)\mathrm{d}\alpha \\
&= \frac{1}{2}\int_0^1 \left(\sum_{i=0}^{n-1}\left[\prod_{k=0,k\ne i}^{n-1}\frac{\lambda_\alpha^L + k\mu_\alpha^L}{(k-i)\mu_\alpha^L}\right]e^{-(\lambda_\alpha^L + i\mu_\alpha^L)t} + \sum_{i=0}^{n-1}\left[\prod_{k=0,k\ne i}^{n-1}\frac{\lambda_\alpha^U + k\mu_\alpha^U}{(k-i)\mu_\alpha^U}\right]e^{-(\lambda_\alpha^U + i\mu_\alpha^U)t}\right)\mathrm{d}\alpha \\
&= \frac{1}{2}\sum_{i=0}^{n-1}\int_0^1 \left(\left[\prod_{k=0,k\ne i}^{n-1}\frac{\lambda_\alpha^L + k\mu_\alpha^L}{(k-i)\mu_\alpha^L}e^{-(\lambda_\alpha^L + i\mu_\alpha^L)t}\right] + \left[\prod_{k=0,k\ne i}^{n-1}\frac{\lambda_\alpha^U + k\mu_\alpha^U}{(k-i)\mu_\alpha^U}e^{-(\lambda_\alpha^U + i\mu_\alpha^U)t}\right]\right)\mathrm{d}\alpha.
\end{aligned}$$

The theorem is proved.

Theorem 4.3.2 *The MTTF of the warm standby system is*

$$\text{MTTF} = \sum_{i=0}^{n-1} E\left[\frac{1}{\lambda + i\mu}\right].$$

Proof It follows from Definition 4.2 and Theorem 1.2.23, and we have

$$\begin{aligned}\text{MTTF} &= \int_0^{+\infty} \text{Cr}\{\theta \in \Theta | E[S_n(\theta)] \geq r\}\text{d}r \\ &= \frac{1}{2}\int_0^1 \left(E[S_n(\theta)]_\alpha^L + E[S_n(\theta)]_\alpha^U\right)\text{d}\alpha. \end{aligned} \tag{4.3.6}$$

From the three warm standby systems constructed in the proof of Theorem 4.3.1, we also can see that

$$E[S_n^3] \leq E[S_n^2] \leq E[S_n^1].$$

Since θ_1 and θ_2 are arbitrary points in A_α and B_α, we have

$$E[S_n(\theta)]_\alpha^L = E[S_n^3] \tag{4.3.7}$$

and

$$E[S_n(\theta)]_\alpha^U = E[S_n^1]. \tag{4.3.8}$$

From the result in classical reliability theory, we can arrive at

$$E[S_n^1] = \sum_{i=0}^{n-1} \frac{1}{\lambda_\alpha^L + i\mu_\alpha^L} \tag{4.3.9}$$

and

$$E[S_n^3] = \sum_{i=0}^{n-1} \frac{1}{\lambda_\alpha^U + i\mu_\alpha^U}. \tag{4.3.10}$$

Then by Eqs. (4.3.6)–(4.3.10), we have

$$\begin{aligned}\text{MTTF} &= \frac{1}{2}\int_0^1 \left(E[S_n(\theta)]_\alpha^L + E[S_n(\theta)]_\alpha^U\right)\text{d}\alpha \\ &= \frac{1}{2}\int_0^1 \left(\sum_{i=0}^{n-1} \frac{1}{\lambda_\alpha^L + i\mu_\alpha^L} + \sum_{i=0}^{n-1} \frac{1}{\lambda_\alpha^U + i\mu_\alpha^U}\right)\text{d}\alpha \\ &= \sum_{i=0}^{n-1} \frac{1}{2}\int_0^1 \left(\frac{1}{\lambda_\alpha^L + i\mu_\alpha^L} + \frac{1}{\lambda_\alpha^U + i\mu_\alpha^U}\right)\text{d}\alpha = \sum_{i=0}^{n-1} E\left[\frac{1}{\lambda + i\mu}\right].\end{aligned}$$

The theorem is proved.

Example 4.3.1 Consider a warm standby system composed by two identical components. The lifetime of component in operating has a random fuzzy exponential distribution with parameter λ, and the lifetime of component in warm standby has a random fuzzy exponential distribution with parameter μ. We assume $\lambda = (1, 2, 3)$ and $\mu = (2, 3, 4)$. Then, we can arrive at

$$\lambda_\alpha^L = 1 + \alpha,\ \lambda_\alpha^U = 3 - \alpha,\ \mu_\alpha^L = 2 + \alpha,\ \mu_\alpha^U = 4 - \alpha.$$

It follows from Theorem 4.3.1 that

$$R(t) = \frac{1}{2}\int_0^1 \left(\frac{\lambda_\alpha^L + \mu_\alpha^L}{\mu_\alpha^L} e^{-\lambda_\alpha^L t} + \frac{\lambda_\alpha^U + \mu_\alpha^U}{\mu_\alpha^U} e^{-\lambda_\alpha^U t} - \frac{\lambda_\alpha^L}{\mu_\alpha^L} e^{-(\lambda_\alpha^L + \mu_\alpha^L)t} - \frac{\lambda_\alpha^U}{\mu_\alpha^U} e^{-(\lambda_\alpha^U + \mu_\alpha^U)t} \right) \mathrm{d}\alpha$$

$$= \frac{1}{2}\int_0^1 \left(\frac{1+\alpha+2+\alpha}{2+\alpha} e^{-(1+\alpha)t} + \frac{3-\alpha+4-\alpha}{4-\alpha} e^{-(3-\alpha)t} \right.$$

$$\left. - \frac{1+\alpha}{2+\alpha} e^{-(1+\alpha+2+\alpha)t} - \frac{3-\alpha}{4-\alpha} e^{-(3-\alpha+4-\alpha)t} \right) \mathrm{d}\alpha.$$

Figure 4.3 shows the plot of the reliability of the warm standby system. We can see the reliability decreases with time. It follows from Theorem 4.3.2 that

$$\mathrm{MTTF} = E\left[\frac{1}{\lambda}\right] + E\left[\frac{1}{\lambda + \mu}\right]$$

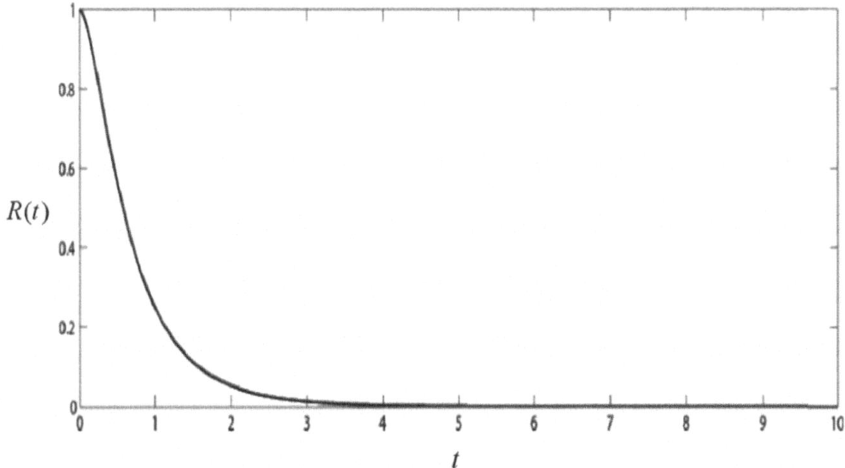

Fig. 4.3 Reliability of the warm standby system

$$= \frac{1}{2}\int_0^1 \left(\frac{1}{1+\alpha} + \frac{1}{3-\alpha}\right)\mathrm{d}\alpha + \frac{1}{2}\int_0^1 \left(\frac{1}{1+\alpha+2+\alpha} + \frac{1}{3-\alpha+4-\alpha}\right)\mathrm{d}\alpha$$
$$= \frac{1}{2}\ln 3 - \frac{1}{4}\ln\frac{7}{3} \approx 0.3375.$$

4.3.2 The Perfect Conversion Switch and Nonidentical Component Case

Let X_1 and X_2 be the lifetimes of component 1 and component 2 in operating. Assume that X_1 and X_2 have random fuzzy exponential distributions with parameters λ_1 and λ_2 defined on the credibility spaces $(\Theta_1, \mathcal{P}(\Theta_1), \mathrm{Cr}_1)$ and $(\Theta_2, \mathcal{P}(\Theta_2), \mathrm{Cr}_2)$, respectively. Initially, the two components are new; component 1 has priority in use, while component 2 is in warm standby. Let Y_2 be the lifetime of component 2 in warm standby which has a random fuzzy exponential distribution with parameter μ defined on the credibility space $(\Theta_3, \mathcal{P}(\Theta_3), \mathrm{Cr}_3)$. We assume X_1, X_2 and Y_2 are independent. We also assume that the conversion switch is completely reliable and the conversion is instantaneous. Let X denote the lifetime of the warm standby system. Then,

$$X = X_1 + X_2 \cdot I_{\{Y_2 > X_1\}},$$

where $I_{\{\cdot\}}$ is the indicator function of a random fuzzy event. Obviously, X is a random fuzzy variable on the product credibility space $(\Theta, \mathcal{P}(\Theta), \mathrm{Cr})$, where $\Theta = \Theta_1 \times \Theta_2 \times \Theta_3$ and $\mathrm{Cr} = \mathrm{Cr}_1 \wedge \mathrm{Cr}_2 \wedge \mathrm{Cr}_3$.

Theorem 4.3.3 *The reliability of the random fuzzy warm standby system is*

$$R(t) = \frac{1}{2}\int_0^1 \left(e^{-\lambda_{1,\alpha}^L t} + \frac{\lambda_{1,\alpha}^L}{\lambda_{1,\alpha}^L - \lambda_{2,\alpha}^L + \mu_\alpha^L}\left[e^{-\lambda_{2,\alpha}^L t} - e^{-(\lambda_{1,\alpha}^L + \mu_\alpha^L)t}\right]\right.$$
$$\left. + e^{-\lambda_{1,\alpha}^U t} + \frac{\lambda_{1,\alpha}^U}{\lambda_{1,\alpha}^U - \lambda_{2,\alpha}^U + \mu_\alpha^U}\left[e^{-\lambda_{2,\alpha}^U t} - e^{-(\lambda_{1,\alpha}^U + \mu_\alpha^U)t}\right]\right)\mathrm{d}\alpha.$$

Proof By Definition 4.1, Definition 1.3.7 and Theorem 1.2.23, we have

$$R(t) = \mathrm{Ch}\{X > t\}$$
$$= \int_0^1 \mathrm{Cr}\{\theta \in \Theta \mid \Pr\{X(\theta) > t\} \geq p\}\mathrm{d}p$$

$$= \frac{1}{2}\int_0^1 \left(\text{Pr}_\alpha^L\{\omega \in \Omega | X(\theta)(\omega) > t\} + \text{Pr}_\alpha^U\{\omega \in \Omega | X(\theta)(\omega) > t\}\right) \mathrm{d}\alpha. \tag{4.3.11}$$

Let $A_\alpha = \{\theta_1 \in \Theta_1 | \mu\{\theta_1\} \geq \alpha\}$, $B_\alpha = \{\theta_2 \in \Theta_2 | \mu\{\theta_2\} \geq \alpha\}$ and $C_\alpha = \{\theta_3 \in \Theta_3 | \mu\{\theta_3\} \geq \alpha\}$. For $\forall \theta_1 \in A_\alpha$, $\forall \theta_2 \in B_\alpha$ and $\forall \theta_3 \in C_\alpha$, we can arrive at $\lambda_{1,\alpha}^L \leq \lambda_1(\theta_1) \leq \lambda_{1,\alpha}^U$, $\lambda_{2,\alpha}^L \leq \lambda_2(\theta_2) \leq \lambda_{2,\alpha}^U$ and $\mu_\alpha^L \leq \mu(\theta_3) \leq \mu_\alpha^U$. Then, we can construct three warm standby systems:

(1) Warm standby system 1: The lifetimes of component 1 and component 2 in operating have exponential distributions with parameter $\lambda_{1,\alpha}^L$ and $\lambda_{2,\alpha}^L$, respectively, and the lifetime of component 2 in warm standby has an exponential distribution with parameter μ_α^L.

(2) Warm standby system 2: The lifetimes of component 1 and component 2 in operating have exponential distributions with parameter $\lambda_1(\theta_1)$ and $\lambda_2(\theta_2)$, respectively, and the lifetime of component 2 in warm standby has an exponential distribution with parameter $\mu(\theta_3)$.

(3) Warm standby system 3: The lifetimes of component 1 and component 2 in operating have exponential distributions with parameter $\lambda_{1,\alpha}^U$ and $\lambda_{2,\alpha}^U$, respectively, and the lifetime of component 2 in warm standby has an exponential distribution with parameter μ_α^U.

It is easy to see that system (1) and system (3) are standard stochastic warm standby systems. For any fixed $\theta_1 \in A_\alpha$, $\theta_2 \in B_\alpha$ and $\theta_3 \in C_\alpha$, system (2) is also a stochastic warm standby system. Let $X^{(1)}$, $X^{(2)}$ and $X^{(3)}$ be the lifetimes of system (1), system (2) and system (3), respectively. Since the warm standby system is a coherent system, then

$$\text{Pr}\{X^{(3)} > t\} \leq \text{Pr}\{X^{(2)} > t\} \leq \text{Pr}\{X^{(1)} > t\}.$$

Since θ_1, θ_2 and θ_3 are arbitrary points in A_α, B_α and C_α, we have

$$\text{Pr}_\alpha^L\{\omega \in \Omega | X(\theta)(\omega) > t\} = \text{Pr}\{X^{(3)} > t\} \tag{4.3.12}$$

and

$$\text{Pr}_\alpha^U\{\omega \in \Omega | X(\theta)(\omega) > t\} = \text{Pr}\{X^{(1)} > t\}. \tag{4.3.13}$$

From the result in classical reliability theory, we can arrive at

$$\text{Pr}\{X^{(1)} > t\} = e^{-\lambda_{1,\alpha}^L t} + \frac{\lambda_{1,\alpha}^L\left[e^{-\lambda_{2,\alpha}^L t} - e^{-(\lambda_{1,\alpha}^L + \mu_\alpha^L)t}\right]}{\lambda_{1,\alpha}^L - \lambda_{2,\alpha}^L + \mu_\alpha^L} \tag{4.3.14}$$

and

$$\Pr\left\{X^{(3)} > t\right\} = e^{-\lambda_{1,\alpha}^{U} t} + \frac{\lambda_{1,\alpha}^{U}\left[e^{-\lambda_{2,\alpha}^{U} t} - e^{-(\lambda_{1,\alpha}^{U}+\mu_{\alpha}^{U})t}\right]}{\lambda_{1,\alpha}^{U} - \lambda_{2,\alpha}^{U} + \mu_{\alpha}^{U}}. \tag{4.3.15}$$

It follows from Eqs. (4.3.11)–(4.3.15) that

$$\begin{aligned} R(t) &= \mathrm{Ch}\{X > t\} \\ &= \frac{1}{2}\int_0^1 \left(\Pr{}_{\alpha}^{L}\{\omega \in \Omega | X(\theta)(\omega) > t\} + \Pr{}_{\alpha}^{U}\{\omega \in \Omega | X(\theta)(\omega) > t\}\right)\mathrm{d}\alpha \\ &= \frac{1}{2}\int_0^1 \left(\Pr\left\{X^{(3)} > t\right\} + \Pr\left\{X^{(1)} > t\right\}\right)\mathrm{d}\alpha \\ &= \frac{1}{2}\int_0^1 \left(e^{-\lambda_{1,\alpha}^{L} t} + \frac{\lambda_{1,\alpha}^{L}\left[e^{-\lambda_{2,\alpha}^{L} t} - e^{-(\lambda_{1,\alpha}^{L}+\mu_{\alpha}^{L})t}\right]}{\lambda_{1,\alpha}^{L} - \lambda_{2,\alpha}^{L} + \mu_{\alpha}^{L}} + e^{-\lambda_{1,\alpha}^{U} t} + \frac{\lambda_{1,\alpha}^{U}\left[e^{-\lambda_{2,\alpha}^{U} t} - e^{-(\lambda_{1,\alpha}^{U}+\mu_{\alpha}^{U})t}\right]}{\lambda_{1,\alpha}^{U} - \lambda_{2,\alpha}^{U} + \mu_{\alpha}^{U}}\right)\mathrm{d}\alpha. \end{aligned}$$

The theorem is proved.

Remark 4.3.1 If X_1, X_2 and Y_2 degenerate to exponentially distributed random variables, i.e., λ_1, λ_2 and μ degenerate to crisp numbers, then the result in Theorem 4.3.3 degenerates to the form

$$R(t) = e^{-\lambda_1 t} + \frac{\lambda_1}{\lambda_1 - \lambda_2 + \mu}\left[e^{-\lambda_2 t} - e^{-(\lambda_1+\mu)t}\right],$$

which is consistent with the result in stochastic case.

Theorem 4.3.4 *The MTTF of the random fuzzy warm standby system is*

$$\mathrm{MTTF} = E\left[\frac{1}{\lambda_1}\right] + \frac{1}{2}\int_0^1 \frac{1}{\lambda_{2,\alpha}^{L}}\left(\frac{\lambda_{1,\alpha}^{L}}{\lambda_{1,\alpha}^{L} + \mu_{\alpha}^{L}}\right)\mathrm{d}\alpha + \frac{1}{2}\int_0^1 \frac{1}{\lambda_{2,\alpha}^{U}}\left(\frac{\lambda_{1,\alpha}^{U}}{\lambda_{1,\alpha}^{U} + \mu_{\alpha}^{U}}\right)\mathrm{d}\alpha.$$

Proof By Definition 4.2 and Theorem 1.2.23, we have

$$\begin{aligned} \mathrm{MTTF} &= \int_0^{+\infty} \mathrm{Cr}\{\theta \in \Theta | E[X(\theta)] \geq r\}\mathrm{d}r \\ &= \frac{1}{2}\int_0^1 \left(E[X(\theta)]_{\alpha}^{L} + E[X(\theta)]_{\alpha}^{U}\right)\mathrm{d}\alpha. \end{aligned} \tag{4.3.16}$$

Based on the three warm standby systems constructed in the proof of Theorem 4.3.3, we also can see that

$$E[X^{(3)}] \le E[X^{(2)}] \le E[X^{(1)}].$$

Since θ_1, θ_2 and θ_3 are arbitrary points in A_α, B_α and C_α, we have

$$E[X(\theta)]_\alpha^L = E[X^{(3)}] \tag{4.3.17}$$

and

$$E[X(\theta)]_\alpha^U = E[X^{(1)}]. \tag{4.3.18}$$

From the result in classical reliability theory, we can arrive at

$$E[X^{(1)}] = \frac{1}{\lambda_{1,\alpha}^L} + \frac{1}{\lambda_{2,\alpha}^L}\left(\frac{\lambda_{1,\alpha}^L}{\lambda_{1,\alpha}^L + \mu_\alpha^L}\right) \tag{4.3.19}$$

and

$$E[X^{(3)}] = \frac{1}{\lambda_{1,\alpha}^U} + \frac{1}{\lambda_{2,\alpha}^U}\left(\frac{\lambda_{1,\alpha}^U}{\lambda_{1,\alpha}^U + \mu_\alpha^U}\right). \tag{4.3.20}$$

It follows from Eqs. (4.3.16)–(4.3.20) that

$$\begin{aligned}
\text{MTTF} &= \frac{1}{2}\int_0^1 \left(E[X(\theta)]_\alpha^L + E[X(\theta)]_\alpha^U\right)\mathrm{d}\alpha \\
&= \frac{1}{2}\int_0^1 \left(E[X^{(3)}] + E[X^{(1)}]\right)\mathrm{d}\alpha \\
&= \frac{1}{2}\int_0^1 \left[\frac{1}{\lambda_{1,\alpha}^L} + \frac{1}{\lambda_{2,\alpha}^L}\left(\frac{\lambda_{1,\alpha}^L}{\lambda_{1,\alpha}^L + \mu_\alpha^L}\right) + \frac{1}{\lambda_{1,\alpha}^U} + \frac{1}{\lambda_{2,\alpha}^U}\left(\frac{\lambda_{1,\alpha}^U}{\lambda_{1,\alpha}^U + \mu_\alpha^U}\right)\right]\mathrm{d}\alpha \\
&= \frac{1}{2}\int_0^1 \left(\frac{1}{\lambda_{1,\alpha}^L} + \frac{1}{\lambda_{1,\alpha}^U}\right)\mathrm{d}\alpha + \frac{1}{2}\int_0^1 \frac{1}{\lambda_{2,\alpha}^L}\left(\frac{\lambda_{1,\alpha}^L}{\lambda_{1,\alpha}^L + \mu_\alpha^L}\right)\mathrm{d}\alpha + \frac{1}{2}\int_0^1 \frac{1}{\lambda_{2,\alpha}^U}\left(\frac{\lambda_{1,\alpha}^U}{\lambda_{1,\alpha}^U + \mu_\alpha^U}\right)\mathrm{d}\alpha \\
&= E\left[\frac{1}{\lambda_1}\right] + \frac{1}{2}\int_0^1 \frac{1}{\lambda_{2,\alpha}^L}\left(\frac{\lambda_{1,\alpha}^L}{\lambda_{1,\alpha}^L + \mu_\alpha^L}\right)\mathrm{d}\alpha + \frac{1}{2}\int_0^1 \frac{1}{\lambda_{2,\alpha}^U}\left(\frac{\lambda_{1,\alpha}^U}{\lambda_{1,\alpha}^U + \mu_\alpha^U}\right)\mathrm{d}\alpha.
\end{aligned}$$

The theorem is proved.

Remark 4.3.2 If X_1, X_2 and Y_2 degenerate to exponentially distributed random variables, i.e., λ_1, λ_2 and μ degenerate to crisp numbers, then the result in Theorem 4.3.4 degenerates to the form

$$\text{MTTF} = \frac{1}{\lambda_1} + \frac{1}{\lambda_2}\left(\frac{\lambda_1}{\lambda_1 + \mu}\right),$$

which is consistent with the result in stochastic case.

Example 4.3.2 If λ_1, λ_2 and μ are triangular fuzzy variables, where $\lambda_1 = (1, 2, 3)$, $\lambda_2 = (2, 3, 4)$ and $\mu = (0, 1, 2)$, we can arrive at

$$\begin{cases} \lambda_{1,\alpha}^L = 1 + \alpha \\ \lambda_{1,\alpha}^U = 3 - \alpha \end{cases}, \begin{cases} \lambda_{2,\alpha}^L = 2 + \alpha \\ \lambda_{2,\alpha}^U = 4 - \alpha \end{cases}, \begin{cases} \mu_\alpha^L = \alpha \\ \mu_\alpha^U = 2 - \alpha \end{cases}.$$

It follows from Theorem 4.3.3 that

$$R(t) = \frac{1}{2}\int_0^1 \left(e^{-(1+\alpha)t} + \frac{1+\alpha}{\alpha - 1}\left[e^{-(2+\alpha)t} - e^{-(1+2\alpha)t}\right]\right.$$
$$\left. + e^{-(3-\alpha)t} + \frac{3-\alpha}{1-\alpha}\left[e^{-(4-\alpha)t} - e^{-(5-2\alpha)t}\right]\right) \mathrm{d}\alpha.$$

Figure 4.4 shows a plot of the reliability of the random fuzzy nonrepairable warm

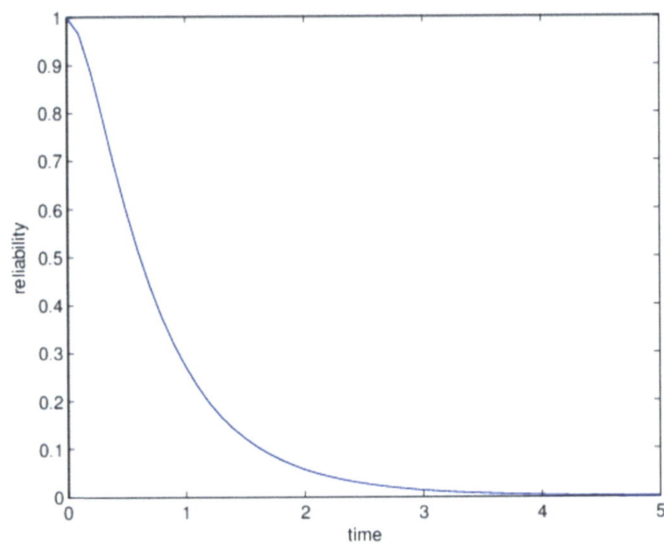

Fig. 4.4 Reliability of the warm standby system

standby system. We can see that the reliability decreases with time. It follows from Theorem 4.3.4 that

$$\begin{aligned}\text{MTTF} &= \frac{1}{2}\int_0^1 \left(\frac{1}{1+\alpha} + \frac{1}{3-\alpha}\right)\mathrm{d}\alpha + \frac{1}{2}\int_0^1 \frac{1}{2+\alpha}\left(\frac{1+\alpha}{1+2\alpha}\right)\mathrm{d}\alpha + \frac{1}{2}\int_0^1 \frac{1}{4-\alpha}\left(\frac{3-\alpha}{5-\alpha}\right)\mathrm{d}\alpha \\ &= \frac{4}{3}\ln 3 + \ln 5 - \frac{1}{12}\ln 2 - \frac{3}{2}\ln 4 \approx 0.9371.\end{aligned}$$

4.3.3 The Imperfect Conversion Switch Case: The Lifetime of the Conversion Switch is Continuous

In this case, we assume the conversion switch can deteriorate and the lifetime of the conversion switch is a continuous positive random fuzzy variable X_K, which has a random fuzzy exponential distribution with parameter λ_K defined on the credibility space $(\Theta_4, \mathcal{P}(\Theta_4), \text{Cr}_4)$. We also assume X_1, X_2, Y_2 and X_K are independent of each other. The warm standby system also can fail as the conversion switch fails immediately. The lifetime of the warm standby system can be expressed by

$$X = \min\left\{X_1 + X_2 \cdot I_{\{Y_2 > X_1\}},\ X_K\right\},$$

where $I_{\{\cdot\}}$ is the indicator function of a random fuzzy event. So, the lifetime of the warm standby system is a random fuzzy variable on the product credibility space $(\Theta,\ \mathcal{P}(\Theta),\ \text{Cr})$, where $\Theta = \Theta_1 \times \Theta_2 \times \Theta_3 \times \Theta_4$ and $\text{Cr} = \text{Cr}_1 \wedge \text{Cr}_2 \wedge \text{Cr}_3 \wedge \text{Cr}_4$.

Theorem 4.3.5 *The reliability of the random fuzzy warm standby system is*

$$\begin{aligned}R(t) &= \frac{1}{2}\int_0^1 e^{-\lambda_{K,\alpha}^L t}\left\{e^{-\lambda_{1,\alpha}^L t} + \frac{\lambda_{1,\alpha}^L\left[e^{-\lambda_{2,\alpha}^L t} - e^{-(\lambda_{1,\alpha}^L + \mu_\alpha^L)t}\right]}{\lambda_{1,\alpha}^L - \lambda_{2,\alpha}^L + \mu_\alpha^L}\right\}\mathrm{d}\alpha \\ &+ \frac{1}{2}\int_0^1 e^{-\lambda_{K,\alpha}^U t}\left\{e^{-\lambda_{1,\alpha}^U t} + \frac{\lambda_{1,\alpha}^U\left[e^{-\lambda_{2,\alpha}^U t} - e^{-(\lambda_{1,\alpha}^U + \mu_\alpha^U)t}\right]}{\lambda_{1,\alpha}^U - \lambda_{2,\alpha}^U + \mu_\alpha^U}\right\}\mathrm{d}\alpha.\end{aligned}$$

Proof By Definition 4.1, Definition 1.3.7 and Theorem 1.2.23, we have

$$\begin{aligned}R(t) &= \text{Ch}\{X > t\} \\ &= \int_0^1 \text{Cr}\{\theta \in \Theta \mid \Pr\{X(\theta) > t\} \geq p\}\mathrm{d}p\end{aligned}$$

$$= \frac{1}{2}\int_{0}^{1}\left(\Pr{}_{\alpha}^{L}\{\omega \in \Omega | X(\theta)(\omega) > t\} + \Pr{}_{\alpha}^{U}\{\omega \in \Omega | X(\theta)(\omega) > t\}\right)\mathrm{d}\alpha. \tag{4.3.21}$$

Let $A_\alpha = \{\theta_1 \in \Theta_1 | \mu\{\theta_1\} \geq \alpha\}$, $B_\alpha = \{\theta_2 \in \Theta_2 | \mu\{\theta_2\} \geq \alpha\}$, $C_\alpha = \{\theta_3 \in \Theta_3 | \mu\{\theta_3\} \geq \alpha\}$ and $D_\alpha = \{\theta_4 \in \Theta_4 | \mu\{\theta_4\} \geq \alpha\}$. For $\forall\theta_1 \in A_\alpha$, $\forall\theta_2 \in B_\alpha$, $\forall\theta_3 \in C_\alpha$ and $\forall\theta_4 \in D_\alpha$, we can arrive at

$$\lambda_{1,\alpha}^{L} \leq \lambda_1(\theta_1) \leq \lambda_{1,\alpha}^{U},\ \lambda_{2,\alpha}^{L} \leq \lambda_2(\theta_2) \leq \lambda_{2,\alpha}^{U}$$

and

$$\mu_{\alpha}^{L} \leq \mu(\theta_3) \leq \mu_{\alpha}^{U},\ \lambda_{K,\alpha}^{L} \leq \lambda_K(\theta_4) \leq \lambda_{K,\alpha}^{U}.$$

Then, we can construct three warm standby systems:

(1) Warm standby system 1: The lifetime of component 1 in operating, the lifetime of component 2 in operating, the lifetime of component 2 in warm standby and the lifetime of conversion switch have exponentially distributions with parameters $\lambda_{1,\alpha}^{L}$, $\lambda_{2,\alpha}^{L}$, μ_{α}^{L} and $\lambda_{K,\alpha}^{L}$, respectively.
(2) Warm standby system 2: The lifetime of component 1 in operating, the lifetime of component 2 in operating, the lifetime of component 2 in warm standby and the lifetime of conversion switch have exponentially distributions with parameters $\lambda_1(\theta_1)$, $\lambda_2(\theta_2)$, $\mu(\theta_3)$ and $\lambda_K(\theta_4)$, respectively.
(3) Warm standby system 3: The lifetime of component 1 in operating, the lifetime of component 2 in operating, the lifetime of component 2 in warm standby and the lifetime of conversion switch have exponentially distributions with parameters $\lambda_{1,\alpha}^{U}$, $\lambda_{2,\alpha}^{U}$, μ_{α}^{U} and $\lambda_{K,\alpha}^{U}$, respectively.

It is easy to see that the system (1) and system (3) are standard stochastic warm standby systems. For any fixed $\theta_1 \in A_\alpha$, $\theta_2 \in B_\alpha$, $\theta_3 \in C_\alpha$ and $\theta_4 \in D_\alpha$, system (2) is also a stochastic warm standby system. Let $X^{(1)}$, $X^{(2)}$ and $X^{(3)}$ be the lifetimes of system (1), system (2) and system (3). Since the warm standby system is a coherent system, we have

$$\Pr\{X^{(3)} > t\} \leq \Pr\{X^{(2)} > t\} \leq \Pr\{X^{(1)} > t\}.$$

Since θ_1, θ_2, θ_3 and θ_4 are arbitrary points in A_α, B_α, C_α and D_α, we have

$$\Pr{}_{\alpha}^{L}\{\omega \in \Omega | X(\theta)(\omega) > t\} = \Pr\{X^{(3)} > t\} \tag{4.3.22}$$

and

$$\Pr{}_{\alpha}^{U}\{\omega \in \Omega | X(\theta)(\omega) > t\} = \Pr\{X^{(1)} > t\}. \tag{4.3.23}$$

From the result in classical reliability theory, we can arrive at

$$\Pr\{X^{(1)} > t\} = e^{-\lambda_{K,\alpha}^L t}\left\{e^{-\lambda_{1,\alpha}^L t} + \frac{\lambda_{1,\alpha}^L\left[e^{-\lambda_{2,\alpha}^L t} - e^{-(\lambda_{1,\alpha}^L + \mu_\alpha^L)t}\right]}{\lambda_{1,\alpha}^L - \lambda_{2,\alpha}^L + \mu_\alpha^L}\right\} \tag{4.3.24}$$

and

$$\Pr\{X^{(3)} > t\} = e^{-\lambda_{K,\alpha}^U t}\left\{e^{-\lambda_{1,\alpha}^U t} + \frac{\lambda_{1,\alpha}^U\left[e^{-\lambda_{2,\alpha}^U t} - e^{-(\lambda_{1,\alpha}^U + \mu_\alpha^U)t}\right]}{\lambda_{1,\alpha}^U - \lambda_{2,\alpha}^U + \mu_\alpha^U}\right\}. \tag{4.3.25}$$

It follows from Eqs. (4.3.21)–(4.3.25) that

$$\begin{aligned} R(t) &= \frac{1}{2}\int_0^1 \left(\Pr_\alpha^L\{\omega \in \Omega | X(\theta)(\omega) > t\} + \Pr_\alpha^U\{\omega \in \Omega | X(\theta)(\omega) > t\}\right)\mathrm{d}\alpha \\ &= \frac{1}{2}\int_0^1 e^{-\lambda_{K,\alpha}^L t}\left\{e^{-\lambda_{1,\alpha}^L t} + \frac{\lambda_{1,\alpha}^L\left[e^{-\lambda_{2,\alpha}^L t} - e^{-(\lambda_{1,\alpha}^L + \mu_\alpha^L)t}\right]}{\lambda_{1,\alpha}^L - \lambda_{2,\alpha}^L + \mu_\alpha^L}\right\}\mathrm{d}\alpha \\ &+ \frac{1}{2}\int_0^1 e^{-\lambda_{K,\alpha}^U t}\left\{e^{-\lambda_{1,\alpha}^U t} + \frac{\lambda_{1,\alpha}^U\left[e^{-\lambda_{2,\alpha}^U t} - e^{-(\lambda_{1,\alpha}^U + \mu_\alpha^U)t}\right]}{\lambda_{1,\alpha}^U - \lambda_{2,\alpha}^U + \mu_\alpha^U}\right\}\mathrm{d}\alpha. \end{aligned}$$

The proof is completed.

Remark 4.3.3 If X_1, X_2, Y_2 and X_K degenerate to exponentially distributed random variables, i.e., λ_1, λ_2, μ and λ_K degenerate to crisp numbers, then the result in Theorem 4.3.5 degenerates to the form

$$R(t) = e^{-\lambda_K t}\left\{e^{-\lambda_1 t} + \frac{\lambda_1\left[e^{-\lambda_2 t} - e^{-(\lambda_1 + \mu)t}\right]}{\lambda_1 - \lambda_2 + \mu}\right\},$$

which is consistent with the result in stochastic case.

Theorem 4.3.6 *The MTTF of the random fuzzy warm standby system is*

$$\begin{aligned} \mathrm{MTTF} = E\left[\frac{1}{\lambda_1 + \lambda_K}\right] &+ \frac{1}{2}\int_0^1 \frac{\lambda_{1,\alpha}^L}{(\lambda_{2,\alpha}^L + \lambda_{K,\alpha}^L)(\lambda_{1,\alpha}^L + \mu_\alpha^L + \lambda_{K,\alpha}^L)}\mathrm{d}\alpha \\ &+ \frac{1}{2}\int_0^1 \frac{\lambda_{1,\alpha}^U}{(\lambda_{2,\alpha}^U + \lambda_{K,\alpha}^U)(\lambda_{1,\alpha}^U + \mu_\alpha^U + \lambda_{K,\alpha}^U)}\mathrm{d}\alpha. \end{aligned}$$

Proof By Definition 4.2 and Theorem 1.2.23, we have

$$\begin{aligned}\text{MTTF} &= \int_0^{+\infty} \text{Cr}\{\theta \in \Theta | E[X(\theta)] \geq r\}\text{d}r \\ &= \frac{1}{2}\int_0^1 \left(E[X(\theta)]_\alpha^L + E[X(\theta)]_\alpha^U\right)\text{d}\alpha. \end{aligned} \tag{4.3.26}$$

Based on the three warm standby systems constructed in the proof of Theorem 3, we also can see that

$$E[X^{(3)}] \leq E[X^{(2)}] \leq E[X^{(1)}].$$

Since $\theta_1,\ \theta_2,\ \theta_3$ and θ_4 are arbitrary points in $A_\alpha,\ B_\alpha,\ C_\alpha$ and D_α, we have

$$E[X(\theta)]_\alpha^L = E[X^{(3)}] \tag{4.3.27}$$

and

$$E[X(\theta)]_\alpha^U = E[X^{(1)}]. \tag{4.3.28}$$

From the result in classical reliability theory, we can arrive at

$$E[X^{(1)}] = \frac{1}{\lambda_{1,\alpha}^L + \lambda_{K,\alpha}^L} + \frac{\lambda_{1,\alpha}^L}{(\lambda_{2,\alpha}^L + \lambda_{K,\alpha}^L)(\lambda_{1,\alpha}^L + \mu_\alpha^L + \lambda_{K,\alpha}^L)} \tag{4.3.29}$$

and

$$E[X^{(3)}] = \frac{1}{\lambda_{1,\alpha}^U + \lambda_{K,\alpha}^U} + \frac{\lambda_{1,\alpha}^U}{(\lambda_{2,\alpha}^L + \lambda_{K,\alpha}^U)(\lambda_{1,\alpha}^U + \mu_\alpha^U + \lambda_{K,\alpha}^U)}. \tag{4.3.30}$$

It follows from Eqs. (4.3.26)–(4.3.30), and we have

$$\begin{aligned}\text{MTTF} &= \frac{1}{2}\int_0^1 \left(E[X(\theta)]_\alpha^L + E[X(\theta)]_\alpha^U\right)\text{d}\alpha \\ &= \frac{1}{2}\int_0^1 \left[\frac{1}{\lambda_{1,\alpha}^L + \lambda_{K,\alpha}^L} + \frac{\lambda_{1,\alpha}^L}{(\lambda_{2,\alpha}^L + \lambda_{K,\alpha}^L)(\lambda_{1,\alpha}^L + \mu_\alpha^L + \lambda_{K,\alpha}^L)}\right. \\ &\quad \left. + \frac{1}{\lambda_{1,\alpha}^U + \lambda_{K,\alpha}^U} + \frac{\lambda_{1,\alpha}^U}{(\lambda_{2,\alpha}^L + \lambda_{K,\alpha}^U)(\lambda_{1,\alpha}^U + \mu_\alpha^U + \lambda_{K,\alpha}^U)}\right]\text{d}\alpha\end{aligned}$$

$$
\begin{aligned}
&= \frac{1}{2}\int_0^1 \left[\frac{1}{\lambda_{1,\alpha}^L + \lambda_{K,\alpha}^L} + \frac{1}{\lambda_{1,\alpha}^U + \lambda_{K,\alpha}^U}\right] d\alpha \\
&+ \frac{1}{2}\int_0^1 \left[\frac{\lambda_{1,\alpha}^L}{(\lambda_{2,\alpha}^L + \lambda_{K,\alpha}^L)(\lambda_{1,\alpha}^L + \mu_\alpha^L + \lambda_{K,\alpha}^L)}\right. \\
&\left.+ \frac{\lambda_{1,\alpha}^U}{(\lambda_{2,\alpha}^U + \lambda_{K,\alpha}^U)(\lambda_{1,\alpha}^U + \mu_\alpha^U + \lambda_{K,\alpha}^U)}\right] d\alpha \\
&= E\left[\frac{1}{\lambda_1 + \lambda_K}\right] + \frac{1}{2}\int_0^1 \frac{\lambda_{1,\alpha}^L}{(\lambda_{2,\alpha}^L + \lambda_{K,\alpha}^L)(\lambda_{1,\alpha}^L + \mu_\alpha^L + \lambda_{K,\alpha}^L)} d\alpha \\
&+ \frac{1}{2}\int_0^1 \frac{\lambda_{1,\alpha}^U}{(\lambda_{2,\alpha}^U + \lambda_{K,\alpha}^U)(\lambda_{1,\alpha}^U + \mu_\alpha^U + \lambda_{K,\alpha}^U)} d\alpha.
\end{aligned}
$$

The theorem is proved.

Remark 4.3.4 If X_1, X_2, Y_2 and X_K degenerate to exponentially distributed random variables, i.e., λ_1, λ_2, μ and λ_K degenerate to crisp numbers, then the result in Theorem 4.3.6 degenerates to the form

$$
\text{MTTF} = \frac{1}{\lambda_1 + \lambda_K} + \frac{\lambda_1}{(\lambda_2 + \lambda_K)(\lambda_1 + \mu + \lambda_K)},
$$

which is consistent with the result in stochastic case.

Example 4.3.3 If λ_1, λ_2, μ and λ_K are triangular fuzzy variables, where $\lambda_1 = (1, 2, 3)$, $\lambda_2 = (2, 3, 4)$, $\mu = (0, 1, 2)$ and $\lambda_K = (0, 1, 2)$. We can arrive at

$$
\lambda_{1,\alpha}^L = 1 + \alpha,\ \lambda_{1,\alpha}^U = 3 - \alpha,\ \lambda_{2,\alpha}^L = 2 + \alpha,\ \lambda_{2,\alpha}^U = 4 - \alpha
$$

and

$$
\mu_\alpha^L = \lambda_{K,\alpha}^L = \alpha,\ \mu_\alpha^U = \lambda_{K,\alpha}^U = 2 - \alpha.
$$

By Theorem 4.3.5, we have

$$
\begin{aligned}
R(t) &= \frac{1}{2}\int_0^1 e^{-\alpha t}\left\{e^{-(1+\alpha)t} + \frac{1+\alpha}{\alpha - 1}\left[e^{-(2+\alpha)t} - e^{-(1+2\alpha)t}\right]\right\} d\alpha \\
&+ \frac{1}{2}\int_0^1 e^{-(2-\alpha)t}\left\{e^{-(3-\alpha)t} + \frac{3-\alpha}{1-\alpha}\left[e^{-(4-\alpha)t} - e^{-(5-2\alpha)t}\right]\right\} d\alpha.
\end{aligned}
$$

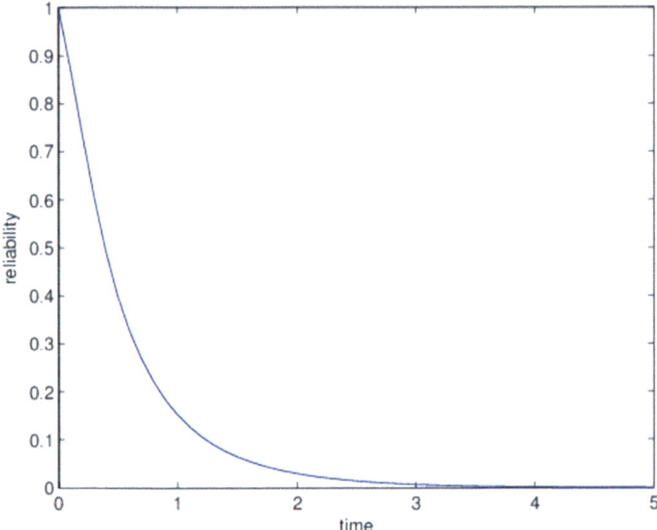

Fig. 4.5 Reliability of the warm standby system

Figure 4.5 shows a plot of the reliability of the nonrepairable warm standby system. We can see the reliability decreases with time. By Theorem 4.3.6, we have

$$\begin{aligned}\text{MTTF} &= \frac{1}{2}\int_0^1 \left(\frac{1}{1+2\alpha} + \frac{1}{5-2\alpha}\right)\mathrm{d}\alpha + \frac{1}{2}\int_0^1 \frac{1+\alpha}{(2+2\alpha)(1+3\alpha)}\mathrm{d}\alpha \\ &\quad + \frac{1}{2}\int_0^1 \frac{3-\alpha}{(6-2\alpha)(7-3\alpha)}\mathrm{d}\alpha \\ &= \frac{1}{12}\ln 3 + \frac{1}{4}\ln 5 + \frac{1}{12}\ln 7 - \frac{1}{12}\ln 4 \approx 0.5405.\end{aligned}$$

4.4 Random Fuzzy Shock Models and the Bivariate Random Fuzzy Exponential Distribution

The random fuzzy nonrepairable systems are composed by independent components in previous sections. In this section, we discuss the case that two components are independent. The random fuzzy shock model and the random fuzzy fatal shock model are proposed, respectively. Then, the bivariate random fuzzy exponential distribution is derived from the random fuzzy fatal shock model. Furthermore, some properties of the bivariate random fuzzy exponential distribution are proposed. Firstly, we introduce two common lemmas.

Lemma 4.4.1 Let X_i, $i = 1, 2, ..., n$ be positive independent random fuzzy variables. Then, we have

$$\mathrm{Ch}\{\min\{X_1, X_2, \ldots, X_n\} > t\} = E\left[\prod_{i=1}^{n} \Pr\{X_i > t\}\right].$$

Lemma 4.4.2 Let X_i, $i = 1, 2, ..., n$ be positive independent random fuzzy variables. Then, we have

$$E[\min\{X_1, X_2, \ldots, X_n\}]$$
$$= \frac{1}{2}\int_0^1\int_0^{+\infty}\left\{\prod_{i=1}^{n} \Pr{}_{\alpha}^{L}\{X_i \geq t\} + \prod_{i=1}^{n} \Pr{}_{\alpha}^{U}\{X_i \geq t\}\right\}\mathrm{d}t\mathrm{d}\alpha.$$

4.4.1 Random Fuzzy Shock Models

Suppose three independent sources of shocks are presented in the environment, see Fig. 4.6. There are three independent random fuzzy Poisson processes $Z_1(t, \lambda_1)$, $Z_2(t, \lambda_2)$ and $Z_{12}(t, \lambda_{12})$ that govern the occurrence of shocks, in which λ_1, λ_2 and λ_{12} are fuzzy variables defined on the credibility spaces $(\Theta_1, \mathcal{P}(\Theta_1), \mathrm{Cr}_1)$, $(\Theta_2, \mathcal{P}(\Theta_2), \mathrm{Cr}_2)$ and $(\Theta_{12}, \mathcal{P}(\Theta_{12}), \mathrm{Cr}_{12})$, respectively. A shock from source $Z_1(t, \lambda_1)$ causes the failure of component 1 with probability q_1, which occurs at the random fuzzy time U_1 and $U_1 \sim \exp(\lambda_1)$. Similarly, a shock from source $Z_2(t, \lambda_2)$ causes the failure of component 2 with probability q_2, which occurs at the random fuzzy time U_2 and $U_2 \sim \exp(\lambda_2)$. Finally, a shock from source $Z_{12}(t, \lambda_{12})$ causes the

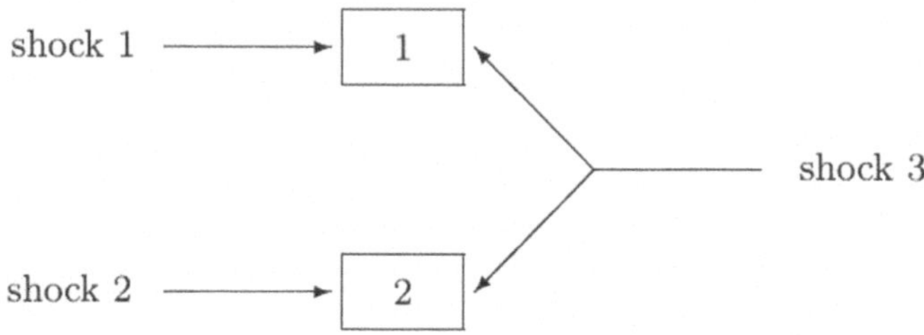

Fig. 4.6 A shock model

failure of both components with probability q_{11}; of component 1 with probability q_{01}; of component 2 with probability q_{10}; and of neither component with probability q_{00}, and the shock occurs at the random fuzzy time U_{12} and $U_{12} \sim \exp(\lambda_{12})$. Obviously, $q_{11}+q_{10}+q_{01}+q_{00} = 1$. The mathematical model is called the random fuzzy shock model.

For convenience, we discuss the reliability of the shock model on the product credibility space $(\Theta, \mathcal{P}(\Theta), \mathrm{Cr})$, where $\Theta = \Theta_1 \times \Theta_2 \times \Theta_{12}$ and $\mathrm{Cr} = \mathrm{Cr}_1 \wedge \mathrm{Cr}_2 \wedge \mathrm{Cr}_{12}$.

We pay attention to the joint survival probability of the lifetime of component 1 (denoted by T_1) and the lifetime of component 2 (denoted by T_2) when θ is fixed. Let $N_1(s,t)$, $N_2(s,t)$ and $N_3(s,t)$ be the numbers of Poisson processes $Z_1(t,\lambda_1)$,$Z_2(t,\lambda_2)$ and $Z_{12}(t,\lambda_{12})$ that occur, respectively.

For any fixed $\theta \in \Theta$, $\theta = (\theta_1, \theta_2, \theta_{12})$ and $0 \le t_1 \le t_2$, we have

$$\begin{aligned}
&\overline{F}(t_1,t_2)(\theta)\\
&= \Pr\{T_1(\theta) > t_1, T_2(\theta) > t_2\}\\
&= \Pr\{N_1(0,t_1)(\theta) = i,\\
&\qquad i \text{ shocks do not cause the failure of component } 1, i = 0, 1, \ldots;\\
&\qquad N_2(0,t_2)(\theta) = j,\\
&\qquad j \text{ shocks do not cause the failure of component } 2, j = 0, 1, \ldots;\\
&\qquad N_3(0,t_1)(\theta) = k,\\
&\qquad k \text{ shocks do not cause the failure of component 1 and } 2, k = 0, 1, \ldots;\\
&\qquad N_3(t_1,t_2)(\theta) = l,\\
&\qquad l \text{ shocks do not cause the failure of component } 2, l = 0, 1, \ldots\}\\
&= \sum_{i=0}^{+\infty} \frac{(\lambda_1(\theta)t_1)^i}{i!} e^{-\lambda_1(\theta)t_1}(1-q_1)^i \cdot \sum_{j=0}^{+\infty} \frac{(\lambda_2(\theta)t_2)^j}{j!} e^{-\lambda_2(\theta)t_2}(1-q_2)^j\\
&\quad \cdot \sum_{k=0}^{+\infty} \frac{(\lambda_{12}(\theta)t_1)^k}{k!} e^{-\lambda_{12}(\theta)t_1} q_{00}^k \cdot \sum_{l=0}^{+\infty} \frac{[\lambda_{12}(\theta)(t_2-t_1)]^l}{l!} e^{-\lambda_{12}(\theta)(t_2-t_1)}(q_{01}+q_{00})^l\\
&= \exp\{-(\lambda_1(\theta)q_1 + \lambda_{12}(\theta)q_{01})t_1 - [\lambda_2(\theta)q_2 + \lambda_{12}(\theta)(q_{11}+q_{01})]t_2\}.
\end{aligned} \tag{4.4.1}$$

Similarly, for $0 \le t_2 \le t_1$, we have

$$\overline{F}(t_1,t_2)(\theta) = \exp\{-[\lambda_1(\theta)q_1 + \lambda_{12}(\theta)(q_{11}+q_{01})]t_1 - (\lambda_2(\theta)q_2 + \lambda_{12}(\theta)q_{10})t_2\}. \tag{4.4.2}$$

Combining Eqs. (4.4.1) and (4.4.2), we can arrive at

$$\begin{aligned}
\overline{F}(t_1,t_2)(\theta) = \exp\{&-(\lambda_1(\theta)q_1 + \lambda_{12}(\theta)q_{01})t_1 - (\lambda_2(\theta)q_2 + \lambda_{12}(\theta)q_{10})t_2\\
&-\lambda_{12}(\theta)q_{11}\max(t_1,t_2)\}.
\end{aligned} \tag{4.4.3}$$

Especially, when $q_1 = q_2 = q_{11} = 1$, each shock causes the destruction of component. Then, the lifetime of component 1 is

$$T_1 = \min\{U_1, U_{12}\} \tag{4.4.4}$$

and the lifetime of component 2 is

$$T_2 = \min\{U_2, U_{12}\}. \tag{4.4.5}$$

Then, Eq. (4.4.3) changes to

$$\overline{F}(t_1, t_2)(\theta) = \exp\{-\lambda_1(\theta)t_1 - \lambda_2(\theta)t_2 - \lambda_{12}(\theta)\max(t_1, t_2)\}. \tag{4.4.6}$$

This mathematical model is called the random fuzzy fatal shock model.

4.4.2 The Bivariate Random Fuzzy Exponential Distribution

In this section, the bivariate random fuzzy exponential distribution is defined. Then, some properties of the bivariate random fuzzy exponential distribution are proposed.

Definition 4.4.1 *Let T_1 and T_2 be random fuzzy variables on the credibility space* $(\Theta, \mathcal{P}(\Theta), \text{Cr})$. *For any $\theta \in \Theta$, if* T_1 *and T_2 has joint survival probability*

$$\begin{aligned}\overline{F}(t_1, t_2)(\theta) &= \Pr\{T_1(\theta) > t_1, T_2(\theta) > t_2\}\\ &= \exp\{-\lambda_1(\theta)t_1 - \lambda_2(\theta)t_2 - \lambda_{12}(\theta)\max(t_1, t_2)\},\ t_1,\ t_2 \geq 0,\end{aligned} \tag{4.4.7}$$

where $\lambda_1,\ \lambda_2$ and λ_{12} are fuzzy variables. (T_1, T_2) is said to have a bivariate random fuzzy exponential distribution with parameters $\lambda_1,\ \lambda_2$ and λ_{12}, denoted by $(T_1, T_2) \sim RFBVE(\lambda_1, \lambda_2, \lambda_{12})$.

Definition 4.4.2 *Let (X, Y) be two-dimensional random fuzzy variable on the credibility space* $(\Theta, \mathcal{P}(\Theta), \text{Cr})$ *and $G(x, y)$ be the bivariate random fuzzy distribution function of (X, Y). $G(x, y)$ is said to have memoryless property, if for any given $\theta \in \Theta$ and $x,\ y,\ t \geq 0$,*

$$G(x + t, y + t)(\theta) = G(x, y)(\theta) \cdot G(t, t)(\theta) \tag{4.4.8}$$

always hold.

Theorem 4.4.1 *Let $(T_1,\ T_2) \sim RFBVE(\lambda_1, \lambda_2, \lambda_{12})$. Then, $\overline{F}(t_1, t_2)$ has memoryless property.*

Proof For any given $\theta \in \Theta$, $(T_1(\theta), T_2(\theta))$ has a bivariate exponential distribution. That is, for any $t_1,\ t_2,\ t \geq 0$, we have

$$\overline{F}(t_1 + t, t_2 + t)(\theta) = \overline{F}(t_1, t_2)(\theta) \cdot \overline{F}(t, t)(\theta).$$

The theorem is proved.

In random fuzzy theory, the average chances of random fuzzy events $\{\ T_1 > t_1\}$ and $\{T_2 > t_2\}$ play the equivalent roles similar to marginal distributions in classical bivariate exponential distributions.

Theorem 4.4.2 *Let* $(T_1,\ T_2) \sim RFBVE(\lambda_1, \lambda_2, \lambda_{12})$. *Then,*

$$\mathrm{Ch}\{T_1 > t_1\} = E[e^{-(\lambda_1+\lambda_{12})t_1}],\ t_1 \geq 0 \tag{4.4.9}$$

and

$$\mathrm{Ch}\{T_2 > t_2\} = E[e^{-(\lambda_2+\lambda_{12})t_2}],\ t_2 \geq 0. \tag{4.4.10}$$

Proof It follows from Lemma 4.4.1 and Eq. (4.4.4) that for any $t_1 \geq 0$,

$$\begin{aligned} \mathrm{Ch}\{T_1 > t_1\} &= E[\Pr\{U_1 > t_1\} \cdot \Pr\{U_{12} > t_1\}] \\ &= E\big[e^{-\lambda_1 t} \cdot e^{-\lambda_{12} t}\big] = E\big[e^{-(\lambda_1+\lambda_{12})t_1}\big]. \end{aligned} \tag{4.4.11}$$

Similarly,

$$\mathrm{Ch}\{T_2 > t_2\} = E\big[e^{-(\lambda_2+\lambda_{12})t_2}\big].$$

The theorem is proved.

Remark 4.4.1 If T_1 and T_2 degenerate to random variables, that is, $(T_1,\ T_2) \sim BVE(\lambda_1, \lambda_2, \lambda_{12})$, the results in Theorem 4.4.2 degenerate to the form

$$\Pr\{T_1 > t_1\} = e^{-(\lambda_1+\lambda_{12})t_1},\ t_1 \geq 0$$

and

$$\Pr\{T_2 > t_2\} = e^{-(\lambda_2+\lambda_{12})t_2},\ t_2 \geq 0,$$

which is consistent with the results in stochastic case.

Theorem 4.4.3 *Let* $(T_1,\ T_2) \sim RFBVE(\lambda_1, \lambda_2, \lambda_{12})$. *Then,*

$$E[T_1] = E\left[\frac{1}{\lambda_1 + \lambda_{12}}\right] \tag{4.4.12}$$

and

$$E[T_2] = E\left[\frac{1}{\lambda_2 + \lambda_{12}}\right] \tag{4.4.13}$$

Proof It follows from Lemma 4.4.2 and Eq. (4.4.4) that

$$\begin{aligned}
E[T_1] &= \frac{1}{2}\int_0^1\int_0^{+\infty} \left\{\Pr{}_{\alpha}^{L}\{U_1 \geq t_1\}\cdot \Pr{}_{\alpha}^{L}\{U_{12} \geq t_1\} + \Pr{}_{\alpha}^{U}\{U_1 \geq t_1\}\cdot \Pr{}_{\alpha}^{U}\{U_{12} \geq t_1\}\right\}\mathrm{d}t_1\mathrm{d}\alpha \\
&= \frac{1}{2}\int_0^1\int_0^{+\infty} \{(e^{-\lambda_1 t_1})_{\alpha}^{L}\cdot (e^{-\lambda_{12} t_1})_{\alpha}^{L} + (e^{-\lambda_1 t_1})_{\alpha}^{U}\cdot (e^{-\lambda_{12} t_1})_{\alpha}^{U}\}\mathrm{d}t_1\mathrm{d}\alpha \\
&= \frac{1}{2}\int_0^1\int_0^{+\infty} \left\{e^{-\lambda_{1,\alpha}^{U} t_1}\cdot e^{-\lambda_{12,\alpha}^{U} t_1} + e^{-\lambda_{1,\alpha}^{L} t_1}\cdot e^{-\lambda_{12,\alpha}^{L} t_1}\right\}\mathrm{d}t_1\mathrm{d}\alpha \\
&= \frac{1}{2}\int_0^1\int_0^{+\infty} \left\{e^{-(\lambda_{1,\alpha}^{U}+\lambda_{12,\alpha}^{U})t_1} + e^{-(\lambda_{1,\alpha}^{L}+\lambda_{12,\alpha}^{L})t_1}\right\}\mathrm{d}t_1\mathrm{d}\alpha \\
&= \frac{1}{2}\int_0^1 \left(\frac{1}{\lambda_{1,\alpha}^{U}+\lambda_{12,\alpha}^{U}} + \frac{1}{\lambda_{1,\alpha}^{L}+\lambda_{12,\alpha}^{L}}\right)\mathrm{d}\alpha \\
&= \frac{1}{2}\int_0^1 \left[\left(\frac{1}{\lambda_1+\lambda_{12}}\right)_{\alpha}^{L} + \left(\frac{1}{\lambda_1+\lambda_{12}}\right)_{\alpha}^{U}\right]\mathrm{d}\alpha \\
&= E\left[\frac{1}{\lambda_1+\lambda_{12}}\right].
\end{aligned}$$

By a similar argument, we see that

$$E[T_2] = E\left[\frac{1}{\lambda_2 + \lambda_{12}}\right].$$

The theorem is proved.

Remark 4.4.3 If T_1 and T_2 degenerate to random variables, that is $(T_1, T_2) \sim BVE(\lambda_1, \lambda_2, \lambda_{12})$, the results in Theorem 4.4.3 degenerate to the form

$$E[T_1] = \frac{1}{\lambda_1 + \lambda_{12}}$$

and

$$E[T_2] = \frac{1}{\lambda_2 + \lambda_{12}}.$$

Theorem 4.4.4 *Let* $(T_1,\ T_2) \sim BVE(\lambda_1, \lambda_2, \lambda_{12})$. *Then,*

$$E[T_1T_2] = E\left[\frac{1}{\lambda_1 + \lambda_2 + \lambda_{12}}\left(\frac{1}{\lambda_1 + \lambda_{12}} + \frac{1}{\lambda_2 + \lambda_{12}}\right)\right]. \tag{4.4.14}$$

Proof It follows from Definition 4.4.3 and Theorem 1.2.23 that

$$\begin{aligned} E[T_1T_2] &= \int_0^{+\infty} Cr\{\theta \in \Theta | E[T_1T_2(\theta)] \geq r\}\mathrm{d}r \\ &= \frac{1}{2}\int_0^1 \left(E[T_1T_2(\theta)]_\alpha^L + E[T_1T_2(\theta)]_\alpha^U\right)\mathrm{d}\alpha. \end{aligned} \tag{4.4.15}$$

Let $A_{i,\alpha} = \{\theta_i \in \Theta_i | \mu\{\theta_i\} \geq \alpha\}$, $i = 1, 2$ and $B_\alpha = \{\theta_{12} \in \Theta_{12} | \mu\{\theta_{12}\} \geq \alpha\}$. Since the α-pessimistic values and the α-optimistic values of fuzzy variables $E[U_i(\theta_i)], E[U_{12}(\theta_{12})]$, $\theta_i \in A_{i,\alpha}$, $\theta_{12} \in B_\alpha$, $i = 1, 2$ are continuous almost everywhere for any α, $\alpha \in (0, 1]$, there at least exist points $\theta_i^{'}$, $\theta_i^{''} \in A_{i,\alpha}$, $i = 1, 2$ and $\theta_{12}^{'}$, $\theta_{12}^{''} \in B_\alpha$ such that

$$\begin{aligned} E[U_i(\theta_i^{'})] &= E[U_i(\theta_i)]_\alpha^L, \\ E[U_i(\theta_i^{''})] &= E[U_i(\theta_i)]_\alpha^U, \\ E[U_{12}(\theta_{12}^{'})] &= E[U_{12}(\theta_{12})]_\alpha^L, \\ E[U_{12}(\theta_{12}^{''})] &= E[U_{12}(\theta_{12})]_\alpha^U. \end{aligned}$$

For $\forall\theta_{i,\alpha} \in A_{i,\alpha}$, $i = 1, 2$ and $\forall\theta_{12,\alpha} \in B_\alpha$, it is clear that

$$E[U_i(\theta_i^{'})] \leq E[U_i(\theta_{i,\alpha})] \leq E[U_i(\theta_i^{''})] \tag{4.4.16}$$

and

$$E[U_{12}(\theta_{12}^{'})] \leq E[U_{12}(\theta_{12,\alpha})] \leq E[U_{12}(\theta_{12}^{''})]. \tag{4.4.17}$$

By Theorem 1.1.21, we have

$$U_i(\theta_i^{'}) \leq_d U_i(\theta_{i,\alpha}) \leq_d U_i(\theta_i^{''}),\ i = 1, 2$$

and

$$U_{12}(\theta_{12}^{'}) \leq_d U_{12}(\theta_{12,\alpha}) \leq_d U_{12}(\theta_{12}^{''}).$$

Hence,

$$\begin{aligned}\min\left\{U_1(\theta_1^{'}), U_{12}(\theta_{12}^{'})\right\} &\leq_d \min\left\{U_1(\theta_{1,\alpha}), U_{12}(\theta_{12,\alpha})\right\}\\ &\leq_d \min\left\{U_1(\theta_1^{''}), U_{12}(\theta_{12}^{''})\right\}\end{aligned} \tag{4.4.18}$$

and

$$\begin{aligned}\min\left\{U_2(\theta_2^{'}), U_{12}(\theta_{12}^{'})\right\} &\leq_d \min\left\{U_2(\theta_{2,\alpha}), U_{12}(\theta_{12,\alpha})\right\}\\ &\leq_d \min\left\{U_2(\theta_2^{''}), U_{12}(\theta_{12}^{''})\right\}.\end{aligned} \tag{4.4.19}$$

It follows from Eqs. (4.4.18) and (4.4.19) that

$$\begin{aligned}&\min\left\{U_1(\theta_1^{'}), U_{12}(\theta_{12}^{'})\right\}\min\left\{U_2(\theta_2^{'}), U_{12}(\theta_{12}^{'})\right\}\\ &\leq_d \min\left\{U_1(\theta_{1,\alpha}), U_{12}(\theta_{12,\alpha})\right\}\min\left\{U_2(\theta_{2,\alpha}), U_{12}(\theta_{12,\alpha})\right\}\\ &\leq_d \min\left\{U_1(\theta_1^{''}), U_{12}(\theta_{12}^{''})\right\}\min\left\{U_2(\theta_2^{''}), U_{12}(\theta_{12}^{''})\right\}.\end{aligned} \tag{4.4.20}$$

That is,

$$T_1T_2(\theta_1^{'}, \theta_2^{'}, \theta_{12}^{'}) \leq_d T_1T_2(\theta_{1,\alpha}, \theta_{2,\alpha}, \theta_{12,\alpha}) \leq_d T_1T_2(\theta_1^{''}, \theta_2^{''}, \theta_{12}^{''}). \tag{4.4.21}$$

From Theorem 1.1.21, we see that

$$\begin{aligned}E\left[T_1T_2(\theta_1^{'}, \theta_2^{'}, \theta_{12}^{'})\right] &\leq E\left[T_1T_2(\theta_{1,\alpha}, \theta_{2,\alpha}, \theta_{12,\alpha})\right]\\ &\leq E\left[T_1T_2(\theta_1^{''}, \theta_2^{''}, \theta_{12}^{''})\right].\end{aligned} \tag{4.4.22}$$

Since $\theta_{i,\alpha}$ and $\theta_{12,\alpha}$ are arbitrary points in $A_{i,\alpha}$, $i = 1, 2$ and B_α, then from the result in classical reliability theory, we have

$$\begin{aligned}&E[T_1T_2(\theta)]_\alpha^L\\ &= E\left[T_1T_2(\theta_1^{'}, \theta_2^{'}, \theta_{12}^{'})\right]\\ &= \frac{1}{\lambda_1(\theta_1^{'}) + \lambda_2(\theta_2^{'}) + \lambda_{12}(\theta_{12}^{'})}\left(\frac{1}{\lambda_1(\theta_1^{'}) + \lambda_{12}(\theta_{12}^{'})} + \frac{1}{\lambda_2(\theta_2^{'}) + \lambda_{12}(\theta_{12}^{'})}\right)\end{aligned} \tag{4.4.23}$$

and

$$\begin{aligned}
&E[T_1T_2(\theta)]_\alpha^U \\
&= E\left[T_1T_2(\theta_1^{''}, \theta_2^{''}, \theta_{12}^{''})\right] \\
&= \frac{1}{\lambda_1(\theta_1^{''}) + \lambda_2(\theta_2^{''}) + \lambda_{12}(\theta_{12}^{''})}\left(\frac{1}{\lambda_1(\theta_1^{''}) + \lambda_{12}(\theta_{12}^{''})} + \frac{1}{\lambda_2(\theta_2^{''}) + \lambda_{12}(\theta_{12}^{''})}\right).
\end{aligned} \tag{4.4.24}$$

On the other hand, it follows from Eqs. (4.4.16) and (4.4.17) that

$$\frac{1}{\lambda_i(\theta_i^{'})} \le \frac{1}{\lambda_i(\theta_{i,\alpha})} \le \frac{1}{\lambda_i(\theta_i^{''})}, \ i = 1, 2 \tag{4.4.25}$$

and

$$\frac{1}{\lambda_{12}(\theta_{12}^{'})} \le \frac{1}{\lambda_{12}(\theta_{12,\alpha})} \le \frac{1}{\lambda_{12}(\theta_{12}^{''})}. \tag{4.4.26}$$

Therefore,

$$\lambda_i(\theta_i^{''}) \le \lambda_i(\theta_{i,\alpha}) \le \lambda_i(\theta_i^{'}), \ i = 1, 2 \tag{4.4.27}$$

and

$$\lambda_{12}(\theta_{12}^{''}) \le \lambda_{12}(\theta_{12,\alpha}) \le \lambda_{12}(\theta_{12}^{'}). \tag{4.4.28}$$

Since $\theta_{i,\alpha}$ and $\theta_{12,\alpha}$ are arbitrary points in $A_{i,\alpha}, \ i = 1, 2$ and B_α, then

$$\lambda_{i,\alpha}^L = \lambda_i(\theta_i^{''}), \ \lambda_{i,\alpha}^U = \lambda_i(\theta_i^{'}), \ i = 1, 2 \tag{4.4.29}$$

and

$$\lambda_{12,\alpha}^L = \lambda_{12}(\theta_{12}^{''}), \ \lambda_{12,\alpha}^U = \lambda_{12}(\theta_{12}^{'}). \tag{4.4.30}$$

From Eqs. (4.4.15), (4.4.23), (4.4.24), (4.4.29) and (4.4.30), we have

$$\begin{aligned}
E[T_1T_2] &= \frac{1}{2}\int_0^1 \left(E[T_1T_2(\theta)]_\alpha^L + E[T_1T_2(\theta)]_\alpha^U\right)\mathrm{d}\alpha \\
&= \frac{1}{2}\int_0^1 \left[\frac{1}{\lambda_1(\theta_1^{'}) + \lambda_2(\theta_2^{'}) + \lambda_{12}(\theta_{12}^{'})}\left(\frac{1}{\lambda_1(\theta_1^{'}) + \lambda_{12}(\theta_{12}^{'})} + \frac{1}{\lambda_2(\theta_2^{'}) + \lambda_{12}(\theta_{12}^{'})}\right)\right] \\
&\quad + \frac{1}{\lambda_1(\theta_1^{''}) + \lambda_2(\theta_2^{''}) + \lambda_{12}(\theta_{12}^{''})}\left(\frac{1}{\lambda_1(\theta_1^{''}) + \lambda_{12}(\theta_{12}^{''})} + \frac{1}{\lambda_2(\theta_2^{''}) + \lambda_{12}(\theta_{12}^{''})}\right)\Bigg]\mathrm{d}\alpha
\end{aligned}$$

$$= \frac{1}{2}\int_0^1 \left[\frac{1}{\lambda_{1,\alpha}^U + \lambda_{2,\alpha}^U + \lambda_{12,\alpha}^U}\left(\frac{1}{\lambda_{1,\alpha}^U + \lambda_{12,\alpha}^U} + \frac{1}{\lambda_{2,\alpha}^U + \lambda_{12,\alpha}^U}\right)\right.$$
$$\left. + \frac{1}{\lambda_{1,\alpha}^L + \lambda_{2,\alpha}^L + \lambda_{12,\alpha}^L}\left(\frac{1}{\lambda_{1,\alpha}^L + \lambda_{12,\alpha}^L} + \frac{1}{\lambda_{2,\alpha}^L + \lambda_{12,\alpha}^L}\right)\right] \mathrm{d}\alpha$$
$$= E\left[\frac{1}{\lambda_1 + \lambda_2 + \lambda_{12}}\left(\frac{1}{\lambda_1 + \lambda_{12}} + \frac{1}{\lambda_2 + \lambda_{12}}\right)\right].$$

The theorem is proved.

Remark 4.4.4 If T_1 and T_2 degenerate to random variables, that is, $(T_1, T_2) \sim BVE(\lambda_1, \lambda_2, \lambda_{12})$, the result in Theorem 4.4.4 degenerates to the form.
$E[T_1 T_2] = \frac{1}{\lambda_1+\lambda_2+\lambda_{12}}\left(\frac{1}{\lambda_1+\lambda_{12}} + \frac{1}{\lambda_2+\lambda_{12}}\right)$.

Theorem 4.4.5 *Let* $(T_1, T_2) \sim RFBVE(\lambda_1, \lambda_2, \lambda_{12})$. *Then,*

$$\begin{aligned} &E\left[e^{-(sT_1+tT_2)}\right] \\ &= E\left[\frac{(\lambda_1+\lambda_{12})(\lambda_2+\lambda_{12})}{(\lambda_1+\lambda_{12}+s)(\lambda_2+\lambda_{12}+t)}\right] \\ &+ \frac{1}{2}\int_0^1 \left[\frac{\lambda_{12,\alpha}^L st}{\left(\lambda_{1,\alpha}^L+\lambda_{2,\alpha}^L+\lambda_{12,\alpha}^L+s+t\right)\left(\lambda_{1,\alpha}^L+\lambda_{12,\alpha}^L+s\right)\left(\lambda_{2,\alpha}^L+\lambda_{12,\alpha}^L+t\right)}\right. \\ &\left. + \frac{\lambda_{12,\alpha}^U st}{\left(\lambda_{1,\alpha}^U+\lambda_{2,\alpha}^U+\lambda_{12,\alpha}^U+s+t\right)\left(\lambda_{1,\alpha}^U+\lambda_{12,\alpha}^U+s\right)\left(\lambda_{2,\alpha}^U+\lambda_{12,\alpha}^U+t\right)}\right]\mathrm{d}\alpha. \end{aligned} \tag{4.4.31}$$

Proof It follows from Definition 4.4.3, Theorem 1.2.23, Eqs. (4.4.4) and (4.4.4) that

$$\begin{aligned} &E\left[e^{-(sT_1+tT_2)}\right] \\ &= \int_0^{+\infty} \mathrm{Cr}\left\{\theta \in \Theta \mid E\left[e^{-(sT_1+tT_2)}\right] \ge r\right\}\mathrm{d}r \\ &= \frac{1}{2}\int_0^1 \left(E\left[e^{-(sT_1+tT_2)}\right]_\alpha^L + E\left[e^{-(sT_1+tT_2)}\right]_\alpha^U\right)\mathrm{d}\alpha \\ &= \frac{1}{2}\int_0^1 \left(E\left[e^{-(s\min\{U_1,U_{12}\}+t\min\{U_2,U_{12}\})}\right]_\alpha^L + E\left[e^{-(s\min\{U_1,U_{12}\}+t\min\{U_2,U_{12}\})}\right]_\alpha^U\right)\mathrm{d}\alpha. \end{aligned} \tag{4.4.32}$$

In Theorem 4.4.4, we have proved that Eqs. (4.4.18) and (4.4.19) hold. Then for any $s,\ t \ge 0$, we have

$$
\begin{aligned}
& s\ \min\left\{U_1(\theta_1'), U_{12}(\theta_{12}')\right\} + t\ \min\left\{U_2(\theta_2'), U_{12}(\theta_{12}')\right\} \\
& \leq_d s\ \min\{U_1(\theta_{1,\alpha}), U_{12}(\theta_{12,\alpha})\} + t\ \min\{U_2(\theta_{1,\alpha}), U_{12}(\theta_{12,\alpha})\} \\
& \leq_d s\ \min\left\{U_1(\theta_1''), U_{12}(\theta_{12}'')\right\} + t\ \min\left\{U_2(\theta_2''), U_{12}(\theta_{12}'')\right\}.
\end{aligned} \tag{4.4.33}
$$

Therefore,

$$
\begin{aligned}
& \exp\left\{s\ \min\left\{U_1(\theta_1'), U_{12}(\theta_{12}')\right\} + t\ \min\left\{U_2(\theta_2'), U_{12}(\theta_{12}')\right\}\right\} \\
& \leq_d \exp\{s\ \min\{U_1(\theta_{1,\alpha}), U_{12}(\theta_{12,\alpha})\} + t\ \min\{U_2(\theta_{1,\alpha}), U_{12}(\theta_{12,\alpha})\}\} \\
& \leq_d \exp\left\{s\ \min\left\{U_1(\theta_1''), U_{12}(\theta_{12}'')\right\} + t\ \min\left\{U_2(\theta_2''), U_{12}(\theta_{12}'')\right\}\right\}.
\end{aligned} \tag{4.4.34}
$$

It follows from Theorem 1.1.21 that

$$
\begin{aligned}
& E\left[\exp\left\{s\ \min\left\{U_1(\theta_1'), U_{12}(\theta_{12}')\right\} + t\ \min\left\{U_2(\theta_2'), U_{12}(\theta_{12}')\right\}\right\}\right] \\
& \leq E[\exp\{s\ \min\{U_1(\theta_{1,\alpha}), U_{12}(\theta_{12,\alpha})\} + t\ \min\{U_2(\theta_{1,\alpha}), U_{12}(\theta_{12,\alpha})\}\}] \\
& \leq E\left[\exp\left\{s\ \min\left\{U_1(\theta_1''), U_{12}(\theta_{12}'')\right\} + t\ \min\left\{U_2(\theta_2''), U_{12}(\theta_{12}'')\right\}\right\}\right].
\end{aligned} \tag{4.4.35}
$$

Since $\theta_{i,\alpha}$ and $\theta_{12,\alpha}$ are arbitrary points in $A_{i,\alpha}$, $i = 1, 2$ and B_α, then by the result in classical reliability theory, we have

$$
\begin{aligned}
& E\left[e^{-(s\min\{U_1,U_{12}\}+t\min\{U_2,U_{12}\})}\right]_\alpha^L \\
& = E\left[\exp\left\{s\ \min\left\{U_1(\theta_1'), U_{12}(\theta_{12}')\right\} + t\ \min\left\{U_2(\theta_2'), U_{12}(\theta_{12}')\right\}\right\}\right] \\
& = \frac{\left(\lambda_1(\theta_1') + \lambda_{12}(\theta_{12}')\right)\left(\lambda_2(\theta_2') + \lambda_{12}(\theta_{12}')\right)}{\left(\lambda_1(\theta_1') + \lambda_{12}(\theta_{12}') + s\right)\left(\lambda_2(\theta_2') + \lambda_{12}(\theta_{12}') + t\right)} \\
& \quad + \frac{\lambda_{12}(\theta_{12}')st}{\left(\lambda_1(\theta_1') + \lambda_2(\theta_2') + \lambda_{12}(\theta_{12}') + s + t\right)\left(\lambda_1(\theta_1') + \lambda_{12}(\theta_{12}') + s\right)\left(\lambda_2(\theta_2') + \lambda_{12}(\theta_{12}') + t\right)}
\end{aligned} \tag{4.4.36}
$$

and

$$
\begin{aligned}
& E\left[e^{-(s\min\{U_1,U_{12}\}+t\min\{U_2,U_{12}\})}\right]_\alpha^U \\
& = E\left[\exp\left\{s\ \min\left\{U_1(\theta_1''), U_{12}(\theta_{12}'')\right\} + t\ \min\left\{U_2(\theta_2''), U_{12}(\theta_{12}'')\right\}\right\}\right] \\
& = \frac{\left(\lambda_1(\theta_1'') + \lambda_{12}(\theta_{12}'')\right)\left(\lambda_2(\theta_2'') + \lambda_{12}(\theta_{12}'')\right)}{\left(\lambda_1(\theta_1'') + \lambda_{12}(\theta_{12}'') + s\right)\left(\lambda_2(\theta_2'') + \lambda_{12}(\theta_{12}'') + t\right)}
\end{aligned}
$$

$$+\frac{\lambda_{12}(\theta_{12}^{''})st}{\left(\lambda_1(\theta_1^{''})+\lambda_2(\theta_2^{''})+\lambda_{12}(\theta_{12}^{''})+s+t\right)\left(\lambda_1(\theta_1^{''})+\lambda_{12}(\theta_{12}^{''})+s\right)\left(\lambda_2(\theta_2^{''})+\lambda_{12}(\theta_{12}^{''})+t\right)}. \tag{4.4.37}$$

By Eqs. (4.4.29), (4.4.30), (4.4.32), (4.4.36) and (4.4.37),

$$\begin{aligned}
&E[e^{-(sT_1+tT_2)}]\\
&=\frac{1}{2}\int_0^1\left(E\left[e^{-(s\min\{U_1,U_{12}\}+t\min\{U_2,U_{12}\})}\right]_\alpha^L+E\left[e^{-(s\min\{U_1,U_{12}\}+t\min\{U_2,U_{12}\})}\right]_\alpha^U\right)\mathrm{d}\alpha\\
&=\frac{1}{2}\int_0^1\left[\frac{\left(\lambda_1(\theta_1^{'})+\lambda_{12}(\theta_{12}^{'})\right)\left(\lambda_2(\theta_2^{'})+\lambda_{12}(\theta_{12}^{'})\right)}{(\lambda_1(\theta_1^{'})+\lambda_{12}(\theta_{12}^{'})+s)(\lambda_2(\theta_2^{'})+\lambda_{12}(\theta_{12}^{'})+t)}\right.\\
&\quad+\frac{\lambda_{12}(\theta_{12}^{'})st}{(\lambda_1(\theta_1^{'})+\lambda_2(\theta_2^{'})+\lambda_{12}(\theta_{12}^{'})+s+t)(\lambda_1(\theta_1^{'})+\lambda_{12}(\theta_{12}^{'})+s)(\lambda_2(\theta_2^{'})+\lambda_{12}(\theta_{12}^{'})+t)}\\
&\quad+\frac{\left(\lambda_1(\theta_1^{''})+\lambda_{12}(\theta_{12}^{''})\right)\left(\lambda_2(\theta_2^{''})+\lambda_{12}(\theta_{12}^{''})\right)}{(\lambda_1(\theta_1^{''})+\lambda_{12}(\theta_{12}^{''})+s)(\lambda_2(\theta_2^{''})+\lambda_{12}(\theta_{12}^{''})+t)}\\
&\quad\left.+\frac{\lambda_{12}(\theta_{12}^{''})st}{(\lambda_1(\theta_1^{''})+\lambda_2(\theta_2^{''})+\lambda_{12}(\theta_{12}^{''})+s+t)(\lambda_1(\theta_1^{''})+\lambda_{12}(\theta_{12}^{''})+s)(\lambda_2(\theta_2^{''})+\lambda_{12}(\theta_{12}^{''})+t)}\right]\mathrm{d}\alpha\\
&=\frac{1}{2}\int_0^1\left[\frac{\left(\lambda_{1,\alpha}^L+\lambda_{12,\alpha}^L\right)\left(\lambda_{2,\alpha}^L+\lambda_{12,\alpha}^L\right)}{\left(\lambda_{1,\alpha}^L+\lambda_{12,\alpha}^L+s\right)\left(\lambda_{2,\alpha}^L+\lambda_{12,\alpha}^L+t\right)}+\frac{\left(\lambda_{1,\alpha}^U+\lambda_{12,\alpha}^U\right)\left(\lambda_{2,\alpha}^U+\lambda_{12,\alpha}^U\right)}{\left(\lambda_{1,\alpha}^U+\lambda_{12,\alpha}^U+s\right)\left(\lambda_{2,\alpha}^U+\lambda_{12,\alpha}^U+t\right)}\right]\mathrm{d}\alpha\\
&\quad+\frac{1}{2}\int_0^1\left[\frac{\lambda_{12,\alpha}^Lst}{\left(\lambda_{1,\alpha}^L+\lambda_{2,\alpha}^L+\lambda_{12,\alpha}^L+s+t\right)\left(\lambda_{1,\alpha}^L+\lambda_{12,\alpha}^L+s\right)\left(\lambda_{2,\alpha}^L+\lambda_{12,\alpha}^L+t\right)}\right.\\
&\quad\left.+\frac{\lambda_{12,\alpha}^Ust}{\left(\lambda_{1,\alpha}^U+\lambda_{2,\alpha}^U+\lambda_{12,\alpha}^U+s+t\right)\left(\lambda_{1,\alpha}^U+\lambda_{12,\alpha}^U+s\right)\left(\lambda_{2,\alpha}^U+\lambda_{12,\alpha}^U+t\right)}\right]\mathrm{d}\alpha\\
&=\frac{1}{2}\int_0^1\left[\left(\frac{(\lambda_1+\lambda_{12})(\lambda_2+\lambda_{12})}{(\lambda_1+\lambda_{12}+s)(\lambda_2+\lambda_{12}+t)}\right)_\alpha^L+\left(\frac{(\lambda_1+\lambda_{12})(\lambda_2+\lambda_{12})}{(\lambda_1+\lambda_{12}+s)(\lambda_2+\lambda_{12}+t)}\right)_\alpha^U\right]\mathrm{d}\alpha\\
&\quad+\frac{1}{2}\int_0^1\left[\frac{\lambda_{12,\alpha}^Lst}{\left(\lambda_{1,\alpha}^L+\lambda_{2,\alpha}^L+\lambda_{12,\alpha}^L+s+t\right)\left(\lambda_{1,\alpha}^L+\lambda_{12,\alpha}^L+s\right)\left(\lambda_{2,\alpha}^L+\lambda_{12,\alpha}^L+t\right)}\right.\\
&\quad\left.+\frac{\lambda_{12,\alpha}^Ust}{\left(\lambda_{1,\alpha}^U+\lambda_{2,\alpha}^U+\lambda_{12,\alpha}^U+s+t\right)\left(\lambda_{1,\alpha}^U+\lambda_{12,\alpha}^U+s\right)\left(\lambda_{2,\alpha}^U+\lambda_{12,\alpha}^U+t\right)}\right]\mathrm{d}\alpha\\
&=E\left[\frac{(\lambda_1+\lambda_{12})(\lambda_2+\lambda_{12})}{(\lambda_1+\lambda_{12}+s)(\lambda_2+\lambda_{12}+t)}\right]\\
&\quad+\frac{1}{2}\int_0^1\left[\frac{\lambda_{12,\alpha}^Lst}{\left(\lambda_{1,\alpha}^L+\lambda_{2,\alpha}^L+\lambda_{12,\alpha}^L+s+t\right)\left(\lambda_{1,\alpha}^L+\lambda_{12,\alpha}^L+s\right)\left(\lambda_{2,\alpha}^L+\lambda_{12,\alpha}^L+t\right)}\right.
\end{aligned}$$

$$+\frac{\lambda_{12,\alpha}^U st}{\left(\lambda_{1,\alpha}^U+\lambda_{2,\alpha}^U+\lambda_{12,\alpha}^U+s+t\right)\left(\lambda_{1,\alpha}^U+\lambda_{12,\alpha}^U+s\right)\left(\lambda_{2,\alpha}^U+\lambda_{12,\alpha}^U+t\right)}\Bigg]\mathrm{d}\alpha.$$

The theorem is proved.

Remark 4.4.5 If T_1 and T_2 degenerate to random variables, that is, $(T_1, T_2) \sim BVE(\lambda_1, \lambda_2, \lambda_{12})$, the result in Theorem 4.4.5 degenerates to the form

$$E\left[e^{-(sT_1+tT_2)}\right]=\frac{(\lambda_1+\lambda_2+\lambda_{12}+s+t)(\lambda_1+\lambda_{12})(\lambda_2+\lambda_{12})+st\lambda_{12}}{(\lambda_1+\lambda_2+\lambda_{12}+s+t)(\lambda_1+\lambda_{12}+s)(\lambda_2+\lambda_{12}+t)}.$$

Theorem 4.4.6 *Let* $(T_1, T_2) \sim RFBVE(\lambda_1, \lambda_2, \lambda_{12})$. *Then,*

$$\mathrm{Ch}\{\min\{T_1, T_2\} \le t\} = 1 - E\left[e^{-(\lambda_1+\lambda_2+\lambda_{12})t}\right],\ t \ge 0.$$

Proof Since $\min\{T_1, T_2\} = \min\{U_1, U_2, U_{12}\}$, by Lemma 4.4.1,

$$\begin{aligned}\mathrm{Ch}\{\min\{T_1, T_2\} \le t\} &= \mathrm{Ch}\{\min\{U_1, U_2, U_{12}\} \le t\}\\ &= 1-\mathrm{Ch}\{\min\{U_1, U_2, U_{12}\} > t\}\\ &= 1-E[\Pr\{U_1 > t\}\Pr\{U_2 > t\}\Pr\{U_{12} > t\}]\\ &= 1-E\left[e^{-\lambda_1 t}e^{-\lambda_2 t}e^{-\lambda_{12} t}\right]\\ &= 1-E\left[e^{-(\lambda_1+\lambda_2+\lambda_{12})t}\right].\end{aligned}$$

The theorem is proved.

Remark 4.4.6 If T_1 and T_2 degenerate to random variables, that is, $(T_1, T_2) \sim BVE(\lambda_1, \lambda_2, \lambda_{12})$, the result in Theorem 4.4.6 degenerates to the form

$$\Pr\{\min\{T_1, T_2\} \le t\} = 1 - e^{-(\lambda_1+\lambda_2+\lambda_{12})t},\ t \ge 0.$$

Example 4.4.1 Consider a series system consisting of two components. Let X_1 and X_2 be the random fuzzy lifetimes of component 1 and component 2, respectively, and $(X_1, X_2) \sim RFBVE(\lambda_1, \lambda_2, \lambda_{12})$, where $\lambda_1 = (0, 1, 2)$, $\lambda_2 = (0, 1, 2)$ and $\lambda_{12} = (1, 2, 3)$. So, we can arrive at

$$\begin{cases}\lambda_{1,\alpha}^L = \alpha\\ \lambda_{1,\alpha}^U = 2-\alpha\end{cases},\ \begin{cases}\lambda_{2,\alpha}^L = \alpha\\ \lambda_{2,\alpha}^U = 2-\alpha\end{cases},\ \begin{cases}\lambda_{12,\alpha}^L = 1+\alpha\\ \lambda_{12,\alpha}^U = 3-\alpha\end{cases}.$$

It follows from Theorem 4.4.2 that

$$\begin{aligned}\mathrm{Ch}\{X_1 > t_1\} &= E\left[e^{-(\lambda_1+\lambda_{12})t_1}\right]\\ &= \frac{1}{2}\int_0^1\left[e^{-(1+2\alpha)t_1}+e^{-(5-2\alpha)t_1}\right]\mathrm{d}\alpha = \frac{1}{4}(e^{-t_1}-e^{-5t_1}),\ t_1 \ge 0\end{aligned}$$

and

$$\begin{aligned}\mathrm{Ch}\{X_2 > t_2\} &= E\left[e^{-(\lambda_2+\lambda_{12})t_2}\right]\\ &= \frac{1}{4}(e^{-t_2} - e^{-5t_2}),\ t_2 \geq 0,\end{aligned}$$

which is just the reliability of component 1 and component 2, respectively. By Theorem 4.4.3, we have

$$E[X_1] = E\left[\frac{1}{\lambda_1+\lambda_2}\right] = \frac{1}{2}\int_0^1 \left[\frac{1}{1+2\alpha} + \frac{1}{5-2\alpha}\right]\mathrm{d}\alpha = \frac{1}{4}\ln 5$$

and

$$E[X_2] = E\left[\frac{1}{\lambda_2+\lambda_{12}}\right] = \frac{1}{4}\ln 5,$$

which is just the MTTF of component 1 and component 2, respectively. Then by Theorem 4.4.4, we have

$$\begin{aligned}E[X_1X_2] &= E\left[\frac{1}{\lambda_1+\lambda_2+\lambda_{12}}\left(\frac{1}{\lambda_1+\lambda_{12}} + \frac{1}{\lambda_2+\lambda_{12}}\right)\right]\\ &= \frac{1}{2}\int_0^1 \left[\frac{1}{(1+3\alpha)}\frac{2}{(1+2\alpha)} + \frac{1}{(7-3\alpha)}\frac{2}{(5-2\alpha)}\right]\mathrm{d}\alpha\\ &= 3\int_0^1 \frac{1}{3\alpha+1}\mathrm{d}\alpha - 2\int_0^1 \frac{1}{2\alpha+1}\mathrm{d}\alpha + 3\int_0^1 \frac{1}{7-3\alpha}\mathrm{d}\alpha - 2\int_0^1 \frac{1}{5-2\alpha}\mathrm{d}\alpha = \ln\frac{10}{7}.\end{aligned}$$

By Theorem 4.4.5, we have

$$\begin{aligned}&E[e^{-(sX_1+tX_2)}]\\ &= \frac{1}{2}\int_0^1 \left[\frac{(1+2\alpha)(1+2\alpha)}{(1+2\alpha+s)(1+2\alpha+t)} + \frac{(5-2\alpha)(5-2\alpha)}{(5-2\alpha+s)(5-2\alpha+t)}\right]\mathrm{d}\alpha\\ &+ \frac{1}{2}\int_0^1 \left[\frac{(1+\alpha)st}{(1+3\alpha+s+t)(1+2\alpha+s)(1+2\alpha+t)} + \frac{(3-\alpha)st}{(7-3\alpha+s+t)(5-2\alpha+s)(5-2\alpha+t)}\right]\mathrm{d}\alpha\\ &= \frac{1}{2} - \frac{s^2}{2(s-t)}\int_0^1 \frac{1}{(1+2\alpha+s)}\mathrm{d}\alpha - \frac{t^2}{2(t-s)}\int_0^1 \frac{1}{(1+2\alpha+t)}\mathrm{d}\alpha + \frac{1}{2} - \frac{s^2}{2(s-t)}\int_0^1 \frac{1}{(5-2\alpha+s)}\mathrm{d}\alpha\\ &- \frac{t^2}{2(t-s)}\int_0^1 \frac{1}{(5-2\alpha+t)}\mathrm{d}\alpha - \frac{3st(t^2-s^2-2t+2s)}{2(2s-t-1)(2t-s-1)(t-s)}\int_0^1 \frac{1}{(1+3\alpha++s+t)}\mathrm{d}\alpha\end{aligned}$$

$$
\begin{aligned}
&+\frac{st(s-1)}{2(2t-s-1)(s-t)}\int_0^1 \frac{1}{(1+2\alpha+s)}\mathrm{d}\alpha+\frac{st(t-1)}{2(2s-t-1)(t-s)}\int_0^1 \frac{1}{(1+2\alpha+t)}\mathrm{d}\alpha\\
&-\frac{3st(t^2-s^2-2t+2s)}{2(2s-t-1)(2t-s-1)(t-s)}\int_0^1 \frac{1}{(7-3\alpha+s+t)}\mathrm{d}\alpha+\frac{st(s-1)}{2(2t-s-1)(s-t)}\int_0^1 \frac{1}{(5-2\alpha+s)}\mathrm{d}\alpha\\
&+\frac{st(t-1)}{2(2s-t-1)(t-s)}\int_0^1 \frac{1}{(5-2\alpha+t)}\mathrm{d}\alpha\\
=&\,1+\frac{s^2}{4(s-t)}\ln\frac{1+s}{5+s}+\frac{t^2}{4(t-s)}\ln\frac{1+t}{5+t}-\frac{st(t^2-s^2-2t+2s)}{4(2s-t-1)(2t-s-1)(t-s)}\ln\frac{7+s+t}{1+s+t}\\
&+\frac{st(s-1)}{2(2t-s-1)(s-t)}\ln\frac{5+s}{1+s}+\frac{st(t-1)}{4(2s-t-1)(t-s)}\ln\frac{5+t}{1+t}.
\end{aligned}
$$

Finally, by Theorem 4.4.6, we can arrive at

$$
\begin{aligned}
&\mathrm{Ch}\{\min\{X_1,X_2\}\le t\}\\
&=1-E\left[e^{-(\lambda_1+\lambda_2+\lambda_{12})t}\right]\\
&=1-\frac{1}{2}\int_0^1\left[e^{-(1+3\alpha)t}+e^{-(7-3\alpha)t}\right]\mathrm{d}\alpha\\
&=1-\frac{1}{6}(e^{-t}-e^{-7t}),\ t\ge 0.
\end{aligned}
$$

Then,

$$
\mathrm{Ch}\{\min\{X_1,X_2\}>t\}=\frac{1}{6}(e^{-t}-e^{-7t}),\ t\ge 0,
$$

which is just the reliability of the series system.

Chapter 5
Repairable Systems with Stochastic Lifetimes and Repair Times

In engineering, maintenance methods are often used to improve systems reliability. Repairable systems usually consist of some components and one or more repair equipment (repairmen). The components can continue to perform their mission after repair. Some important reliability indices for evaluating repairable systems are considered in this chapter, such as the distribution of time to first system failure, mean time to first system failure, system availability, average number of system failures in $(0, t]$, mean up time and mean down time.

Let X_1 be the time to first system failure. Then, the distribution of time to first system failure is

$$F_1(t) = \Pr\{X_1 \le t\}.$$

The mean time to first system failure, denoted by MTTFF, can be expressed by

$$\text{MTTFF} = E[X_1] = \int_0^{+\infty} t \mathrm{d}F_1(t).$$

Consider a repairable system with only functioning state and failed state; we use a binary function to describe the state of the repairable system at time t. For $t \ge 0$, let

$$X(t) = \begin{cases} 1, & \text{the system is in functioning state at time } t \\ 0, & \text{the system is in failed state at time } t. \end{cases}$$

The availability at time t, denoted by $A(t)$, is defined as

$$A(t) = \Pr\{X(t) = 1\},$$

Y. Liu, *Reliability Theory Based on Uncertain Lifetimes*,
https://doi.org/10.1007/978-981-16-0995-4_5

which is the probability that the system is in functioning state at time t. The availability at time t only relates to whether the system is functioning at time t, and it does not care whether the system has failed before time t. Then, the average availability in $[0, t]$ is defined by

$$\tilde{A}(t) = \frac{1}{t}\int_0^t A(u)\mathrm{d}u.$$

If

$$\tilde{A} = \lim_{t\to+\infty} \tilde{A}(t)$$

exists, then $\tilde{A}$ is called the long-run average availability. If

$$A = \lim_{t\to+\infty} A(t)$$

exists, then A is called the limiting availability (steady-state availability). Obviously, if the limiting availability A exists, then the long-run average availability $\tilde{A}$ exists and $\tilde{A} = A$.

In engineering applications, engineers are more interested in limiting availability, which means that the time proportion A is in functioning state as the system has been running for a long time.

The progress of repairable system over time is a series of processes that alternate between functioning and failed states. For $t > 0$, the number of system failures in $(0, t]$, denoted by $N(t)$, is a nonnegative random variable. The distribution of the number of system failures in $(0, t]$ is

$$P_k(t) = \Pr\{N(t) = k\},\ \ k = 0, 1, 2, \ldots$$

The average number of system failures in $(0, t]$ is

$$M(t) = E[N(t)] = \sum_{k=1}^{+\infty} k P_k(t).$$

If the derivative of $M(t)$ exists,

$$m(t) = \frac{\mathrm{d}}{\mathrm{d}t} M(t)$$

is called the instantaneous failure frequency. In engineering applications, engineers are more interested in steady-state failure frequency, which is defined as

$$M = \lim_{t \to +\infty} \frac{M(t)}{t}$$

if the limit exists.

$M(t)$ and M are important reliability indices, which can help us determine how many redundant components we need to prepare in replacement problems.

The mean up time (MUT) is defined as

$$\text{MUT} = \lim_{n \to +\infty} \frac{1}{n} \sum_{i=1}^{n} E[X_i]$$

and the mean down time (MDT) is defined as

$$\text{MDT} = \lim_{n \to +\infty} \frac{1}{n} \sum_{i=1}^{n} E[Y_i].$$

For different practical problems, the reliability indices that people are interested in are different. The reliability indices mentioned above reflect the performance of repairable systems from different aspects. The main mathematical tool for studying repairable systems is stochastic process theory. If lifetimes and repair times of components are exponentially distributed, and the state of the system is properly defined, such a system can be described by a Markov process. However, in practice, it is often have the case that lifetimes or repair times of components do not have exponential distributions. At this moment, the stochastic process constituted by the repairable system is not a Markov process. Therefore, other mathematical tools are needed to study this kind of more general system. In this chapter, we mainly discuss non-Markov repairable systems. Firstly, we introduce the main tools used in this chapter, i.e., renewal processes and Markov renewal processes. Then, the reliability mathematical models of repairable single component systems, series systems, parallel systems, cold standby systems and coherent systems are established, respectively. Furthermore, the expressions of reliability indices are given for each repairable system.

5.1 Renewal Processes and Markov Renewal Processes

5.1.1 Renewal Processes

Let $X_1, X_2, \ldots$ be a sequence of independent and identically distributed nonnegative random variables, their distribution functions are $F(t)$, their expected values are μ, and $\Pr\{X_n = 0\} < 1$. Let

$$S_0 = 0,$$
$$S_n = X_1 + X_2 + \cdots + X_n, \ n = 1, 2, \ldots$$

They are the partial sums of random variables $X_1, X_2, \ldots$ Obviously,

$$\Pr\{S_n \leq t\} = F^{(n)}(t), \ n = 0, 1, 2, \ldots$$

where $F^{(n)}(t)$ is the n-fold convolution of $F(t)$ and

$$F^{(0)}(t) = \begin{cases} 1, \ t \geq 0 \\ 0, \ t < 0. \end{cases}$$

Let

$$N(t) = \sup\{n: S_n \leq t\}. \tag{5.1.1}$$

$\{N(t), \ t \geq 0\}$ is a stochastic process that takes nonnegative integer values, we call it the renewal process generated by independent and identically distributed random variables $X_1, X_2, \ldots$ Then, X_n is called the renewal lifetime, S_n is the moment when the nth renewal occurs (regeneration point), and $N(t)$ is the total number of renewals that occurs in $(0, t]$.

For example, in places where continuous lighting is required, we install a new lamp. When the lamp fails, we replace it with a new one immediately. Suppose that lifetimes of lamps (denoted by $U_n, \ n = 1, 2, \ldots$) are independent and identically distributed, and the replacement times of corresponding lamps (denoted by $V_n, \ n = 1, 2, \ldots$) are also independent and identically distributed. Let $X_n = U_n + V_n, \ n = 1, 2, \ldots$ A renewal process $\{N(t), \ t \geq 0\}$ is generated by Eq. (5.1.1), where $N(t)$ represents the total number of lamp renewals that occurs in $(0, t]$ and S_n represents the moment when the nth lamp renewal occurs.

Obviously, random events $\{N(t) \geq k\}$ and $\{S_k \leq t\}$ are equivalent. Therefore, the partial sum process $\{S_n, \ n \geq 0\}$ is also called a renewal process.

Due to

$$\{N(t) = k\} = \{S_k \leq t < S_{k+1}\}, \ k = 0, 1, \ldots$$

Then,

$$\begin{aligned} \Pr\{N(t) = k\} &= \Pr\{S_k \leq t\} - \Pr\{S_{k+1} \leq t\} \\ &= F^{(k)}(t) - F^{(k+1)}(t), \ k = 0, 1, \ldots \end{aligned} \tag{5.1.2}$$

By (5.1.2), the average number of renewals in $(0, t]$ is

$$M(t) = E\{N(t)\} = \sum_{k=1}^{+\infty} k \Pr\{N(t) = k\} = \sum_{k=1}^{+\infty} F^{(k)}(t). \tag{5.1.3}$$

We call $M(t)$ the renewal function. The Laplace–Stieltjes transform of $M(t)$ is

$$\hat{M}(s) = \int_0^{+\infty} e^{-st} \mathrm{d}M(t) = \sum_{k=1}^{+\infty} [\hat{F}(s)]^k = \frac{\hat{F}(s)}{1 - \hat{F}(s)}, \quad s > 0. \tag{5.1.4}$$

It can be proved that under the condition of $\Pr\{X_n = 0\} < 1$, for any $t \geq 0$, we have $M(t) < +\infty$. However, if $\Pr\{X_n = 0\} = 1$, the renewal process obtained is meaningless. Therefore, when discussing the renewal process, it is always assumed that $\Pr\{X_n = 0\} < 1$ holds. If $F(t)$ has a density function $f(t)$, then $M(t)$ is differentiable and

$$m(t) = \frac{\mathrm{d}}{\mathrm{d}t} M(t) = \sum_{k=1}^{+\infty} f^{(k)}(t), \tag{5.1.5}$$

where $f^{(k)}(t)$ is the k-fold convolution of the density function $f(t)$, i.e.,

$$\begin{cases} f^{(2)}(t) = \displaystyle\int_0^t f(t-u) f(u) \mathrm{d}u \\ f^{(k)}(t) = \displaystyle\int_0^t f^{(k-1)}(t-u) f(u) \mathrm{d}u, \quad k = 3, 4, \ldots \end{cases}$$

$m(t)$ is called the renewal density.

In practical problems, the following renewal equation is often used

$$h(t) = g(t) + \int_0^t h(t-u) \mathrm{d}F(u),$$

that is,

$$h(t) = g(t) + F(t) * h(t) \tag{5.1.6}$$

where $h(t)$ and $g(t)$ are nonnegative and bounded in any finite interval. When $F(t)$ and $g(t)$ are given functions, the solution of renewal Eq. (5.1.6) is

$$\begin{aligned} h(t) &= g(t) + \int_0^t g(t-u) \mathrm{d}M(u) \\ &= g(t) + M(t) * g(t). \end{aligned} \tag{5.1.7}$$

In fact, by Laplace transform on both sides of (5.1.6), we get

$$h^*(s) = g^*(s) + \hat{F}(s)h^*(s), \quad s > 0.$$

The solution is

$$\begin{aligned} h^*(s) &= \frac{g^*(s)}{1 - \hat{F}(s)} = \left[1 + \frac{\hat{F}(s)}{1 - \hat{F}(s)}\right] g^*(s) \\ &= g^*(s) + \hat{M}(s) g^*(s), \end{aligned}$$

in which the last equation is obtained by (5.1.4). Then, Eq. (5.1.7) is arrived by inversion of the above formula.

We call the distribution function of a nonnegative random variable X has a lattice distribution, if there has a $\delta > 0$ such that

$$\Pr\{X = n\delta, \ n = 0, 1, 2, \ldots\} = 1.$$

The largest δ that satisfies the above condition is called the period of lattice distribution.

We quote the following limit theorems without proof.

Theorem 5.1.1 $\lim\limits_{t \to +\infty} \frac{M(t)}{t} = \frac{1}{\mu}$.

Theorem 5.1.2 *If $F(t)$ is not lattice, then*

$$\lim_{t \to +\infty} [M(t + a) - M(t)] = \frac{a}{\mu}.$$

Theorem 5.1.3 *If $F(t)$ is not lattice, $g(t)$ is nonnegative and nonincreasing on $t \geq 0$, and*

$$\int_0^{+\infty} g(t) \mathrm{d}t < +\infty.$$

Then,

$$\lim_{t \to +\infty} \int_0^t g(t - u) \mathrm{d}M(u) = \frac{1}{\mu} \int_0^{+\infty} g(t) \mathrm{d}t.$$

When $\mu = +\infty$, the limits in the above three theorems are all zero.

If $\{N(t), \ t \geq 0\}$ is the renewal process generated by nonnegative random variables $X_1, X_2, \ldots$ Suppose Y_n is the reward of the nth renewal lifetime $X_n, \ n = 1, 2, \ldots,$ and $(X_n, Y_n), \ n = 1, 2, \ldots$ are independent and identically distributed. Let

$$Y(t) = \sum_{n=1}^{N(t)} Y_n$$

be the total reward in $(0, t]$. Then, we have the following theorem.

Theorem 5.1.4 *If $E|Y_n|$ and EX_n are finite, then*

$$\lim_{t \to +\infty} \frac{Y(t)}{t} = \frac{EY_n}{EX_n} \text{ with probability 1}$$

and

$$\lim_{t \to +\infty} \frac{EY(t)}{t} = \frac{EY_n}{EX_n}.$$

Suppose that $X_1, X_2, \ldots$ are independent and follow the same discrete probability distribution

$$\Pr\{X_k = l\} = \mathrm{P}_l, \ l = 1, 2, \ldots$$

Their partial sums $S_k = X_1 + X_2 + \cdots + X_k$, $k = 1, 2, \ldots$ follow the probability distributions

$$\Pr\{S_k = l\} = \mathrm{P}_l^{(k)}, \ l = k, k+1, \ldots, \ \ k = 1, 2, \ldots$$

where $\{\mathrm{P}_l^{(k)}, \ l = k, k+1, \ldots\}$ are the k-fold convolution of $\{\mathrm{P}_l, \ l = 1, 2, \ldots\}$, i.e.,

$$\begin{cases} \mathrm{P}_l^{(1)} = \mathrm{P}_l, & l \geq 1 \\ \mathrm{P}_l^{(k)} = \sum\limits_{j=1}^{l-1} \mathrm{P}_j \mathrm{P}_{l-j}^{(k-1)}, & 2 \leq k \leq l < +\infty \\ \mathrm{P}_l^{(k)} = 0, & k > l. \end{cases}$$

Let $\Phi(z) = \sum_{l=1}^{+\infty} \mathrm{P}_l z^l$, $|z| \leq 1$ be the generating function. The generating functions of $\{\mathrm{P}_l^{(k)}, \ l = k, k+1, \ldots\}$ are

$$\sum_{l=k}^{+\infty} \mathrm{P}_l^{(k)} z^l = [\Phi(z)]^k, \ k = 1, 2, \ldots$$

Let $N(n) = \sup\{k\colon S_k \leq n\}$ be the number of renewals in $(0, n]$. $\{N(n), \ n = 1, 2, \ldots\}$ is called the renewal process generated by $X_1, X_2, \ldots$ The renewal function is

$$M(n) = E[N(n)]$$

$$
\begin{aligned}
&= \sum_{k=1}^{n} k \Pr\{N(n) = k\} \\
&= \sum_{k=1}^{n} k \Pr\{S_k \le n < S_{k+1}\} \\
&= \sum_{k=1}^{n} k \left[\sum_{j=k}^{n} \mathrm{P}_j^{(k)} - \sum_{j=k+1}^{n} \mathrm{P}_j^{(k+1)} \right] \\
&= \sum_{k=1}^{n} \left[\sum_{j=k}^{n} \mathrm{P}_j^{(k)} \right] = \sum_{j=1}^{n} \left[\sum_{k=1}^{j} \mathrm{P}_j^{(k)} \right],
\end{aligned}
$$

in which $m(n) = \sum_{k=1}^{n} \mathrm{P}_n^{(k)}$ is called the renewal density, it represents the probability of exactly one renewal occurs at time n. Then, $M(n)$ can be expressed by

$$
M(n) = \sum_{j=1}^{n} m(j). \tag{5.1.8}
$$

The generating function of renewal density $m(j)$ is

$$
\begin{aligned}
\Phi_m(z) &= \sum_{j=1}^{+\infty} m(j) z^j \\
&= \sum_{j=1}^{+\infty} \sum_{k=1}^{j} \mathrm{P}_j^{(k)} z^j \\
&= \sum_{k=1}^{+\infty} \sum_{j=k}^{+\infty} \mathrm{P}_j^{(k)} z^j \\
&= \sum_{k=1}^{+\infty} [\Phi(z)]^k = \frac{\Phi(z)}{1 - \Phi(z)}.
\end{aligned}
$$

5.1.2 Markov Renewal Processes

Consider random variable Z_n takes values on $E = \{0, 1, \ldots, K\}$ and random variable T_n takes values on $[0, +\infty)$, where $0 = T_0 \le T_1 \le T_2 \le \ldots,\ n = 0, 1, \ldots$ The stochastic process $(\mathbf{Z}, \mathbf{T}) = \{Z_n, T_n,\ n = 0, 1, \ldots\}$ is called the Markov renewal process on the state space E, if for all $n = 0, 1, \ldots,\ j \in E,\ t \ge 0$, we have

$$\begin{aligned}&\Pr\{Z_{n+1}=j,\ T_{n+1}-T_n\le t|Z_0, Z_1,\ldots,Z_n,T_0,T_1,\ldots,T_n\}\\&\quad=\Pr\{Z_{n+1}=j,\ T_{n+1}-T_n\le t|Z_n\}.\end{aligned}\tag{5.1.9}$$

If for any $i, j \in E,\ t \ge 0$,

$$\Pr\{Z_{n+1}=j,\ T_{n+1}-T_n\le t|Z_n=i\}=Q_{ij}(t)\tag{5.1.10}$$

is irrelevant to n, $(\mathbf{Z}, \mathbf{T})$ is said to be homogeneous. $\{Q_{ij}(t),\ i, j \in E\}$ is called the semi-Markov kernel, which can form a matrix $\mathbf{Q}(t) = (Q_{ij}(t))$. Obviously,

$$\lim_{t\to+\infty} Q_{ij}(t)=\Pr\{Z_{n+1}=j|Z_n=i\}=\mathrm{P}_{ij}\tag{5.1.11}$$

is less than 1, and $Q_{ij}(t),\ i, j \in E$ have the properties of distribution functions. It is easy to prove that

$$\mathrm{P}_{ij}\ge 0,\ \sum_{j\in E}\mathrm{P}_{ij}=1.$$

Denote $\mathbf{P} = (\mathrm{P}_{ij})$. If $(\mathbf{Z}, \mathbf{T})$ is a homogeneous Markov renewal process, we can arrive at the following theorems.

Theorem 5.1.5 *If $\{Z_n,\ n \ge 0\}$ is a Markov chain with transition probability matrix* **P** *on the state space E.*

Let

$$G_{ij}(t)=\frac{Q_{ij}(t)}{\mathrm{Pr}_{ij}},\quad i, j\in E,\ t\ge 0.$$

(When $\mathrm{P}_{ij} = 0$, we set $G_{ij}(t) = 1$.) For each $i, j \in E$, $G_{ij}(t)$ is a distribution function and

$$G_{ij}(t)=\Pr\{T_{n+1}-T_n\le t|Z_n=i,\ Z_{n+1}=j\}.$$

Theorem 5.1.6 *For any $n \ge 1,\ t_1, t_2, \ldots, t_n \ge 0$, we have*

$$\begin{aligned}&\Pr\{T_1-T_0\le t_1,\ldots,T_n-T_{n-1}\le t_n|Z_0,Z_1,\ldots,Z_n\}\\&\quad=G_{Z_0Z_1}(t_1)G_{Z_1Z_2}(t_2)\cdots G_{Z_{n-1}Z_n}(t_n).\end{aligned}\tag{5.1.12}$$

That is, for a given Markov chain $Z_0, Z_1, \ldots$, the increments $T_1 - T_0, T_2 - T_1, \ldots$ are conditionally independent, and the distribution of $T_{n+1} - T_n$ depends only on Z_n and Z_{n+1}.

Let $X(t) = Z_n,\ T_n \le t \le T_{n+1}$. $\{X(t),\ t \ge 0\}$ is called a semi-Markov process associated with Markov renewal process $(\mathbf{Z}, \mathbf{T})$. $X(t)$ can be regarded as the state of the process at time t. The state transition occurs at time $T_1, T_2, \ldots$ At time T_n, the state transfers to state Z_n, and the sojourn time in state Z_n is $T_{n+1} - T_n$, whose distribution depends on the present state Z_n and the next state Z_{n+1}. The successive states $\{Z_n, n \geq 0\}$ form a Markov chain. For given Z_n, $n = 0, 1, 2, \ldots$, the consecutive sojourn times of states are conditionally independent.

In a Markov process, the sojourn time in each state follows an exponential distribution. Due to the memoryless property of the exponential distribution, any time t is the regeneration point of the process, which means that it has Markov property at any time t. The regeneration point means that under the condition that the present state of the process is known, the probability law of the future development of the process is irrelevant to the past states. In the case of a semi-Markov process, the distributions of sojourn times of states have general distributions. Therefore, not any time t is the regeneration point of the process, only the state transition moments are the regeneration points and have Markov property.

When all $G_{ij}(t)$ have exponential distributions, the semi-Markov process $\{X(t), t \geq 0\}$ is called the Markov process. When there is only one state in the state space of the process, i.e., $E = \{0\}$, then $\{T_{n+1} - T_n, n = 0, 1, 2, \ldots\}$ is a sequence of independent and identically distributed random variables, so $\{T_n, n = 0, 1, \ldots\}$ is the partial sum process discussed in Sect. 5.1.1. In this case, the Markov renewal process is called the renewal process.

In practical problems, we often use the following Markov renewal equation

$$h_i(t) = g_i(t) + \sum_{j \in E} \int_0^t h_j(t-u) \mathrm{d}Q_{ij}(u), \quad i \in E,$$

i.e.,

$$h_i(t) = g_i(t) + \sum_{j \in E} Q_{ij}(t) * h_j(t), \quad i \in E, \tag{5.1.13}$$

in which both $h_i(t)$ and $g_i(t)$ defined on $[0, +\infty)$ are nonnegative and bounded functions on any finite interval. Then, Eq. (5.1.13) can be abbreviated as the form of vector

$$\mathbf{h}(t) = \mathbf{g}(t) + \mathbf{Q}(t) * \mathbf{h}(t), \tag{5.1.14}$$

where $\mathbf{h}(t) = (h_0(t), h_1(t), \ldots, h_K(t))^T$ and $\mathbf{g}(t) = (g_0(t), g_1(t), \ldots, g_K(t))^T$ are column vectors, $\mathbf{Q}(t)$ is a matrix composed of semi-Markov kernels, $\mathbf{Q}(t) * \mathbf{h}(t)$ has the rules of matrix multiplication but the usual multiplication is replaced by convolution according to the operation on the second term on the right side of Eq. (5.1.13). If $\mathbf{g}(t)$ and $\mathbf{Q}(t)$ are given, Laplace transform and Laplace–Stieltjes transform are usually used to solve Eqs. (5.1.13) and (5.1.14).

Then, we introduce the limit behavior of $\mathbf{h}(t)$ without proof. Let v_i be a constant measure of $\{Z_n, n = 0, 1, \ldots\}$, that is

$$\sum_{i \in E} v_i \mathrm{P}_{ij} = v_j, \ \ j \in E. \tag{5.1.15}$$

Then, the mean sojourn time in state i is

$$\begin{aligned} m(i) &= E\{T_1 | Z_0 = i\} \\ &= \int_0^{+\infty} \left[1 - \sum_{j \in E} Q_{ij}(t) \right] \mathrm{d}t \\ &= -\sum_{j \in E} \hat{Q}'_{ij}(0), \ i \in E. \end{aligned} \tag{5.1.16}$$

We call a nonnegative function $Q(t)$ defined on $[0, +\infty)$ is lattice, if there exist $\alpha \geq 0$ and $\delta > 0$ such that $Q(t)$ is a jumping step function on the set $\{\alpha, \ \alpha + \delta, \ \alpha + 2\delta, \ \ldots\}$.

Theorem 5.1.7 *If all states in $\{Z_n, \ n = 0, 1, \ldots\}$ can reach each other and there exists a pair of states $(i, \ j)$ such that $Q_{ij}(t)$ is not lattice, $g_j(t)$ is nonnegative and nonincreasing and $\int_0^{+\infty} g_j(t)\mathrm{d}t < +\infty$, $j \in E$. Then, for all $k \in E$, we have*

$$\lim_{t \to +\infty} h_k(t) = \frac{1}{\sum_{j \in E} v_j m(j)} \sum_{j \in E} v_j \int_0^{+\infty} g_j(t)\mathrm{d}t$$

is irrelevant to state k.

5.2 The Single Component System

Suppose that the system is composed of one component, and its lifetime X follows a general distribution $F(t)$. If the component (system) fails, it is repaired by the repair equipment immediately, and the repair time Y follows a general distribution $G(t)$. When the repair is completed, the component transfers to the functioning state immediately. Denote

$$\frac{1}{\lambda} = \int_0^{+\infty} t\mathrm{d}F(t), \ \frac{1}{\mu} = \int_0^{+\infty} t\mathrm{d}G(t),$$

where $\lambda, \ \mu > 0$. The component is like a new one after repair. We also assume that X and Y are independent.

For convenience, we assume that the component is new at initial time, that is, the component enters the functioning state at $t = 0$. Then, we consider a renewal process in which the operating times and repair times occur alternately. Denote $Z_i = X_i + Y_i$, where X_i and Y_i are the lifetime and repair time of the component in the ith cycle, $i = 1, 2, \ldots$ Then, $\{Z_i, i = 1, 2, \ldots\}$ is a sequence of independent and identically distributed random variables, which can constitute a renewal process and the distribution of the renewal lifetime is

$$\begin{aligned} Q(t) &= \Pr\{Z_i \le t\} \\ &= \Pr\{X_i + Y_i \le t\} \\ &= \int_0^t G(t-u)\mathrm{d}F(u) \\ &= F * G(t), \quad i = 1, 2, \ldots \end{aligned}$$

The time to first system failure is denoted by X_1, then the reliability of the series system is

$$R(t) = \Pr\{X_1 > t\} = 1 - F(t). \tag{5.2.1}$$

The mean time to first system failure is

$$\mathrm{MTTFF} = E[X_1] = \frac{1}{\lambda}.$$

To obtain the instantaneous availability of the system, we introduce a stochastic process $\{X(t), t \ge 0\}$, where

$$X(t) = \begin{cases} 1, & \text{the system is in functioning state at time } t \\ 0, & \text{the system is in repair state at time } t. \end{cases}$$

Then, the system availability at time t is

$$A(t) = \Pr\{X(t) = 1 | \text{the system is new at } t = 0\}.$$

By using the total probability formula, we have

$$\begin{aligned} A(t) = {} & \Pr\{X_1 > t,\ X(t) = 1 | \text{the system is new at } t = 0\} \\ & + \Pr\{X_1 \le t < X_1 + Y_1,\ X(t) = 1 | \text{the system is new at } t = 0\} \\ & + \Pr\{X_1 + Y_1 \le t,\ X(t) = 1 | \text{the system is new at } t = 0\}. \end{aligned}$$

In the first term on the right side of the above formula, if $X_1 > t$, the system must be in functioning state at time t, i.e., $X(t) = 1$, so the first term is $\Pr\{X_1 > t\}$; in the second term, if $X_1 \le t < X_1 + Y$, the system must be in repair state at time t, which cannot occur with $X(t) = 1$ at the same time, so the second term is equal to zero; the third term is

$$
\begin{aligned}
&\Pr\{X_1 + Y_1 \le t,\ X(t) = 1 | \text{the system is new at } t = 0\} \\
&= \int_0^t \Pr\{X(t) = 1 | \text{the system is new at } t = 0,\ X_1 + Y_1 = u\} \mathrm{d}\Pr\{X_1 + Y_1 \le u\} \\
&= \int_0^t \Pr\{X(t) = 1 | \text{the system is new at } t = u\} \mathrm{d}\Pr\{X_1 + Y_1 \le u\} \\
&= \int_0^t \Pr\{X(t-u) = 1 | \text{the system is new at } t = 0\} \mathrm{d}Q(u) \\
&= \int_0^t A(t-u)\mathrm{d}Q(u) = Q(t) * A(t).
\end{aligned}
$$

The first equation in above formula is by using the total probability formula. Since $X_1 + Y_1 = u$ is the regeneration point, then $A(t)$ satisfies the renewal equation

$$
A(t) = 1 - F(t) + Q(t) * A(t). \tag{5.2.2}
$$

The following equation is obtained from Laplace transform of the above formula,

$$
A^*(s) = \frac{1 - \hat{F}(s)}{s} + \hat{F}(s)\hat{G}(s)A^*(s),
$$

in which

$$
A^*(s) = \int_0^{+\infty} e^{-st} A(t)\mathrm{d}t,
$$

$$
\hat{F}(s) = \int_0^{+\infty} e^{-st} \mathrm{d}F(t)
$$

and

$$
\hat{G}(\mathrm{s}) = \int_0^{+\infty} e^{-st} \mathrm{d}G(t).
$$

Thus, the Laplace transform of $A(t)$ is

$$
A^*(s) = \frac{1}{s} \cdot \frac{1 - \hat{F}(s)}{1 - \hat{F}(s)\hat{G}(\mathrm{s})}. \tag{5.2.3}
$$

According to the Tauberian theorem and L'Hospital's rule of Laplace transform, the long-run average availability of the system is

$$\begin{aligned}
\tilde{A} &= \lim_{t\to+\infty}\frac{1}{t}\int\limits_0^t A(u)\mathrm{d}u \\
&= \lim_{s\to 0} sA^*(s) \\
&= \lim_{s\to 0}\frac{1-\hat{F}(s)}{1-\hat{F}(s)\hat{G}(s)} \\
&= \lim_{s\to 0}\frac{-\hat{F}'(s)}{-\hat{F}'(s)\hat{G}(s)-\hat{F}(s)\hat{G}'(s)} \\
&= \frac{\mu}{\lambda+\mu}.
\end{aligned} \tag{5.2.4}$$

By Eq. (5.1.7), the solution of renewal Eq. (5.2.2) is

$$A(t) = \overline{F}(t) + \tilde{M}(t) * \overline{F}(t), \tag{5.2.5}$$

where $\tilde{M}(t) = \sum_{k=1}^{+\infty} F^{(k)}(t) * G^{(k)}(t)$ is the renewal function. If at least one of $F(t)$ and $G(t)$ does not have a lattice distribution, the renewal lifetime distribution $Q(t)$ does not have a lattice distribution. By Theorem 5.1.3, the limiting availability of the system is

$$\begin{aligned}
A &= \lim_{t\to+\infty} A(t) \\
&= \lim_{t\to+\infty} \tilde{M}(t) * \overline{F}(t) \\
&= \frac{1}{E[X+Y]}\int\limits_0^{+\infty}\overline{F}(t)\mathrm{d}t \\
&= \frac{\frac{1}{\lambda}}{\frac{1}{\lambda}+\frac{1}{\mu}} = \frac{\mu}{\lambda+\mu}.
\end{aligned} \tag{5.2.6}$$

Assume the system is new at $t = 0$. Let $N(t)$ be the number of system failures in $(0, t]$, and $M(t) = E[N(t)]$. By the total probability formula, we have

$$\begin{aligned}
M(t) &= E[N(t)] \\
&= E[N(t)|X_1 > t]\Pr\{X_1 > t\} \\
&\quad + E[N(t)|X_1 \le t < X_1 + Y_1]\Pr\{X_1 \le t < X_1 + Y_1\}
\end{aligned}$$

$$+\int_0^t E[N(t)|X_1+Y_1=u]\mathrm{d}\Pr\{X_1+Y_1\le u\}.$$

In the first term on the right side of the above formula, under the condition of $X_1 > t$, the number of system failures in $(0, t]$ is zero, i.e., $E[N(t)|X_1 > t] = 0$; in the second term, under the condition of $X_1 \le t < X_1 + Y_1$, the system fails only once in $(0, t]$, i.e., $E[N(t)|X_1 \le t < X_1 + Y_1] = 1$; in the third term,

$$E[N(t)|X_1+Y_1=u]=E[N(t-u)]+1=M(t-u)+1,$$

which implies that the system fails only once in $(0, u]$ under the condition of $X_1 + Y_1 = u$; therefore, the average number of system failures in $(0, t]$ is equal to the average number of system failures in $(u, t]$ plus 1. Since $X_1 + Y_1 = u$ is the regeneration point, the average number of system failures in $(u, t]$ is equal to the average number of failures in $(0, t-u]$. Therefore, $M(t)$ satisfies the renewal equation

$$\begin{aligned}M(t)&=\Pr\{X_1\le t<X_1+Y_1\}+\int_0^t[M(t-u)+1]\mathrm{d}\Pr\{X_1+Y_1\le u\}\\&=F(t)-F(t)*\mathrm{G}(t)+Q(t)*[M(t)+1],\end{aligned}$$

that is,

$$M(t)=F(t)+Q(t)*M(t).$$

The solution can be arrived by Laplace–Stieltjes transform on the above formula, then

$$\hat{M}(s)=\frac{\hat{F}(s)}{1-\hat{F}(s)\hat{G}(s)},\tag{5.2.7}$$

where $\hat{M}(s)=\int_0^{+\infty}e^{-st}\mathrm{d}M(t)$. The steady-state failure frequency of the system is

$$M=\lim_{t\to+\infty}\frac{M(t)}{t}=\lim_{s\to0}s\hat{M}(s)=\lim_{s\to0}\frac{\hat{F}(s)}{\frac{1-\hat{F}(s)\hat{G}(s)}{s}}=\frac{\lambda\mu}{\lambda+\mu}.\tag{5.2.8}$$

In particular, if the lifetime and the repair time of the component have exponential distributions, that is,

$$F(t)=1-e^{-\lambda t},\quad G(t)=1-e^{-\mu t},\quad t\ge0.$$

At this moment, we have

$$\hat{F}(s) = \frac{\lambda}{s+\lambda} \text{ and } \hat{G}(s) = \frac{\mu}{s+\mu}.$$

The system reliability is obtained by (5.2.1), namely

$$R(t) = e^{-\lambda t}. \tag{5.2.9}$$

Formula (5.2.3) changes to

$$A^*(s) = \frac{s+\mu}{s(s+\lambda+\mu)} = \frac{\mu}{\lambda+\mu} \cdot \frac{1}{s} + \frac{\lambda}{\lambda+\mu} \cdot \frac{1}{s+\lambda+\mu}.$$

Inverse the above formula, we have

$$A(t) = \frac{\mu}{\lambda+\mu} + \frac{\lambda}{\lambda+\mu} e^{-(\lambda+\mu)t}. \tag{5.2.10}$$

Formula (5.2.7) changes to

$$\hat{M}(s) = \frac{\lambda(s+\mu)}{s(s+\lambda+\mu)} = \frac{\lambda\mu}{\lambda+\mu} \cdot \frac{1}{s} + \frac{\lambda^2}{\lambda+\mu} \cdot \frac{1}{s+\lambda+\mu}.$$

Inverse the above formula, we have

$$M(t) = \frac{\lambda\mu}{\lambda+\mu} t + \frac{\lambda^2}{(\lambda+\mu)^2}[1 - e^{-(\lambda+\mu)t}]. \tag{5.2.11}$$

Then, the instantaneous failure frequency of the system is

$$m(t) = M'(t) = \frac{\lambda\mu}{\lambda+\mu} + \frac{\lambda^2}{\lambda+\mu} e^{-(\lambda+\mu)t}. \tag{5.2.12}$$

We can arrive at A and M by Formulas (5.2.10) and (5.2.11), which are consistent with Formulas (5.2.5) and (5.2.8), respectively.

5.3 The Series System

Assume that the series system is composed of n components, where the lifetime of the ith component X_i follows an exponential distribution $F_i(t) = 1 - e^{-\lambda_i t}$, $t \geq 0$. The ith component is repaired by the repair equipment immediately after failure and the corresponding repair time Y_i follows a general distribution $G_i(t)$ which has mean value $\frac{1}{\mu_i} = \int_0^{+\infty} t dG_i(t)$, $(\lambda_i, \mu_i > 0)$, $i = 1, 2, \ldots, n$. We assume the component is like a new one after repair. The lifetimes and the repair times of components are also independent of each other. When a component fails and enters the repair state, the remaining components stop working immediately. At this moment, the series system is in repair state. When the failed component is repaired, all components start working at the same time, and the series system enters the functioning state.

For convenience, we assume all components are in functioning state at $t = 0$, that is, the series system is in functioning state. Obviously, the lifetime of the series system is

$$\zeta = \min_{1 \le i \le n} X_i.$$

The repair time χ depends on ζ. If $\zeta = \min_{1 \le i \le n} X_i = X_j$, that is, if the jth component fails, the corresponding repair time $\chi = Y_j$. When the failed component is repaired, the series system enters the functioning state again. Due to the memoryless property of the exponential distribution, the series system returns to the situation at $t = 0$. Therefore, the moments when the failed components finish repair are the regeneration points of the series system. Let ζ_k and χ_k be the kth lifetime and the repair time of the series system. Then, $\{\zeta_k + \chi_k,\ k = 1, 2, \ldots\}$ is a sequence of independent and identically distributed random variables and the distribution function is

$$\begin{aligned}
Q(t) &= \Pr\{\zeta_k + \chi_k \le t\} \\
&= \Pr\left\{\min_{1 \le i \le n} X_i + \chi \le t\right\} \\
&= \sum_{j=1}^{n} \Pr\left\{X_j = \min_{1 \le i \le n} X_i,\ X_j + Y_j \le t\right\} \\
&= \sum_{j=1}^{n} \Pr\{X_j \le X_1, X_2, \ldots, X_{j-1}, X_{j+1}, \ldots, X_n, X_j + Y_j \le t\} \\
&= \sum_{j=1}^{n} \int_0^t \Pr\{u \le X_1, X_2, \ldots, X_{j-1}, X_{j+1}, \ldots, X_n, Y_j \le t - u\} \mathrm{d}\Pr\{X_j \le u\} \\
&= \sum_{j=1}^{n} \int_0^t G_j(t-u) e^{-\sum_{i \ne j} \lambda_i u} \mathrm{d}(1 - e^{-\lambda_j u}) \\
&= \sum_{j=1}^{n} \frac{\lambda_j}{\Lambda} \int_0^t G_j(t-u) \mathrm{d}(1 - e^{-\Lambda u}),\ k = 1, 2, \ldots
\end{aligned} \tag{5.3.1}$$

where $\Lambda = \sum_{i=1}^{n} \lambda_i$. It is easy to see that $Q(t)$ is irrelevant to k. Then, a renewal process $\{\zeta_k + \chi_k,\ k = 1, 2, \ldots\}$ is constituted, and $Q(t)$ is the distribution of the renewal lifetime. By Laplace–Stieltjes transform on both sides of Eq. (5.3.1), we can arrive at

$$\hat{Q}(s) = \sum_{j=1}^{n} \frac{\lambda_j}{s+\Lambda} \hat{G}_j(s). \tag{5.3.2}$$

The time to first system failure is $\zeta_1 = \min_{1 \le i \le n} X_i$, so the reliability of the series system is

$$R(t) = \Pr\left\{\min_{1<i<n} X_i > t\right\} = \prod_{i=1}^{n} \Pr\{X_i > t\} = e^{-\Lambda t}. \tag{5.3.3}$$

The mean time to first system failure is

$$\text{MTTFF} = \frac{1}{\Lambda}.$$

Let

$$X(t) = \begin{cases} 1, & \text{the system is in functioning state at time } t \\ 0, & \text{the system is in repair state at time } t. \end{cases}$$

The availability at time t is

$$A(t) = \Pr\{X(t) = 1 | \text{the system is in functioning state at } t = 0\}.$$

Then, we can arrive at the renewal equation

$$A(t) = e^{-\Lambda t} + Q(t) * A(t). \tag{5.3.4}$$

By Laplace transform of the above formula, we have

$$A^*(s) = \frac{1}{s+\Lambda} + \hat{Q}(s) A^*(s).$$

The solution is

$$A^*(s) = \frac{\frac{1}{s+\Lambda}}{1-\hat{Q}(s)} = \frac{1}{s+\Lambda-\sum_{j=1}^{n}\lambda_j \hat{G}_j(s)}. \tag{5.3.5}$$

From Eq. (5.1.7), it is easy to obtain the solution of the renewal Eq. (5.3.4), that is,

$$A(t) = e^{-\Lambda t} + \tilde{M}(t) * e^{-\Lambda t}, \tag{5.3.6}$$

where $\tilde{M}(t) = \sum_{k=1}^{+\infty} Q^{(k)}(t)$ is the renewal function. By Eq. (5.2.1), the distribution of the renewal lifetime $Q(t)$ does not have lattice distribution. From Theorem 5.1.3, the limiting availability of the series system is

$$
\begin{aligned}
A &= \lim_{t\to+\infty} A(t) \\
&= \lim_{t\to+\infty} \{e^{-\Lambda t} + \tilde{M}(t) * e^{-\Lambda t}\} \\
&= \frac{1}{E\{\zeta_k + \chi_k\}} \int_0^{+\infty} e^{-\Lambda t}\mathrm{d}t \\
&= \frac{\frac{1}{\Lambda}}{-\hat{Q}'(0)} = \frac{1}{1+\sum_{j=1}^{n} \frac{\lambda_j}{\mu_j}}. \qquad (5.3.7)
\end{aligned}
$$

Similarly, $M(t)$ satisfies the renewal equation

$$M(t) = 1 - e^{-\Lambda t} + Q(t) * M(t). \qquad (5.3.8)$$

By Laplace transform of the above formula, we have

$$\hat{M}(s) = \frac{\Lambda}{s+\Lambda} + \hat{Q}(s)\hat{M}(s).$$

The solution is

$$\hat{M}(s) = \frac{\Lambda}{s+\Lambda - \sum_{j=1}^{n} \lambda_j \hat{G}_j(s)}. \qquad (5.3.9)$$

The steady-state failure frequency of the series system is

$$
\begin{aligned}
M &= \lim_{t\to+\infty} \frac{M(t)}{t} \\
&= \lim_{s\to 0} s\hat{M}(s) = \frac{\Lambda}{1+\sum_{j=1}^{n} \frac{\lambda_j}{\mu_j}}. \qquad (5.3.10)
\end{aligned}
$$

5.4 The Parallel System

Consider a parallel system consisting of two different types of components and a repair equipment. Assume the lifetime of component i (denoted by X_i) follows an exponential distribution $1 - e^{-\lambda_i t}$, $t \geq 0$, $\lambda_i > 0$, and the corresponding repair time of component i (denoted by Y_i) follows a general distribution $G_i(t)$ with mean value $1/\mu_i$, $(\mu_i > 0)$, $i = 1, 2$. We also assume that X_1, X_2, Y_1 and Y_2 are independent of each other. The component is like a new one after the repair is finished.

We define the three states of the parallel system:

State 0: both components are in functioning state.
State 1: component 2 is in functioning state, and component 1 is in repair state.
State 2: component 1 is in functioning state, and component 2 is in repair state.

Due to the memoryless property of the exponential distribution, the moments that the system enters states 0, 1, 2 are the regeneration points. Denote the parallel system in state j at time t by $X(t) = j$, $j = 0, 1, 2$. Let T_n be the nth state transition moment of the parallel system and $T_0 = 0$, and let $Z_n = X(T_n)$ be the state that the system enters at the nth state transition moment. It is easy to verify that $\{Z_n, T_n;\ n = 0, 1, 2, \dots\}$ is a homogeneous Markov renewal process, and $\{X(t),\ t \geq 0\}$ is a semi-Markov process.

For convenience, two virtual states are introduced:

State 3: component 1 is under repair, and component 2 is to be repaired.
State 4: component 2 is under repair, and component 1 is to be repaired.

Since the repair times have general distributions, the moment when the system enters state 3 or state 4 is not the regeneration point. Obviously, state 3 and state 4 are the failed state of the parallel system.

The semi-Markov kernels can be calculated by

$$Q_{01}(t) = \Pr\{X_1 \leq t,\ X_1 < X_2\} = \frac{\lambda_1}{\lambda_1 + \lambda_2}[1 - e^{-(\lambda_1+\lambda_2)t}],$$

$$Q_{02}(t) = \Pr\{X_2 \leq t,\ X_2 < X_1\} = \frac{\lambda_2}{\lambda_1 + \lambda_2}[1 - e^{-(\lambda_1+\lambda_2)t}],$$

$$Q_{10}(t) = \Pr\{Y_1 \leq t,\ Y_1 < X_2\} = \int_0^t e^{-\lambda_2 u} \mathrm{d}G_1(u),$$

$$Q_{20}(t) = \Pr\{Y_2 \leq t,\ Y_2 < X_1\} = \int_0^t e^{-\lambda_1 u} \mathrm{d}G_2(u),$$

$$\begin{aligned} Q_{12}(t) &= Q_{12}^{(3)}(t) \\ &= \Pr\{Y_1 \leq t,\ X_2 < Y_1\} \\ &= \int_0^t (1 - e^{-\lambda_2 u}) \mathrm{d}G_1(u) \\ &= G_1(t) - \int_0^t e^{-\lambda_2 u} \mathrm{d}G_1(u) \end{aligned}$$

and

$$
\begin{aligned}
Q_{21}(t) &= Q_{21}^{(4)}(t) \\
&= \Pr\{Y_2 \le t,\ X_1 < Y_2\} \\
&= G_2(t) - \int_0^t e^{-\lambda_1 u} \mathrm{d}G_2(u).
\end{aligned}
$$

To obtain the reliability indices of the parallel system, we need to use

$$
Q_{13}(t) = \Pr\{X_2 \le t,\ X_2 < Y_1\} = \int_0^t \overline{G}_1(u) \mathrm{d}(1 - e^{-\lambda_2 u})
$$

and

$$
Q_{24}(t) = \Pr\{X_1 \le t,\ X_1 < Y_2\} = \int_0^t \overline{G}_2(u) \mathrm{d}(1 - e^{-\lambda_1 u}).
$$

By Laplace–Stieltjes transform on the above formulas, we get

$$
\begin{cases}
\hat{Q}_{01}(s) = \dfrac{\lambda_1}{s + \lambda_1 + \lambda_2} \\
\hat{Q}_{02}(s) = \dfrac{\lambda_2}{s + \lambda_1 + \lambda_2} \\
\hat{Q}_{10}(s) = \hat{G}_1(s + \lambda_2) \\
\hat{Q}_{20}(s) = \hat{G}_2(s + \lambda_1) \\
\hat{Q}_{12}(s) = \hat{G}_1(s) - \hat{G}_1(s + \lambda_2) \\
\hat{Q}_{21}(s) = \hat{G}_2(s) - \hat{G}_2(s + \lambda_1) \\
\hat{Q}_{13}(s) = \dfrac{\lambda_2}{s + \lambda_2}[1 - \hat{G}_1(s + \lambda_2)] \\
\hat{Q}_{24}(s) = \dfrac{\lambda_1}{s + \lambda_1}[1 - \hat{G}_2(s + \lambda_1)].
\end{cases}
\tag{5.4.1}
$$

Let

$$
\begin{aligned}
\Phi_i(t) = \Pr\{&\text{the time to first system failure} \le t \,|\, \text{the} \\
&\text{system enters state } i \text{ at } t = 0\},\quad i = 0, 1, 2.
\end{aligned}
$$

By the relationship between state transition of the parallel system, $\Phi_i(t)$, $i = 0, 1, 2$ satisfy the Markov renewal equations

$$\begin{cases} \Phi_0(t) = Q_{01}(t) * \Phi_1(t) + Q_{02}(t) * \Phi_2(t) \\ \Phi_1(t) = Q_{10}(t) * \Phi_0(t) + Q_{13}(t) \\ \Phi_2(t) = Q_{20}(t) * \Phi_0(t) + Q_{24}(t). \end{cases}$$

By Laplace–Stieltjes transform on the above formulas, we get

$$\begin{cases} \hat{\Phi}_0(s) = \hat{Q}_{01}(s)\hat{\Phi}_1(s) + \hat{Q}_{02}(s)\hat{\Phi}_2(s) \\ \hat{\Phi}_1(s) = \hat{Q}_{10}(s)\hat{\Phi}_0(s) + \hat{Q}_{13}(s) \\ \hat{\Phi}_2(s) = \hat{Q}_{20}(s)\hat{\Phi}_0(s) + \hat{Q}_{24}(s). \end{cases} \tag{5.4.2}$$

The solution is

$$\begin{cases} \hat{\Phi}_0(s) = \dfrac{\hat{Q}_{01}\hat{Q}_{13} + \hat{Q}_{02}\hat{Q}_{24}}{1 - \hat{Q}_{01}\hat{Q}_{10} - \hat{Q}_{02}\hat{Q}_{20}} \\ \hat{\Phi}_1(s) = \dfrac{\hat{Q}_{10}\hat{Q}_{02}\hat{Q}_{24} + \hat{Q}_{13}(1 - \hat{Q}_{02}\hat{Q}_{20})}{1 - \hat{Q}_{01}\hat{Q}_{10} - \hat{Q}_{02}\hat{Q}_{20}} \\ \hat{\Phi}_1(s) = \dfrac{\hat{Q}_{20}\hat{Q}_{01}\hat{Q}_{13} + \hat{Q}_{24}(1 - \hat{Q}_{01}\hat{Q}_{10})}{1 - \hat{Q}_{01}\hat{Q}_{10} - \hat{Q}_{02}\hat{Q}_{20}}, \end{cases} \tag{5.4.3}$$

where $\hat{Q}_{ij} = \hat{Q}_{ij}(s)$ is given by (5.4.1).

To calculate the mean time to first system failure $T_i,\ i = 0, 1, 2$, a more convenient method is needed; i.e., taking the derivative of both sides of Formula (5.4.2) with respect to s and substituting $-\hat{\Phi}_i'(0) = T_i$, we can obtain $T_i,\ i = 0, 1, 2$ such that

$$\begin{cases} T_0 = \dfrac{\lambda_1}{\lambda_1 + \lambda_2} T_1 + \dfrac{\lambda_2}{\lambda_1 + \lambda_2} T_2 + \dfrac{1}{\lambda_1 + \lambda_2} \\ T_1 = \hat{G}_1(\lambda_2) T_0 + \dfrac{1}{\lambda_2}[1 - \hat{G}_1(\lambda_2)] \\ T_2 = \hat{G}_2(\lambda_1) T_0 + \dfrac{1}{\lambda_1}[1 - \hat{G}_2(\lambda_1)]. \end{cases} \tag{5.4.4}$$

Then, we can arrive at

$$T_0 = \frac{1 + \frac{\lambda_1}{\lambda_2}[1 - \hat{G}_1(\lambda_2)] + \frac{\lambda_2}{\lambda_1}[1 - \hat{G}_2(\lambda_1)]}{\lambda_1[1 - \hat{G}_1(\lambda_2)] + \lambda_2[1 - \hat{G}_2(\lambda_1)]}. \tag{5.4.5}$$

Substituting (5.4.5) to (5.4.4), we can obtain the expressions of T_1 and T_2, respectively.

The availabilities $A_i(t),\ i = 0, 1, 2$ satisfy Markov renewal equations, namely

$$\begin{cases} A_0(t) = Q_{01}(t) * A_1(t) + Q_{02}(t) * A_2(t) + [1 - Q_{01}(t) - Q_{02}(t)] \\ A_1(t) = Q_{10}(t) * A_0(t) + Q_{12}(t) * A_2(t) + [1 - Q_{10}(t) - Q_{13}(t)] \\ A_2(t) = Q_{20}(t) * A_0(t) + Q_{21}(t) * A_1(t) + [1 - Q_{20}(t) - Q_{24}(t)]. \end{cases} \tag{5.4.6}$$

By Laplace transform on the above formulas, we get

$$\begin{cases} A_0^*(s) = \hat{Q}_{01}(s)A_1^*(s) + \hat{Q}_{02}(s)A_2^*(s) + \dfrac{1}{s}[1 - \hat{Q}_{01}(s) - \hat{Q}_{02}(s)] \\ A_1^*(s) = \hat{Q}_{10}(s)A_0^*(s) + \hat{Q}_{12}(s)A_2^*(s) + \dfrac{1}{s}[1 - \hat{Q}_{10}(s) - \hat{Q}_{13}(s)] \\ A_2^*(s) = \hat{Q}_{20}(s)A_0^*(s) + \hat{Q}_{21}(s)A_1^*(s) + \dfrac{1}{s}[1 - \hat{Q}_{20}(s) - \hat{Q}_{24}(s)]. \end{cases} \tag{5.4.7}$$

It follows from Formula (5.4.1) that

$$\frac{1}{s}[1 - \hat{Q}_{01}(s) - \hat{Q}_{02}(s)] = \frac{1}{s + \lambda_1 + \lambda_2},$$

$$\frac{1}{s}[1 - \hat{Q}_{10}(s) - \hat{Q}_{13}(s)] = \frac{1}{s + \lambda_2}[1 - \hat{G}_1(s + \lambda_2)]$$

and

$$\frac{1}{s}[1 - \hat{Q}_{20}(s) - \hat{Q}_{24}(s)] = \frac{1}{s + \lambda_1}[1 - \hat{G}_2(s + \lambda_1)].$$

Solving Eq. (5.4.7), we have

$$A_0^*(s) = \frac{\begin{vmatrix} \frac{1}{s+\lambda_1+\lambda_2} & -\hat{Q}_{01}(s) & -\hat{Q}_{02}(s) \\ \frac{1-\hat{G}_1(s+\lambda_2)}{s+\lambda_2} & 1 & -\hat{Q}_{12}(s) \\ \frac{1-\hat{G}_2(s+\lambda_1)}{s+\lambda_1} & -\hat{Q}_{21}(s) & 1 \end{vmatrix}}{\begin{vmatrix} 1 & -\hat{Q}_{01}(s) & -\hat{Q}_{02}(s) \\ -\hat{Q}_{10}(s) & 1 & -\hat{Q}_{12}(s) \\ -\hat{Q}_{20}(s) & -\hat{Q}_{21}(s) & 1 \end{vmatrix}}. \tag{5.4.8}$$

The expressions of $A_1^*(s)$ and $A_2^*(s)$ can be obtained similarly.

Obviously, $Q_{01}(t)$ in Formula (5.4.1) is not lattice; by limit theorem of Markov renewal process, it is easy to see that $\lim_{t\to+\infty} A_i(t) = A$ exists and is irrelevant to the initial state i. Therefore, from Tauberian theorem, the limiting availability of the parallel system is

$$A = \lim_{t\to+\infty} A_i(t) = \lim_{t\to+\infty} \frac{1}{t}\int_0^t A_i(u)\mathrm{d}u = \lim_{s\to 0} sA_i^*(s).$$

Substitute (5.4.8) into the right side of the above formula to get the specific expression.

We can use Theorem 5.1.7 to calculate the limiting availability A of the parallel system directly. From Formula (5.1.16), we have

$$\begin{cases} m(0) = \dfrac{1}{\lambda_1 + \lambda_2} \\ m(1) = \dfrac{1}{\mu_1} \\ m(2) = \dfrac{1}{\mu_2}. \end{cases} \tag{5.4.9}$$

By Eq. (5.4.1), we can arrive at $P_{ij} = \lim_{t\to+\infty} Q_{ij}(t)$. Then, solving Eq. (5.1.15), we can arrive at

$$\begin{cases} v_0 = 1 - \overline{\hat{G}}_1(\lambda_2)\overline{\hat{G}}_2(\lambda_1) \\ v_1 = 1 - \dfrac{\lambda_2}{\lambda_1 + \lambda_2}\hat{G}_2(\lambda_1) \\ v_2 = 1 - \dfrac{\lambda_1}{\lambda_1 + \lambda_2}\hat{G}_1(\lambda_2). \end{cases} \tag{5.4.10}$$

It follows from (5.1.16) that

$$A = \frac{1}{\sum_{j=0}^{2} v_j m(j)}\left\{\frac{v_0}{\lambda_1 + \lambda_2} + \frac{v_1}{\lambda_2}\left[1 - \hat{G}_1(\lambda_2)\right] + \frac{v_2}{\lambda_1}\left[1 - \hat{G}_2(\lambda_1)\right]\right\}. \tag{5.4.11}$$

The average numbers of system failures $M_j(t)$, $j = 0, 1, 2$ satisfy the Markov renewal equations

$$\begin{cases} M_0(t) = Q_{01}(t) * M_1(t) + Q_{02}(t) * M_2(t) \\ M_1(t) = Q_{10}(t) * M_0(t) + Q_{12}(t) * [M_2(t) + 1] + [Q_{13}(t) - Q_{12}(t)] \\ M_2(t) = Q_{20}(t) * M_0(t) + Q_{21}(t) * [M_1(t) + 1] + [Q_{24}(t) - Q_{21}(t)]. \end{cases} \tag{5.4.12}$$

By Laplace–Stieltjes transform on the above formulas, we get

$$\begin{cases} \hat{M}_0(s) = \hat{Q}_{01}(s)\hat{M}_1(s) + \hat{Q}_{02}(s)\hat{M}_2(s) \\ \hat{M}_1(s) = \hat{Q}_{10}(s)\hat{M}_0(s) + \hat{Q}_{12}(s)\hat{M}_2(s) + \hat{Q}_{13}(s) \\ \hat{M}_2(s) = \hat{Q}_{20}(s)\hat{M}_0(s) + \hat{Q}_{21}(s)\hat{M}_1(s) + \hat{Q}_{24}(s). \end{cases} \tag{5.4.13}$$

Solving the above equations, we have

$$\hat{M}_0(s) = \frac{\begin{vmatrix} 0 & -\hat{Q}_{01}(s) & -\hat{Q}_{02}(s) \\ \hat{Q}_{13}(s) & 1 & -\hat{Q}_{12}(s) \\ \hat{Q}_{24}(s) & -\hat{Q}_{21}(s) & 1 \end{vmatrix}}{\begin{vmatrix} 1 & -\hat{Q}_{01}(s) & -\hat{Q}_{02}(s) \\ -\hat{Q}_{10}(s) & 1 & -\hat{Q}_{12}(s) \\ -\hat{Q}_{20}(s) & -\hat{Q}_{21}(s) & 1 \end{vmatrix}}. \tag{5.4.14}$$

The expressions of $\hat{M}_1(s)$ and $\hat{M}_2(s)$ can be obtained similarly.

The steady-state failure frequency of the parallel system can be obtained by

$$M = \lim_{t\to+\infty} \frac{M_i(t)}{t} = \lim_{s\to 0} s\hat{M}_i(s), \ i = 0, 1, 2.$$

Substituting (5.4.14) into the above equations, it is easy to see that the above equations are irrelevant to the initial state i.

Then, we can use another method to calculate M. Note that (5.4.7) and (5.4.13) have the same form; it is easy to see that $\lim_{s\to 0} sA_i^*(s)$ can be obtained by (5.4.11). Then, we have the similar results for (5.4.13), namely

$$M = \lim_{s\to 0} s\hat{M}_i(s) = \frac{1}{\sum_{j=0}^{2} m(j)v_j}\left\{v_1\left[1 - \hat{G}_1(\lambda_2)\right] + v_2\left[1 - \hat{G}_2(\lambda_1)\right]\right\},$$

which is irrelevant to the initial state i, $m(j)$ and v_j are given by (5.4.9) and (5.4.10), respectively.

5.5 Cold Standby Systems

5.5.1 The Cold Standby System with Two Identical Components

Consider a cold standby system consisting of two identical components and two repair equipment (the case of one repair equipment can be discussed in a similar method). Assume that the lifetime of each component X follows a general distribution $F(t)$ and $\frac{1}{\lambda} = \int_0^{+\infty} t dF(t)$, $(\lambda > 0)$, and the repair time of component Y follows an exponential distribution with parameter μ. We also assume that

(1) Lifetimes and repair times of components are independent.
(2) The conversion switch is completely reliable, and the transition is instantaneous.
(3) The component is like a new one after repair.
(4) The standby component is neither failed nor deteriorated.

The cold standby system has three states:

State 0: one component is in operating and another component is in cold standby. The moment when the new component starts to work is defined as the moment when the system enters state 0.
State 1: one component is in operating, and another component is being repaired. The moment when the component starts to work is defined as the moment when the system enters state 1.
State 2: one component is failed when another component is being repaired. The moment when the component fails is defined as the moment when the system enters state 2.

Obviously, state 2 is the failed state of the cold standby system. Let $E = \{0, 1, 2\}$ and $X(t)$ be the state of the system at time t. Denote the nth state transition moment of the system by T_n and $T_0 = 0$. Let $Z_n = X(T_n)$ be the state that the system enters at the nth state transition moment.

By the definition of the above three states and the memoryless property of the exponential distribution, it is easy to see that all state transition moments T_n, $n = 1, 2, \ldots$ are regeneration points. That is, as long as the state Z_n that the system enters at time T_n is given, the probability law of system development after T_n is irrelevant to the history before T_n. Therefore, $\{Z_n, T_n;\ n = 0, 1, \ldots\}$ is a Markov renewal process. Then, $\{X(t), t \geq 0\}$ is a semi-Markov process. By the definition of semi-Markov kernel, we have

$$Q_{ij}(t) = \Pr\{Z_{n+1} = j,\ T_{n+1} - T_n \leq t | Z_n = i\},\ i, j = 0, 1, 2,\ n = 0, 1, 2, \ldots$$

It follows from the relationship between state transition that

$$\begin{cases} Q_{01}(t) = \Pr\{X \leq t\} = F(t) \\ Q_{11}(t) = \Pr\{X \leq t,\ Y < X\} = \int\limits_0^t (1 - e^{-\mu u}) \mathrm{d}F(u) \\ Q_{12}(t) = \Pr\{X \leq t,\ X < Y\} = \int\limits_0^t e^{-\mu u} \mathrm{d}F(u) \\ Q_{21}(t) = 1 - e^{-2\mu t} \\ Q_{ij}(t) = 0,\ \text{otherwise}. \end{cases}$$

By Laplace–Stieltjes transform on the above formulas, we get

$$\begin{cases} \hat{Q}_{01}(s) = \hat{F}(s) \\ \hat{Q}_{11}(s) = \hat{F}(s) - \hat{F}(s+\mu) \\ \hat{Q}_{12}(s) = \hat{F}(s+\mu) \\ \hat{Q}_{21}(s) = \dfrac{2\mu}{s+2\mu}. \end{cases} \tag{5.5.1}$$

Let

$$\Phi_i(t) = \Pr\{\text{the time to first system failure} \le t | \text{the system enters state } i \text{ at } t = 0\}, \quad i = 0, 1.$$

From the relationship between state transition of the system, $\Phi_i(t)$, $i = 0, 1$ satisfy the Markov renewal equations

$$\begin{cases} \Phi_0(t) = Q_{01}(t) * \Phi_1(t) \\ \Phi_1(t) = Q_{11}(t) * \Phi_1(t) + Q_{12}(t). \end{cases} \tag{5.5.2}$$

Actually,

$$\begin{aligned} \Phi_0(t) &= \Pr\{\text{the time to first system failure } \tau \le t | Z_0 = 0\} \\ &= \Pr\{\tau \le t,\ T_1 > t | Z_0 = 0\} + \Pr\{\tau \le t,\ T_1 \le t | Z_0 = 0\}. \end{aligned}$$

In the first term on the right side of above equation, the event $\{T_1 > t\}$ indicates that the system is still in state 0 at time t and it cannot happen at the same time as $\tau \le t$, so this term is zero. The second term on the right side can be obtained by using the full probability formula, namely

$$\begin{aligned} &\Pr\{\tau \le t,\ T_1 \le t | Z_0 = 0\} \\ &= \sum_{j \in E} \int_0^t \Pr\{\tau \le t | Z_1 = j,\ T_1 = u,\ Z_0 = 0\} \mathrm{d}\Pr\{T_1 \le u,\ Z_1 = j | Z_0 = 0\}. \end{aligned}$$

Starting from state 0, the next step can only transfer to state 1; hence,

$$\begin{aligned} \Phi_0(t) &= \int_0^t \Pr\{\tau \le t | Z_1 = 1,\ T_1 = u,\ Z_0 = 0\} \mathrm{d}Q_{01}(u) \\ &= \int_0^t \Phi_1(t-u) \mathrm{d}Q_{01}(u) = Q_{01}(t) * \Phi_1(t). \end{aligned}$$

Similarly,

$$\begin{aligned}\Phi_1(t) &= \Pr\{\tau \le t | Z_0 = 1\} \\ &= \Pr\{\tau \le t,\ T_1 > t | Z_0 = 1\} + \Pr\{\tau \le t,\ T_1 \le t | Z_0 = 1\} \\ &= \Pr\{\tau \le t,\ T_1 \le t | Z_0 = 1\} \\ &= \sum_{j \in E} \int_0^t \Pr\{\tau \le t | Z_1 = j,\ T_1 = u,\ Z_0 = 1\} \mathrm{d}Q_{1j}(u) \\ &= Q_{11}(t) * \Phi_1(t) + Q_{12}(t) * 1.\end{aligned}$$

By Laplace–Stieltjes transform on both sides of (5.5.2), we get

$$\begin{cases} \hat{\Phi}_0(s) = \hat{Q}_{01}(s)\hat{\Phi}_1(s) \\ \hat{\Phi}_1(s) = \hat{Q}_{11}(s)\hat{\Phi}_1(s) + \hat{Q}_{12}(s). \end{cases}$$

The solution is

$$\begin{cases} \hat{\Phi}_0(s) = \dfrac{\hat{Q}_{01}(s)\hat{Q}_{12}(s)}{1 - \hat{Q}_{11}(s)} = \dfrac{\hat{F}(s)\hat{F}(s+\mu)}{1 - \hat{F}(s) + \hat{F}(s+\mu)} \\ \hat{\Phi}_1(s) = \dfrac{\hat{Q}_{12}(s)}{1 - \hat{Q}_{11}(s)} = \dfrac{\hat{F}(s+\mu)}{1 - \hat{F}(s) + \hat{F}(s+\mu)}. \end{cases} \tag{5.5.3}$$

Let $T_i = \int_0^{+\infty} t\mathrm{d}\Phi_i(t)$ be the mean time to first system failure starting from state i at time t, $i = 0, 1$. Then, by Eq. (5.5.3) and $\hat{F}'(0) = -\frac{1}{\lambda}$, we have

$$\begin{cases} T_0 = -\dfrac{\mathrm{d}}{\mathrm{d}s}\hat{\Phi}_0(s)\Big|_{s=0} = \dfrac{1 + \hat{F}(\mu)}{\lambda \hat{F}(\mu)} \\ T_1 = -\dfrac{\mathrm{d}}{\mathrm{d}s}\hat{\Phi}_1(s)\Big|_{s=0} = \dfrac{1}{\lambda \hat{F}(\mu)}. \end{cases}$$

Let

$$\begin{aligned}A_i(t) = \Pr\{&\text{the system is functioning at time } t | \text{the system enters} \\ &\text{state } i \text{ at } t = 0\},\quad i = 0, 1, 2.\end{aligned}$$

From the relationship between state transition, $A_i(t)$, $i = 0, 1, 2$ satisfy the Markov renewal equations

$$\begin{cases} A_0(t) = Q_{01}(t) * A_1(t) + [1 - Q_{01}(t)] \\ A_1(t) = Q_{11}(t) * A_1(t) + Q_{12}(t) * A_2(t) + [1 - Q_{11}(t) - Q_{12}(t)] \\ A_2(t) = Q_{21}(t) * A_1(t). \end{cases} \tag{5.5.4}$$

In fact, the second formula in Eq. (5.5.4) is

$$\begin{aligned} A_1(t) &= \Pr\{\text{the system is functioning at time } t|Z_0 = 1\} \\ &= \Pr\{\text{the system is functioning at time } t,\ T_1 > t|Z_0 = 1\} \\ &\quad + \Pr\{\text{the system is functioning at time } t,\ T_1 \le t|Z_0 = 1\}. \end{aligned}$$

In the first term on the right side of the above equation, the system is in state 1, $T_1 > t$ indicates that no state transition occurs at time t and the system remains in state 1, then the system must be in functioning state at time t, so the first term on the right side is equal to

$$\Pr\{T_1 > t|Z_0 = 1\} = 1 - Q_{11}(t) - Q_{12}(t).$$

The second term on the right side is equal to

$$\begin{aligned} &\sum_{j\in E}\int_0^t \Pr\{\text{the system is functioning at time } t|\mathrm{Z}_1 = j,\ T_1 = u,\ \mathrm{Z}_0 = 1\}\mathrm{d}Q_{1j}(u) \\ &= \int_0^t A_1(t-u)\mathrm{d}Q_{11}(u) + \int_0^t A_2(t-u)\mathrm{d}Q_{12}(u). \end{aligned}$$

The first formula in Eq. (5.5.4) can be proved similarly. The third formula in Eq. (5.5.4) is

$$\begin{aligned} A_2(t) &= \Pr\{\text{system is functioning at time } t|Z_0 = 2\} \\ &= \Pr\{\text{system is functioning at time } t,\ T_1 > t|Z_0 = 2\} \\ &\quad + \Pr\{\text{system is functioning at time } t,\ T_1 \le t|Z_0 = 2\}. \end{aligned}$$

In the first term on the right side of above equation, the system starts from state 2 at time 0 and remains in state 2 until time t; therefore, the system at time t cannot be in functioning state, so this term is zero. The second term on the right side is $Q_{21}(t) * A_1(t)$.

By Laplace transform on both sides of (5.5.4), we get

$$\begin{cases} A_0^*(s) = \hat{Q}_{01}(s)A_1^*(s) + \dfrac{1}{s}[1 - \hat{Q}_{01}(s)] \\ A_1^*(s) = \hat{Q}_{11}(s)A_1^*(s) + \hat{Q}_{12}(s)A_2^*(s) + \dfrac{1}{s}[1 - \hat{Q}_{11}(s) - \hat{Q}_{12}(s)] \\ A_2^*(s) = \hat{Q}_{21}(s)A_1^*(s). \end{cases}$$

Solving this equations and substituting (5.5.1) into the solutions, we have

$$\begin{cases} A_1^*(s) = \dfrac{\frac{1}{s}[1-\hat{F}(s)]}{1-\hat{F}(s)+\hat{F}(s+\mu)\frac{s}{s+2\mu}} \\ A_2^*(s) = \dfrac{\frac{1}{s}[1-\hat{F}(s)]\frac{2\mu}{s+2\mu}}{1-\hat{F}(s)+\hat{F}(s+\mu)\frac{s}{s+2\mu}} \\ A_0^*(s) = \dfrac{\frac{1}{s}[1-\hat{F}(s)]\left[1+\hat{F}(s+\mu)\frac{s}{s+2\mu}\right]}{1-\hat{F}(s)+\hat{F}(s+\mu)\frac{s}{s+2\mu}}. \end{cases} \tag{5.5.5}$$

According to Tauberian theorem of Laplace transform, the long-run average availability of the system is

$$\tilde{A}_i = \lim_{t\to+\infty}\frac{1}{t}\int_0^t A_i(u)\mathrm{d}u = \lim_{s\to 0} sA_i^*(s) = \frac{2\mu}{2\mu+\lambda\hat{F}(\mu)}, \quad i=0,1,2.$$

The right side of the above formula is irrelevant to the initial state i. Obviously, $Q_{21}(t)$ is not lattice. It follows from Theorem 5.1.7 that

$$\lim_{t\to+\infty} A_i(t) = A$$

exists and is irrelevant to state i. Therefore, the limiting availability of the system is

$$A = \lim_{t\to+\infty} A_i(t) = \lim_{t\to+\infty}\frac{1}{t}\int_0^t A_i(u)\mathrm{d}u = \frac{2\mu}{2\mu+\lambda\hat{F}(u)}. \tag{5.5.6}$$

Let $N(t)$ denote the number of system failures in $(0, t]$. Then,

$$M_i(t) = E[N(t)|\mathrm{Z}_0 = i]$$

represents the average number of system failures in $(0, t]$ starting from state i at initial time, $i = 0, 1, 2$, which satisfy the following Markov renewal equations

$$\begin{cases} M_0(t) = Q_{01}(t) * M_1(t) \\ M_1(t) = Q_{11}(t) * M_1(t) + Q_{12}(t) * [M_2(t) + 1] \\ M_2(t) = Q_{21}(t) * M_1(t). \end{cases} \tag{5.5.7}$$

Actually,

$$M_0(t) = E[N(t)|Z_0 = 0]$$

$$= \sum_{j \in E} \int_0^t E[N(t)|Z_1 = j, \ T_1 = u, \ Z_0 = 0] \mathrm{d}Q_{0j}(u)$$
$$+ E[N(t)|T_1 > t, \ Z_0 = 0] \Pr\{T_1 > t | Z_0 = 0\}.$$

If the system starts from state 0, it can only transfer to state 1, then there is only term $j = 1$ in the sum on the right side, and state 1 is the functioning state of the system. Then, the first term on the right side of the above equation is equal to

$$\int_0^t E[N(t)|Z_1 = 1, \ T_1 = u, \ Z_0 = 0] \mathrm{d}Q_{01}(u)$$
$$= \int_0^t M_1(t-u) \mathrm{d}Q_{01}(u) = Q_{01}(t) * M_1(t).$$

In the second term on the right side, under the conditions of $Z_0 = 0$ and $T_1 > t$, the system always stays in functioning state 0 in $(0, t]$, so

$$E[N(t)|Z_0 = 0, \ T_1 > t] = 0.$$

The third equation in (5.5.7) can be proved similarly. The left side of the second equation in (5.5.7) is

$$M_1(t) = E[N(t)|Z_0 = 1]$$
$$= \sum_{j \in E} \int_0^t E[N(t)|Z_1 = j, \ T_1 = u, \ Z_0 = 1] \mathrm{d}Q_{1j}(u)$$
$$+ E[N(t)|T_1 > t, \ Z_0 = 1] \Pr\{T_1 > t | Z_0 = 1\}.$$

The second term on the right side is zero, and there are only two cases $j = 1, 2$ in the sum of the first term on the right side; hence,

$$M_1(t) = \int_0^t E[N(t)|Z_1 = 1, \ T_1 = u, \ Z_0 = 1] \mathrm{d}Q_{11}(u)$$
$$+ \int_0^t E[N(t)|Z_1 = 2, \ T_1 = u, \ Z_0 = 1] \mathrm{d}Q_{12}(u)$$

$$= \int_0^t M_1(t-u)\mathrm{d}Q_{11}(u) + \int_0^t [M_2(t-u)+1]\mathrm{d}Q_{12}(u).$$

The latter term is arrived since the fact that under the conditions of $Z_0 = 1,\ Z_1 = 2$ and $T_1 = u$, the system transfers from the functioning state 1 to the failed state 2 at time u. Then, the number of system failures in $(0, t]$ is equal to one failure occurs before time u plus the number of failures in $(u, t]$ starting from state 2 at time u.

$$E[N(t)|Z_1 = 2,\ T_1 = u,\ Z_0 = 1] = M_2(t-u) + 1.$$

Therefore, the second formula in (5.5.7) is obtained. By Laplace–Stieltjes transform on both sides of (5.5.7), we get

$$\begin{cases} \hat{M}_0(s) = \hat{Q}_{01}(s)\hat{M}_1(s) \\ \hat{M}_1(s) = \hat{Q}_{11}(s)\hat{M}_1(s) + \hat{Q}_{12}(s)\hat{M}_2(s) + \hat{Q}_{12}(s) \\ \hat{M}_2(s) = \hat{Q}_{21}(s)\hat{M}_1(s). \end{cases}$$

Solving the above equations and substituting (5.5.1) into the solution, we have

$$\begin{cases} \hat{M}_1(s) = \dfrac{\hat{F}(s+\mu)}{1-\hat{F}(s)+\hat{F}(s+\mu)\frac{s}{s+2\mu}} \\ \hat{M}_2(s) = \dfrac{2\mu}{s+2\mu}\hat{M}_1(s) \\ \hat{M}_0(s) = \hat{F}(s)\hat{M}_1(s). \end{cases} \tag{5.5.8}$$

According to Tauberian theorem and Formula (5.5.8), the steady-state failure frequency of the cold standby system is

$$M = \lim_{t\to+\infty}\frac{M_i(t)}{t} = \lim_{s\to 0} s\hat{M}_i(s) = \frac{2\lambda\mu\hat{F}(\mu)}{2\mu+\lambda\hat{F}(\mu)},$$

which is irrelevant to the initial state i.

5.5.2 The Cold Standby System with Two Nonidentical Components

Consider a cold standby system consisting of two nonidentical components and a repair equipment. Suppose that the lifetime X_i and the repair time Y_i of component i

follow general distributions $F_i(t)$ and $G_i(t)$, respectively, and their expected values are $1/\lambda_i$ and $1/\mu_i(\lambda_i, \mu_i > 0)$, $i = 1, 2$, respectively. We also assume that

(1) X_1, X_2, Y_1 and Y_2 are mutually independent.
(2) The standby component is neither failed nor deteriorated.
(3) The conversion switch is completely reliable, and the state transition is instantaneous.
(4) The component works like a new one after repair.

We define three states of the system:

State -1: the moment that the system enters state -1 is the moment when component 1 starts to operation and component 2 is in standby, in which component 1 is new at the start of operation.
State 0: the moment that the system enters state 0 is the moment when component 2 starts to operation and component 1 starts to repair.
State 1: the moment that the system enters state 1 is the moment when component 1 starts to operation and component 2 starts to repair.

The initial state -1 is a transient state; that is, after the system leaves state -1, it is impossible to return state -1. It is easy to verify that the moments when the system enters state 0 and state 1 are the regeneration points. There are no other regeneration points in the system. Let

$$X(t) = j, \text{ if the system is in state } j, \ \ j = -1, 0, 1,$$

and denote the moment of the nth state transition by T_n and $T_0 = 0$. Let $Z_n = X(T_n)$ be the state that the system enters when the nth state transition occurs. It is easy to verify that $\{Z_n, T_n;\ n = 0, 1, \ldots\}$ is a Markov renewal process with state space $E = \{-1, 0, 1\}$, and $\{X(t),\ t \geq 0\}$ is a semi-Markov process. Let $Q_{ij}(t)$ be the semi-Markov kernel, $i = -1, 0, 1,\ j = 0, 1$.

For convenience, we introduce two virtual states:

State 2: the moment that the system enters state 2 is the moment when one component is in operation and the repair of another component is finished.
State 3: the moment that the system enters state 3 is the moment when one component is under repair and another component fails.

Since lifetimes and repair times of components are general distributions, the moment the system enters state 2 or state 3 is not the regeneration point. During the period from state 0 to state 1, the Markov renewal process may go through virtual state 2 or virtual state 3. Similarly, during the period from state 1 to state 0, it may go through virtual state 2 or virtual state 3. Let $Q_{ij}^{(k)}(t)$ denote the probability that the system starts from the state i, goes through the virtual state k, then transfers to the state j, and the sojourn time in the state i or state k is less than or equal to t. All semi-Markov kernels can be calculated by

$$Q_{-10}(t) = \Pr\{X_1 \leq t\} = F_1(t),$$

$$\begin{aligned}Q_{01}(t) &= Q_{01}^{(2)}(t) + Q_{01}^{(3)}(t)\\ &= \Pr\{X_2 \le t,\ X_2 > Y_1\} + \Pr\{Y_1 \le t,\ Y_1 > X_2\}\\ &= \int_0^t G_1(u)\mathrm{d}F_2(u) + \int_0^t F_2(u)\mathrm{d}G_1(u)\\ &= F_2(t)G_1(t)\end{aligned}$$

and

$$\begin{aligned}Q_{10}(t) &= Q_{10}^{(2)}(t) + Q_{10}^{(3)}(t)\\ &= \Pr\{X_1 \le t, X_1 > Y_2\} + \Pr\{Y_2 \le t, Y_2 > X_1\}\\ &= \int_0^t G_2(u)\mathrm{d}F_1(u) + \int_0^t F_1(u)\mathrm{d}G_2(u)\\ &= F_1(t)G_2(t).\end{aligned}$$

To obtain various reliability indices of the system, we also need to use

$$Q_{01}^{(2)}(t) = \int_0^t G_1(u)\mathrm{d}F_2(u),$$

$$Q_{10}^{(2)}(t) = \int_0^t G_2(u)\mathrm{d}F_1(u),$$

$$Q_{03}(t) = \Pr\{X_2 \le t, X_2 < Y_1\} = \int_0^t \overline{G}_1(u)\mathrm{d}F_2(u),$$

$$Q_{13}(t) = \Pr\{X_1 \le t, X_1 < Y_2\} = \int_0^t \overline{G}_2(u)\mathrm{d}F_1(u).$$

The definitions of the latter two formulas are similar with the semi-Markov kernel, but the state 3 is a virtual state which is not the regeneration point of the system. It is easy to see that only virtual state 3 is the failed state of the system. By Laplace–Stieltjes transform on the above formulas, we get

$$\begin{cases} \hat{Q}_{-10}(s) = \hat{F}_1(s) \\ \hat{Q}_{01}(s) = \int\limits_0^{+\infty} e^{-st} G_1(t) \mathrm{d}F_2(t) + \int\limits_0^{+\infty} e^{-st} F_2(t) \mathrm{d}G_1(t) \\ \hat{Q}_{10}(s) = \int\limits_0^{+\infty} e^{-st} G_2(t) \mathrm{d}F_1(t) + \int\limits_0^{+\infty} e^{-st} F_1(t) \mathrm{d}G_2(t) \\ \hat{Q}_{01}^{(2)}(s) = \int\limits_0^{+\infty} e^{-st} G_1(t) \mathrm{d}F_2(t) \\ \hat{Q}_{10}^{(2)}(s) = \int\limits_0^{+\infty} e^{-st} G_2(t) \mathrm{d}F_1(t) \\ \hat{Q}_{03}(s) = \int\limits_0^{+\infty} e^{-st} \overline{G}_1(t) \mathrm{d}F_2(t) \\ \hat{Q}_{13}(s) = \int\limits_0^{+\infty} e^{-st} \overline{G}_2(t) \mathrm{d}F_1(t). \end{cases}$$

Let $\Phi_i(t)$ denote the distribution of time to first system failure when the system enters state i at time 0, that is

$$\Phi_i(t) = \Pr\{\text{the time to first system failure} \le t | Z_0 = i\}, \ \ i = -1, 0, 1.$$

Similar with Eq. (5.5.3), it can be proved that $\Phi_i(t)$, $i = -1, 0, 1$ satisfy the Markov renewal equations

$$\begin{cases} \Phi_{-1}(t) = Q_{-10}(t) * \Phi_0(t) \\ \Phi_0(t) = Q_{01}^{(2)}(t) * \Phi_1(t) + Q_{03}(t) \\ \Phi_1(t) = Q_{10}^{(2)}(t) * \Phi_0(t) + Q_{13}(t). \end{cases}$$

By Laplace–Stieltjes transform on the above formula, the solution is

$$\begin{cases} \hat{\Phi}_0(s) = \dfrac{\hat{Q}_{01}^{(2)}(s)\hat{Q}_{13}(s) + \hat{Q}_{03}(s)}{1 - \hat{Q}_{01}^{(2)}(s)\hat{Q}_{10}^{(2)}(s)} \\ \hat{\Phi}_1(s) = \dfrac{\hat{Q}_{10}^{(2)}(s)\hat{Q}_{03}(s) + \hat{Q}_{13}(s)}{1 - \hat{Q}_{10}^{(2)}(s)\hat{Q}_{01}^{(2)}(s)} \\ \hat{\Phi}_{-1}(s) = \hat{F}_1(s)\hat{\Phi}_0(s). \end{cases} \tag{5.5.9}$$

Let

$$T_i = \int_0^{+\infty} t\mathrm{d}\Phi_i(t),\ i = -1, 0, 1.$$

If the system starts from entering state -1 at time 0, the mean time to first system failure is

$$T_{-1} = -\hat{\Phi}'_{-1}(0) = \frac{1}{\lambda_1} + \frac{\frac{1}{\lambda_2} + \frac{\hat{Q}_{01}^{(2)}(0)}{\lambda_1}}{1 - \hat{Q}_{01}^{(2)}(0)\hat{Q}_{10}^{(2)}(0)}. \tag{5.5.10}$$

The expressions of T_0 and T_1 can be arrived similarly.

Let

$$A_i(t) = \Pr\{\text{the system is functioning at time } t | Z_0 = i\},\ i = -1, 0, 1.$$

Similar with Formula (5.5.4), it is easy to prove that $A_i(t),\ i = -1, 0, 1$ satisfy the Markov renewal equations

$$\begin{cases} A_{-1}(t) = Q_{-10}(t) * A_0(t) + \left[1 - Q_{-10}(t)\right] \\ A_0(t) = Q_{01}(t) * A_1(t) + \left[1 - Q_{03}(t) - Q_{01}^{(2)}(t)\right] \\ A_1(t) = Q_{10}(t) * A_0(t) + \left[1 - Q_{13}(t) - Q_{10}^{(2)}(t)\right]. \end{cases} \tag{5.5.11}$$

The second term at the right side of the second formula indicates that the system starts from state 0 at time 0 and stays in state 0 or in virtual state 2 in $(0, t]$; that is, the system does not transfer to the next regeneration point state 1 and failed state in $(0, t]$. At this moment, the system is functioning at time t. The second term at the right side of the third formula can be obtained similarly. Due to

$$\begin{aligned} 1 - Q_{03}(t) - Q_{01}^{(2)}(t) &= 1 - \Pr\{X_2 \le t,\ X_2 < Y_1\} - \Pr\{X_2 \le t,\ X_2 > Y_1\} \\ &= 1 - \Pr\{X_2 \le t\} \\ &= 1 - F_2(t) \end{aligned}$$

and

$$1 - Q_{13}(t) - Q_{10}^{(2)}(t) = 1 - F_1(t),$$

by Laplace transform on both sides of (5.5.11), we can obtain

$$\begin{cases} A_{-1}^*(s) = \hat{Q}_{-10}(s)A_0^*(s) + \dfrac{1}{s}\Big[1 - \hat{F}_1(s)\Big] \\ A_0^*(s) = \hat{Q}_{01}(s)A_1^*(s) + \dfrac{1}{s}\Big[1 - \hat{F}_2(s)\Big] \\ A_1^*(s) = \hat{Q}_{10}(s)A_0^*(s) + \dfrac{1}{s}\Big[1 - \hat{F}_1(s)\Big]. \end{cases}$$

Solving the above equations, we have

$$\begin{cases} A_0^*(s) = \dfrac{1}{s} \cdot \dfrac{\Big[1 - \hat{F}_1(s)\Big]\hat{Q}_{01}(s) + \Big[1 - \hat{F}_2(s)\Big]}{1 - \hat{Q}_{01}(s)\hat{Q}_{10}(s)} \\ A_1^*(s) = \dfrac{1}{s} \cdot \dfrac{\Big[1 - \hat{F}_2(s)\Big]\hat{Q}_{10}(s) + \Big[1 - \hat{F}_1(s)\Big]}{1 - \hat{Q}_{01}(s)\hat{Q}_{10}(s)} \\ A_{-1}^*(s) = \hat{F}_1(s)A_0^*(s) + \dfrac{1}{s}\Big[1 - \hat{F}_1(s)\Big]. \end{cases} \tag{5.5.12}$$

From the Tauberian theorem of Laplace transform, the long-run limiting availability of the system can be obtained by

$$\tilde{A}_i = \lim_{t\to+\infty} \frac{1}{t}\int_0^t A_i(u)\mathrm{d}u = \lim_{s\to 0} sA_i^*(s) = \frac{\frac{1}{\lambda_1} + \frac{1}{\lambda_2}}{l}, \; i = -1, 0, 1, \tag{5.5.13}$$

which is irrelevant to the initial state i, where

$$l = \frac{1}{\lambda_1} + \frac{1}{\lambda_2} + \frac{1}{\mu_1} + \frac{1}{\mu_2} - \int_0^{+\infty} \overline{F}_1(t)\overline{G}_2(t)\mathrm{d}t - \int_0^{+\infty} \overline{F}_2(t)\overline{G}_1(t)\mathrm{d}t. \tag{5.5.14}$$

If at least one of $F_1(t)$, $F_2(t)$, $G_1(t)$ and $G_2(t)$ does not have a lattice distribution, then at least one of $Q_{01}(t)$ and $Q_{10}(t)$ is not lattice. From the limit theorem of the Markov renewal process, it is easy to see that $\lim_{t\to+\infty} A_i(t) = A$ exists and is irrelevant to the initial state i. Therefore, the limiting availability of the system is

$$A = \lim_{t\to+\infty} A_i(t) = \lim_{t\to+\infty} \frac{1}{t}\int_0^t A_i(u)\mathrm{d}u = \frac{\lambda_1 + \lambda_2}{\lambda_1\lambda_2 l}, \tag{5.5.15}$$

where l is given by (5.5.14).

Let $N(t)$ be the number of system failures in $(0, t]$. Then,

$$M_i(t) = E[N(t)|Z_0 = i], \; i = -1, 0, 1.$$

Similar with Eq. (5.5.7), $M_i(t)$, $i = -1, 0, 1$ satisfy the Markov renewal equations

$$\begin{cases} M_{-1}(t) = Q_{-10}(t) * M_0(t) \\ M_0(t) = Q_{01}^{(2)}(t) * M_1(t) + Q_{01}^{(3)}(t) * [M_1(t) + 1] + \left[Q_{03}(t) - Q_{01}^{(3)}(t)\right] \\ M_1(t) = Q_{10}^{(2)}(t) * M_0(t) + Q_{10}^{(3)}(t) * [M_0(t) + 1] + \left[Q_{13}(t) - Q_{10}^{(3)}(t)\right]. \end{cases} \tag{5.5.16}$$

By Laplace–Stieltjes transform on the above equations, we get

$$\begin{cases} \hat{M}_{-1}(s) = \hat{Q}_{-10}(s)\hat{M}_0(s) \\ \hat{M}_0(s) = \hat{Q}_{01}(s)\hat{M}_1(s) + \hat{Q}_{03}(s) \\ \hat{M}_1(s) = \hat{Q}_{10}(s)\hat{M}_0(s) + \hat{Q}_{13}(s). \end{cases}$$

Solving the above equations, we have

$$\begin{cases} \hat{M}_0(s) = \dfrac{\hat{Q}_{01}(s)\hat{Q}_{13}(s) + \hat{Q}_{03}(s)}{1 - \hat{Q}_{01}(s)\hat{Q}_{10}(s)} \\ \hat{M}_1(s) = \dfrac{\hat{Q}_{10}(s)\hat{Q}_{03}(s) + \hat{Q}_{13}(s)}{1 - \hat{Q}_{01}(s)\hat{Q}_{10}(s)} \\ \hat{M}_{-1}(s) = \hat{F}_1(s)\hat{M}_0(s). \end{cases} \tag{5.5.17}$$

According to Tauberian theorem, the steady-state failure frequency of the cold standby system is

$$M = \lim_{t\to+\infty} \frac{M_i(t)}{t} = \lim_{s\to 0} s\hat{M}_i(s) = \frac{\hat{Q}_{03}(0) + \hat{Q}_{13}(0)}{l}, \tag{5.5.18}$$

which is irrelevant to the initial state i, where l is given by (5.5.14).

5.6 The Coherent System

Consider a repairable coherent system composed of n components and sufficient repair equipment. That is, each component can be repaired immediately after failure, and there is no waiting for repair. We assume that when the system fails, the operating components may fail, and their failure rules are not affected by the system failures. Let the distribution of the lifetime of the ith component be $F_i(t)$ and $\frac{1}{\lambda_i} = \int_0^{+\infty} t\mathrm{d}F_i(t)$, $0 < \lambda_i < +\infty$, and let the distribution of the corresponding repair time of the ith component be $G_i(t)$ and $\frac{1}{\mu_i} = \int_0^{+\infty} t\mathrm{d}G_i(t)$, $0 < \mu_i < +\infty$, $i = 1, 2, \ldots, n$. We also assume that lifetimes and repair times of all components are independent with each other, and the component will behave like a new one after repair.

Let the structure function of the coherent system be

$$\phi = \phi(x_1, x_2, \ldots, x_n) \tag{5.6.1}$$

and let the reliability function be

$$R = h(R_1, R_2, \ldots, R_n), \tag{5.6.2}$$

where $x_1, x_2, \ldots, x_n$ are the states of components and $R_1, R_2, \ldots, R_n$ are the reliabilities of components. Since there have sufficient repair equipment, all components operate independently. The state of each component is a process of alternating functioning state and failed state over time. In Sect. 5.2, we have discussed the single component replacement case, then $A_i(t)$, A_i, $m_i(t)$ and M_i of component i, $i = 1, 2, \ldots, n$ can be obtained, respectively. In particular, the formulas of limiting availability and steady-state failure frequency of each component are

$$\begin{cases} A_i = \dfrac{\mu_i}{\lambda_i + \mu_i} \\ M_i = \dfrac{\lambda_i \mu_i}{\lambda_i + \mu_i}, \quad i = 1, 2, \ldots, n. \end{cases} \tag{5.6.3}$$

From the above discussion, we can arrive at the availability at time t and the limiting availability of the repairable coherent system, i.e.,

$$\begin{cases} A(t) = h(A_1(t), A_2(t), \ldots, A_n(t)) \\ A = h(A_1, A_2, \ldots, A_n). \end{cases} \tag{5.6.4}$$

In this section, we assume that lifetimes and repair times of components have finite density functions. Before discussing the failure frequency of the coherent system, we review some lemmas first.

When a component transfers from the functioning state to the failed state or from the failed state to the functioning state, we call an event occurs in the component. In a specified period, the total number of events that all components occur is called the number of events that occur in the system during this period.

Lemma 5.6.1 *For any $i = 1, 2, \ldots, n$, $t \geq 0$ and sufficiently small $\Delta t > 0$, there exist finite $G_i(t) > 0$ and $\Lambda_i(t) > 0$ such that*

$$\begin{aligned} &\Pr\{\text{the } i\text{th component occurs at least } j \text{ events in } (t, t + \Delta t]\} \\ &\quad \leq G_i(t)[\Lambda_i(t)\Delta t]^j, \quad j = 1, 2, \ldots \end{aligned}$$

Proof Since the lifetime and repair time of the ith component have finite density functions, for any $t \geq 0$, there at least exists a finite $\Lambda_i(t) > 0$ such that

$$\begin{cases} \max\limits_{0 \leq u \leq t} [F_i(u + \Delta t) - F_i(u)] \leq \Lambda_i(t)\Delta t \\ \max\limits_{0 \leq u \leq t} [G_i(u + \Delta t) - G_i(u)] \leq \Lambda_i(t)\Delta t. \end{cases} \tag{5.6.5}$$

For convenience, we assume that the ith component is new at $t = 0$. Let $X_1, X_2, \ldots$ and $Y_1, Y_2, \ldots$ be the lifetimes and the repair times of the ith component alternately starting from $t = 0$, and let

$$\begin{cases} S_0 = 0 \\ S_k = \sum\limits_{r=1}^{k} (X_r + Y_r), \ k = 1, 2, \ldots \end{cases}$$

Then, we have

$$\begin{aligned} &\Pr\{\text{the } i\text{th component occurs at least one event in } (t, t + \Delta t]\} \\ &= \sum_{k=0}^{+\infty} \Pr\{S_k \leq t < S_k + X_{k+1} \leq t + \Delta t\} \\ &\quad + \sum_{k=1}^{+\infty} \Pr\{S_k + X_{k+1} \leq t < S_{k+1} \leq t + \Delta t\} \end{aligned}$$

and

$$\begin{aligned} &\Pr\{\text{the } i\text{th component occurs at least two events in } (t, t + \Delta t]\} \\ &= \sum_{k=0}^{+\infty} \Pr\{S_k \leq t < S_k + X_{k+1},\ S_{k+1} \leq t + \Delta t\} \\ &\quad + \sum_{k=0}^{+\infty} \Pr\{S_k + X_{k+1} \leq t < S_{k+1},\ S_{k+1} + X_{k+2} \leq t + \Delta t\}. \end{aligned}$$

If j is an odd number, that is, $j = 2l + 1$, $l = 0, 1, \ldots$, there have

$$\begin{aligned} &\Pr\{\text{the } i\text{th component occurs at least } j \text{ events in } (t, t + \Delta t]\} \\ &= \sum_{k=0}^{+\infty} \Pr\{S_k \leq t < S_k + X_{k+1},\ S_{k+l} + X_{k+l+1} \leq t + \Delta t\} \\ &\quad + \sum_{k=0}^{+\infty} \Pr\{S_k + X_{k+1} \leq t < S_{k+1},\ S_{k+l+1} \leq t + \Delta t\} \end{aligned}$$

$$= \sum_{k=0}^{+\infty} \int_0^t \Pr\{t - u < X_{k+1},\ X_{k+1} + Y_{k+1} + \cdots + X_{k+l+1}$$

$$\leq t - u + \Delta t\} \mathrm{d}\Pr\{S_k \leq u\}$$

$$+ \sum_{k=0}^{+\infty} \int_0^t \Pr\{t - u < Y_{k+1},\ Y_{k+1} + X_{k+2} + \cdots + Y_{k+l+1}$$

$$\leq t - u + \Delta t\} \mathrm{d}\Pr\{S_k + X_{k+1} \leq u\}$$

$$\leq \sum_{k=0}^{+\infty} \int_0^t \Pr\{t - u < X_{k+1} \leq t - u + \Delta t,\ Y_{k+1} \leq \Delta t, \ldots,$$

$$X_{k+l+1} \leq \Delta t\} \mathrm{d}\Pr\{S_k \leq u\}$$

$$+ \sum_{k=0}^{+\infty} \int_0^t \Pr\{t - u < Y_{k+1} \leq t - u + \Delta t,\ X_{k+2} \leq \Delta t, \ldots,$$

$$Y_{k+l+1} \leq \Delta t\} \mathrm{d}\Pr\{S_k + X_{k+1} \leq u\}. \tag{5.6.6}$$

Since the events in the integrand function at the right side of the above formula are independent of each other, by Formula (5.6.5), we can arrive at

$$\Pr\{\text{the } i\text{th component occurs at least } j \text{ events in } (t, t + \Delta t]\}$$

$$\leq \sum_{k=0}^{+\infty} \int_0^t [\Lambda_i(t)\Delta t]^j \mathrm{d}\Pr\{S_k \leq u\} + \sum_{k=0}^{+\infty} \int_0^t [\Lambda_i(t)\Delta t]^j \mathrm{d}\Pr\{S_k + X_{k+1} \leq u\}$$

$$= [\Lambda_i(t)\Delta t]^j \left[\sum_{k=0}^{+\infty} \Pr\{S_k \leq t\} + \sum_{k=0}^{+\infty} \Pr\{S_k + X_{k+1} \leq t\} \right]$$

$$= [\Lambda_i(t)\Delta t]^j \left\{ [1 + \tilde{M}(t)] + F(t) * [1 + \tilde{M}(t)] \right\},$$

in which

$$\tilde{M}(t) = \sum_{k=1}^{+\infty} \Pr\{S_k \leq t\} = \sum_{k=1}^{+\infty} F_i^{(k)}(t) * G_i^{(k)}(t)$$

is the renewal function of the renewal process $X_k + Y_k,\ k = 1, 2, \ldots$ Since $\tilde{M}(t) < +\infty$, there exists a finite $G_i(t) > 0$ such that

$$\Pr\{\text{the } i\text{th component occurs at least } j \text{ events in } (t, t + \Delta t]\} \leq C_i(t)[\Lambda_i(t)\Delta t]^j.$$

If j is an even number, that is, $j = 2l,\ l = 1, 2, \ldots$, there have

$$\begin{aligned}&\Pr\{\text{the } i\text{th component occurs at least } j \text{ events in } (t, t+\Delta t]\}\\&=\sum_{k=0}^{+\infty}\Pr\{S_k \le t < S_k + X_{k+1},\ S_{k+l} \le t+\Delta t\}\\&+\sum_{k=0}^{+\infty}\Pr\{S_k + X_{k+1} \le t < S_{k+1},\ S_{k+l} + X_{k+l+1} \le t+\Delta t\},\end{aligned}$$

which is similar with Formula (5.6.6). The similar method can be used to prove the result of the lemma.

For the case that the ith component is not new at $t = 0$, it starts from a residual lifetime or residual repair time at time 0. In this case, the similar method can be used to prove the lemma. The proof is complete.

Obviously, for any component $i,\ i = 1, 2, \dots, n$, there have

$$\Pr\{\text{the } i\text{th component occurs at least 0 event in } (t, t+\Delta t]\} = 1.$$

By the appropriate choice of $C_i(t) \ge 1$, we have

$$\Pr\{\text{the } i\text{th component occurs at least 0 event in } (t, t+\Delta t]\} \le C_i(t). \tag{5.6.7}$$

Therefore, Lemma 5.6.1 holds for $j = 0$.

Lemma 5.6.2 *For any $t \ge 0$ and sufficiently small $\Delta t > 0$, there exist finite $C(t) > 0$ and $\Lambda(t) > 0$ such that*

$$\Pr\{\text{the system occurs at least } j \text{ events in } (t, t+\Delta t]\} \le C(t)[n\Lambda(t)\Delta t]^j,\quad j = 1, 2, \dots$$

Proof For any $t \ge 0$, let

$$\begin{cases}\Lambda(t) = \max\limits_{1\le i\le n} \Lambda_i(t)\\ C(t) = \prod\limits_{i=1}^{n} C_i(t).\end{cases} \tag{5.6.8}$$

Obviously, both $\Lambda(t) > 0$ and $C(t) > 0$ are finite. It is easy to see that

$$\begin{aligned}&\Pr\{\text{the system occurs at least } j \text{ events in } (t, t+\Delta t]\}\\&=\sum_{\substack{j_1+j_2+\cdots+j_n=j\\0\le j_i\le j}}\prod_{i=1}^{n}\Pr\{\text{the } i\text{th component occurs at least } j_i \text{ events in } (t, t+\Delta t]\}.\end{aligned}$$

Substituting the result of Lemma 5.6.1 and Formula (5.6.7) into the right side of the above formula and using the symbols in (5.6.8), we have

$$\Pr\{\text{the system occurs at least } j \text{ events in } (t, t+\Delta t]\}$$
$$\leq \sum_{\substack{j_1+\cdots+j_n=j \\ 0\leq j_i\leq j}} \prod_{i=1}^{n} C_i(t)[\Lambda_i(t)\Delta t]^{j_i}$$
$$\leq C(t) \sum_{\substack{j_1+\cdots+j_n=j \\ 0\leq j_i\leq j}} [\Lambda(t)\Delta t]^{j}$$
$$\leq C(t) \sum_{\substack{j_1+\cdots+j_n=j \\ 0\leq j_i\leq j}} \frac{j!}{j_1!j_2!\cdots j_n!}[\Lambda(t)\Delta t]^{j}$$
$$= C(t)[n\Lambda(t)\Delta t]^{j}.$$

The proof is complete.

Lemma 5.6.3 *For any $t \geq 0$ and sufficiently small $\Delta t > 0$, there has*

$$M_i(t+\Delta t) - M_i(t)$$
$$= \Pr\{\text{the } i\text{th component is functioning at time } t, \text{ occurs exactly one event in } (t, t+\Delta t], \text{ and transfers from the functioning state to the failed state}\} + o(\Delta t),\ i = 1, 2, \ldots, n.$$

Proof It follows from definition that

$$M_i(t+\Delta t) - M_i(t) = E[N_i(t+\Delta t) - N_i(t)]$$
$$= \sum_{k=1}^{+\infty} k\Pr\{N_i(t+\Delta t) - N_i(t) = k\}.$$

However,

$$\sum_{k=2}^{+\infty} k\Pr\{N_i(t+\Delta t) - N_i(t) = k\}$$
$$= \sum_{k=2}^{+\infty}\sum_{j=1}^{k} \Pr\{N_i(t+\Delta t) - N_i(t) = k\}$$
$$= \sum_{k=2}^{+\infty} \Pr\{N_i(t+\Delta t) - N_i(t) = k\} + \sum_{j=2}^{+\infty}\sum_{k=j}^{+\infty} \Pr\{N_i(t+\Delta t) - N_i(t) = k\}$$
$$= \Pr\{N_i(t+\Delta t) - N_i(t) \geq 2\} + \sum_{j=2}^{+\infty} \Pr\{N_i(t+\Delta t) - N_i(t) \geq j\}$$
$$\leq \Pr\{\text{the } i\text{th component occurs at least two events in } (t, t+\Delta t]\}$$

$$+\sum_{j=2}^{+\infty} \Pr\{\text{the } i\text{th component occurs at least } j \text{ events in } (t, t+\Delta t]\}$$

$$\leq C_i(t)[\Lambda_i(t)\Delta t]^2 + \sum_{j=2}^{+\infty} C_i(t)[\Lambda_i(t)\Delta t]^j = o(\Delta t).$$

The last inequality is due to Lemma 5.6.1. Since the probability that the ith component occurs at least two events in $(t, t+\Delta t]$ is $o(\Delta t)$, only one event may occur since the ith component fails only once in $(t, t+\Delta t]$ and transfers from the functioning state to the failed state. Therefore,

$$\begin{aligned}
&\Pr\{N_i(t+\Delta t) - N_i(t) = 1\} \\
&\quad = \Pr\{\text{the } i\text{th component is functioning at time } t, \text{ occurs exactly one event} \\
&\qquad \text{in } (t, t+\Delta t], \text{ and transfers from the functioning state to the failed state}\} \\
&\qquad + o(\Delta t),\ i = 1, 2, \ldots, n.
\end{aligned}$$

The proof is complete.

Let $M(t)$ be the average number of system failures in $(0, t]$. Similarly with Lemma 5.6.3, by the result of Lemma 5.6.2, we can prove the following Lemma.

Lemma 5.6.4 *For any $t \geq 0$ and sufficiently small $\Delta t > 0$, we have*

$$\begin{aligned}
&M(t+\Delta t) - M(t) \\
&\quad = \Pr\{\text{the system is fuctioning at time } t, \text{ occurs only one event in } (t+\Delta t], \\
&\qquad \text{and transfers from the functioning state to the failed state}\} + o(\Delta t).
\end{aligned}$$

The following theorem gives the results of system failure frequency.

Theorem 5.6.1 *For any $t \geq 0$, we have*

$$m(t) = \sum_{j=1}^{n} \left[h(1_j, \mathbf{A}(t)) - h(0_j, \mathbf{A}(t))\right] m_j(t) \tag{5.6.9}$$

and

$$M = \sum_{j=1}^{n} \left[h(1_j, \mathbf{A}) - h(0_j, \mathbf{A})\right] \frac{\lambda_j \mu_j}{\lambda_j + \mu_j}, \tag{5.6.10}$$

where $h(p_1, p_2, \ldots, p_n)$ is the reliability function of the coherent system, $\mathbf{A}(t) = (A_1(t), A_2(t), \ldots, A_n(t))$ and $\mathbf{A} = (A_1, A_2, \ldots, A_n)$.

Proof Let

$$X_i(t) = \begin{cases} 1, & \text{if the } i\text{th component is functioning at time } t \\ 0, & \text{if the } i\text{th component is failed at time } t \end{cases}$$

and

$$\mathbf{X}(t) = (X_1(t), X_2(t), \ldots, X_n(t)).$$

It follows from Lemma 5.6.4 that

$$\begin{aligned}
&M(t+\Delta t) - M(t) \\
&= \Pr\left\{\begin{array}{l}\text{the system is functioning at time } t, \text{ occurs only one event in } (t, t+\Delta t], \\ \text{and the system transfers from the functioning state to the failed state}\end{array}\right\} \\
&\quad + o(\Delta t) \\
&= \sum_{j=1}^{n} \Pr\left\{\begin{array}{l}\text{the system is functioning at time } t, \text{ and fails since the } j\text{th component} \\ \text{transfers from the functioning state to the failed state in } (t, t+\Delta t]\end{array}\right\} \\
&\quad + o(\Delta t) \\
&= \sum_{j=1}^{n} \Pr\left\{\begin{array}{l}(\cdot_j, \mathbf{X}(t)) \text{ is the critical path vector of the } j\text{th component}, \\ \text{the } j\text{th component occurs only one event in } (t, t+\Delta t] \text{ and} \\ \text{transfers from the functioning state to the failed state}\end{array}\right\} \\
&\quad + o(\Delta t).
\end{aligned}$$

By the independence of all components and Lemma 5.6.3, we have

$$\begin{aligned}
&M(t+\Delta t) - M(t) \\
&= \sum_{j=1}^{n} \Pr\{(\cdot_j, \mathbf{X}(t)) \text{ is the critical path vector of the } j\text{th component}\} \\
&\quad \times \Pr\left\{\begin{array}{l}\text{the } j\text{th component is functioning at time } t, \text{ occurs only one event in} \\ (t, t+\Delta t] \text{ and transfers from the functioning state to the failed state}\end{array}\right\} \\
&\quad + o(\Delta t) \\
&= \sum_{j=1}^{n} \Pr\{\phi(1_j, \mathbf{X}(t)) - \phi(0_j, \mathbf{X}(t)) = 1\}[M_j(t+\Delta t) - M_j(t)] + o(\Delta t) \\
&= \sum_{j=1}^{n} E\{\phi(1_j, \mathbf{X}(t)) - \phi(0_j, \mathbf{X}(t))\} m_j(t)\Delta t + o(\Delta t) \\
&= \sum_{j=1}^{n} [h(1_j, \mathbf{A}(t)) - h(0_j, \mathbf{A}(t))] m_j(t)\Delta t + o(\Delta t).
\end{aligned}$$

Divide the two sides of the above equation by Δt and let $\Delta t \to 0$, $M(t)$ is differentiable since the limit on the right side exists, so the instantaneous failure frequency of the system $m(t)$ is obtained, which is just Eq. (5.6.9). Let $t \to +\infty$ on both sides of (5.6.9). It is easy to see that the limit on the right side exists, then the steady-state failure frequency of the system M is obtained, which is just formula (5.6.10). The proof is complete.

The value space of system state vector $\mathbf{X}(t) = (X_1(t), X_2(t), \ldots, X_n(t))$ is

$$\Omega = \{(k_1, k_2, \ldots, k_n): k_i = 0 \text{ or } 1,\ i = 1, 2, \ldots, n\}.$$

The set of all functioning states of the system is a subset of Ω, that is,

$$W = \{(k_1, k_2, \ldots, k_n): \phi(k_1, k_2, \ldots, k_n) = 1\}.$$

For any $\mathbf{K} = (k_1, k_2, \ldots, k_n) \in \Omega$, let

$$C_0(\mathbf{K}) = \{i: k_i = 0,\ i = 1, 2, \ldots, n\}$$

and

$$C_1(\mathbf{K}) = \{i: k_i = 1,\ i = 1, 2, \ldots, n\}$$

be the subscript sets of state 0 and state 1 in state vector $\mathbf{K}$, respectively.

Theorem 5.6.2 *The steady-state failure frequency of the coherent system is*

$$M = \sum_{\mathbf{K}\in W} \left[\prod_{i=1}^{n} A_i^{k_i} \overline{A}_i^{1-k_i}\right] \left[\sum_{j\in C_1(\mathbf{K})} \lambda_j - \sum_{j\in C_0(\mathbf{K})} \mu_j\right]. \tag{5.6.11}$$

Proof It follows from (5.6.10) and $A_j = \frac{\mu_j}{\lambda_j+\mu_j}$ that

$$\begin{aligned} M &= \sum_{j=1}^{n} \left[h(1_j, \mathbf{A}) - h(0_j, \mathbf{A})\right] \frac{\lambda_j \mu_j}{\lambda_j + \mu_j} \\ &= \sum_{j=1}^{n} \sum_{(\cdot_j, \mathbf{K})} \left[\phi(1_j, \mathbf{K}) - \phi(0_j, \mathbf{K})\right] \left[\prod_{\substack{i=1 \\ i\neq j}}^{n} A_i^{k_i} \overline{A}_i^{1-k_i}\right] \times \frac{\lambda_j \mu_j}{\lambda_j + \mu_j} \\ &= \sum_{j=1}^{n} \sum_{(\cdot_j, \mathbf{K})} \phi(1_j, \mathbf{K}) \left[\prod_{\substack{i=1 \\ i\neq j}}^{n} A_i^{k_i} \overline{A}_i^{1-k_i}\right] A_j \lambda_j \end{aligned}$$

$$+\sum_{j=1}^{n}\sum_{(\cdot_j,\mathbf{K})}\phi(0_j,\mathbf{K})\left[\prod_{\substack{i=1\\ i\neq j}}^{n}A_i^{k_i}\overline{A}_i^{1-k_i}\right]\overline{A}_j(-\mu_j)$$

$$=\sum_{j=1}^{n}\sum_{\mathbf{K}}\phi(\mathbf{K})\left[\prod_{i=1}^{n}A_i^{k_i}\overline{A}_i^{1-k_i}\right]\left[\lambda_j^{k_j}(-\mu_j)^{1-k_j}\right]$$

$$=\sum_{\mathbf{K}}\phi(\mathbf{K})\left[\prod_{i=1}^{n}A_i^{k_i}\overline{A}_i^{1-k_i}\right]\left[\sum_{j=1}^{n}\lambda_j^{k_j}(-\mu_j)^{1-k_j}\right]$$

$$=\sum_{\mathbf{K}\in W}\left[\prod_{i=1}^{n}A_i^{k_i}\overline{A}_i^{1-k_i}\right]\left[\sum_{j\in C_1(\mathbf{K})}\lambda_j-\sum_{j\in C_0(\mathbf{K})}\mu_j\right].$$

The proof is complete.

The limiting availability of the system can be obtained by the state enumeration method

$$A=h(\mathbf{A})=\sum_{\mathbf{K}\in W}\left[\prod_{i=1}^{n}A_i^{k_i}\overline{A}_i^{1-k_i}\right]. \tag{5.6.12}$$

Comparing Formulas (5.6.11) and (5.6.12), we can see that the calculation of system failure frequency only needs to multiply each term in (5.6.12) by a corresponding factor. However, due to the large amount of calculation for the state enumeration method, it is impractical to calculate A and M by Formulas (5.6.11) and (5.6.12), respectively. We should use more effective methods to calculate the system reliability. For example, we can obtain the system reliability by finding all the minimum path sets of the system. Let P_k denote the event that the kth smallest path is normal in the system, and $P_1, P_2, \ldots, P_m$ be the events that all the smallest paths are normal; then, the event that the system is normal can be expressed as

$$S=\bigcup_{k=1}^{m}P_k=\sum_{k=1}^{r}B_k,$$

where $\sum_{k=1}^{r}B_k$ is the sum of mutually exclusive terms (disjoint sum) of $\bigcup_{k=1}^{m}P_k$. Therefore, the limiting availability of the coherent system is

$$A=\sum_{k=1}^{r}\Pr\{B_k\}. \tag{5.6.13}$$

In general, (5.6.13) is much less computation than (5.6.12). Similar with Eq. (5.6.13), the following theorem gives the expression of steady-state failure frequency of the coherent system.

Theorem 5.6.3 *The steady-state failure frequency of the system is*

$$M = \sum_{k=1}^{r} \Pr\{B_k\} \left[\sum_{j \in C_1(B_k)} \lambda_j - \sum_{j \in C_0(B_k)} \mu_j \right], \tag{5.6.14}$$

where $C_0(B_k)$ represents the subscript set of the failed components in B_k, and $C_1(B_k)$ represents the subscript set of the functioning components in B_k.

We prove a lemma first.

Lemma 5.6.5 *If I is a proper subset of $\{1, 2, \ldots, n\}$, for any $l \overline{\in} I$ $(1 \le l \le n)$, we have*

$$\prod_{i \in I} A_i^{k_i} \overline{A}_i^{1-k_i} = A_l \prod_{i \in I} A_i^{k_i} \overline{A}_i^{1-k_i} + \overline{A}_l \prod_{i \in I} A_i^{k_i} \overline{A}_i^{1-k_i}. \tag{5.6.15}$$

Correspondingly, we have

$$\begin{aligned} &\left[\prod_{i \in I} A_i^{k_i} \overline{A}_i^{1-k_i} \right] \left[\sum_{j \in C_1(I)} \lambda_j - \sum_{j \in C_0(I)} \mu_j \right] \\ &= \left[A_l \prod_{i \in I} A_i^{k_i} \overline{A}_i^{1-k_i} \right] \left[\sum_{j \in C_1(I)} \lambda_j + \lambda_l - \sum_{j \in C_0(I)} \mu_j \right] \\ &\quad + \left[\overline{A}_l \prod_{i \in I} A_i^{k_i} \overline{A}_i^{1-k_i} \right] \left[\sum_{j \in C_1(I)} \lambda_j - \sum_{j \in C_0(I)} \mu_j - \mu_l \right], \end{aligned} \tag{5.6.16}$$

in which

$$C_0(I) = \{ j : k_j = 0,\ j \in I \}$$

and

$$C_1(I) = \{ j : k_j = 1,\ j \in I \}.$$

Proof Since $A_l + \overline{A}_l = 1$, Formula (5.6.15) can be arrived immediately. The right side of (5.6.16) is equal to

$$\begin{aligned} &\left[\prod_{i \in I} A_i^{k_i} \overline{A}_i^{1-k_i} \right] \left[\sum_{j \in C_1(I)} \lambda_j - \sum_{j \in C_0(I)} \mu_j \right] \\ &\quad + \left[\prod_{i \in I} A_i^{k_i} \overline{A}_i^{1-k_i} \right] A_l \lambda_l - \left[\prod_{i \in I} A_i^{k_i} \overline{A}_i^{1-k_i} \right] \overline{A}_l \mu_l. \end{aligned}$$

Since $A_l\lambda_l = \overline{A}_l\mu_l$, the above formula is just the left side of (5.6.16). The proof is complete.

We use a simple example to explain the proof idea of Theorem 5.6.3.

Example 5.6.1 Consider a repairable $2/3(G)$ system consisting of three components. By the state enumeration method, the limiting availability of the system is

$$A = A_1A_2A_3 + A_1A_2\overline{A}_3 + A_1\overline{A}_2A_3 + \overline{A}_1A_2A_3. \tag{5.6.17}$$

By searching all the smallest paths and combining the first two terms, we get

$$A = A_1A_2 + A_1\overline{A}_2A_3 + \overline{A}_1A_2A_3. \tag{5.6.18}$$

Obviously, by Formula (5.6.15), each term in Formula (5.6.18) can be decomposed into several terms in Formula (5.6.12) by the state enumeration method. Here, the first term of Formula (5.6.18) can be decomposed into the first two terms of Formula (5.6.17). According to Theorem 5.6.2 and Formula (5.6.17), the steady-state failure frequency of the system is

$$\begin{aligned} M = {} & A_1A_2A_3(\lambda_1 + \lambda_2 + \lambda_3) + A_1A_2\overline{A}_3(\lambda_1 + \lambda_2 - \mu_3) \\ & + A_1\overline{A}_2A_3(\lambda_1 + \lambda_3 - \mu_2) + \overline{A}_1A_2A_3(\lambda_2 + \lambda_3 - \mu_1). \end{aligned} \tag{5.6.19}$$

The first two terms of (5.6.19) are combined into one term, then we get

$$M = A_1A_2(\lambda_1 + \lambda_2) + A_1\overline{A}_2A_3(\lambda_1 + \lambda_3 - \mu_2) + \overline{A}_1A_2A_3(\lambda_2 + \lambda_3 - \mu_1),$$

which is just the result of Theorem 5.6.3.

Generally, to prove Theorem 5.6.3, we start from Formula (5.6.13) and repeat the decomposition by Formula (5.6.15), then we can arrive at the limiting availability of the system, which is just Formula (5.6.12). It follows from Theorem 5.6.2 that the corresponding steady-state failure frequency is just Formula (5.6.11). Then, start from the Formula (5.6.11), repeat the Formula (5.6.16), and combine some terms; Formula (5.6.11) is changed to Formula (5.6.14).

In more complex systems, (5.6.13) is simpler than (5.6.12) for calculating the limiting availability. Therefore, (5.6.14) is simpler than (5.6.11) for calculating the steady-state failure frequency.

Chapter 6
Repairable Systems with Random Fuzzy Lifetimes and Repair Times

In real life, there are more cases that randomness and fuzziness coexist in one repairable system. Then in this chapter, lifetimes and repair times of components are considered as independent random fuzzy variables. On this basis, the reliability mathematical models of repairable systems are established, including repairable series systems, repairable parallel systems, repairable cold standby systems and repairable coherent systems. Then the reliability indices such as limiting availability and steady-state failure frequency of the above repairable systems are given accordingly.

Firstly, the limiting availability and the steady-state failure frequency are redefined for random fuzzy repairable systems. Let up times ξ_i and down times η_i of the repairable system be random fuzzy variables defined on the credibility space $(\Theta_i, \mathcal{P}(\Theta_i), \text{Cr}_i)$ and $(\tau_i, \mathcal{P}(\tau_i), \text{Cr}_i')$, $i = 1, 2, \ldots$, respectively. The up times and the down times occur alternatively. We assume $S(t) = 1$ if the system is in functioning state at time t and $S(t) = 0$ if the system is in repair state at time t. For convenience, an infinite credibility product space $(\Theta, \mathcal{P}(\Theta), \text{Cr})$ is defined, where $\theta \in \Theta$, $\Theta = \prod_{i=1}^{+\infty} (\Theta_i, \tau_i)$ and $\text{Cr} = \text{Cr}_1 \wedge \text{Cr}_1' \wedge \text{Cr}_2 \wedge \text{Cr}_2' \cdots$

Definition 6.1 Let $S(t)$ be the state of the random fuzzy repairable system at time t. The limiting availability A is given by

$$A = \lim_{t \to +\infty} \text{Ch}\{S(t) = 1\}$$

when it exists.

Lemma 6.1 *The limiting availability of the random fuzzy repairable system is*

$$A = \frac{1}{2}\left[\int_0^1 \left(\lim_{t \to +\infty} \Pr\{S(t)(\theta) = 1\}\right)_\alpha^L + \left(\lim_{t \to +\infty} \Pr\{S(t)(\theta) = 1\}\right)_\alpha^U\right] \text{d}\alpha.$$

Y. Liu, *Reliability Theory Based on Uncertain Lifetimes*,
https://doi.org/10.1007/978-981-16-0995-4_6

Proof By Definition 6.1, Definition 1.3.7 and Theorem 1.2.23, we have

$$\begin{aligned} A &= \lim_{t\to+\infty} \mathrm{Ch}\{S(t) = 1\} \\ &= \lim_{t\to+\infty} \int_0^1 \mathrm{Cr}\{\theta \in \Theta | \Pr\{S(t)(\theta) = 1\} \geq p\}\mathrm{d}p \\ &= \frac{1}{2} \lim_{t\to+\infty} \int_0^1 \left(\Pr{}_\alpha^L\{S(t)(\theta) = 1\} + \Pr{}_\alpha^U\{S(t)(\theta) = 1\}\right)\mathrm{d}\alpha. \end{aligned} \tag{6.1}$$

At any time t, we have

$$0 \leq \Pr{}_\alpha^L\{S(t)(\theta) = 1\} + \Pr{}_\alpha^U\{S(t)(\theta) = 1\} \leq 2.$$

Obviously, 2 is an integrable function of α, $\alpha \in (0, 1]$. By dominated convergence theorem, Eq. (6.1) can be written as

$$A = \frac{1}{2} \int_0^1 \left(\lim_{t\to+\infty} \Pr{}_\alpha^L\{S(t)(\theta) = 1\} + \lim_{t\to+\infty} \Pr{}_\alpha^U\{S(t)(\theta) = 1\} \right)\mathrm{d}\alpha. \tag{6.2}$$

Since $\Pr\{S(t) = 1\}$ is continuous almost everywhere, then

$$\left(\lim_{t\to+\infty} \Pr\{S(t)(\theta) = 1\}\right)_\alpha^L = \lim_{t\to+\infty} \Pr{}_\alpha^L\{S(t)(\theta) = 1\} \tag{6.3}$$

and

$$\left(\lim_{t\to+\infty} \Pr\{S(t)(\theta) = 1\}\right)_\alpha^U = \lim_{t\to+\infty} \Pr{}_\alpha^U\{S(t)(\theta) = 1\}. \tag{6.4}$$

It follows from Eqs. (6.2) (6.3) and (6.4) that

$$A = \frac{1}{2} \int_0^1 \left[\left(\lim_{t\to+\infty} \Pr\{S(t)(\theta) = 1\}\right)_\alpha^L + \left(\lim_{t\to+\infty} \Pr\{S(t)(\theta) = 1\}\right)_\alpha^U \right]\mathrm{d}\alpha.$$

The lemma is proved.

Definition 6.2 Let $N(t)$ denote the number of failures that have occurred in $(0, t]$. The steady-state failure frequency, denoted by M, is defined as

$$M = \lim_{t\to+\infty} \frac{E[N(t)]}{t},$$

if $\lim_{t\to+\infty}\frac{E[N(t)]}{t}$ exists.

Lemma 6.2 *The steady-state failure frequency of the random fuzzy repairable system is*

$$M=\frac{1}{2}\int_0^1\left\{\left(\lim_{t\to+\infty}\frac{E[N(t)(\theta)]}{t}\right)_\alpha^L+\left(\lim_{t\to+\infty}\frac{E[N(t)(\theta)]}{t}\right)_\alpha^U\right\}\mathrm{d}\alpha.$$

Proof By Definition 6.2, Definition 1.3.9 and Theorem 1.2.23, we have

$$\begin{aligned}M&=\lim_{t\to+\infty}\frac{E[N(t)]}{t}\\&=\lim_{t\to+\infty}\int_0^{+\infty}\mathrm{Cr}\left\{\theta\in\Theta|\frac{E[N(t)(\theta)]}{t}\geq r\right\}\mathrm{d}r\\&=\frac{1}{2}\int_0^1\left\{\lim_{t\to+\infty}\frac{E[N(t)(\theta)]_\alpha^L}{t}+\lim_{t\to+\infty}\frac{E[N(t)(\theta)]_\alpha^U}{t}\right\}\mathrm{d}\alpha.\end{aligned}\tag{6.5}$$

At any time t, there at least exist a finite crisp number q sufficiently large so that

$$0\leq\frac{E[N(t)(\theta)]_\alpha^L}{t}+\frac{E[N(t)(\theta)]_\alpha^U}{t}\leq q.$$

Obviously, q is an integrable function of α, $\alpha\in(0,1]$. By dominated convergence theorem, Eq. (6.5) can be written as

$$M=\frac{1}{2}\int_0^1\left\{\lim_{t\to+\infty}\frac{E[N(t)(\theta)]_\alpha^L}{t}+\lim_{t\to+\infty}\frac{E[N(t)(\theta)]_\alpha^U}{t}\right\}\mathrm{d}\alpha.\tag{6.6}$$

On the other hand, since $\frac{E[N(t)(\theta)]}{t}$ is continuous almost everywhere, then

$$\left(\lim_{t\to+\infty}\frac{E[N(t)(\theta)]}{t}\right)_\alpha^L=\lim_{t\to+\infty}\frac{E[N(t)(\theta)]_\alpha^L}{t}.\tag{6.7}$$

and

$$\left(\lim_{t\to+\infty}\frac{E[N(t)(\theta)]}{t}\right)_\alpha^U=\lim_{t\to+\infty}\frac{E[N(t)(\theta)]_\alpha^U}{t}.\tag{6.8}$$

It follows from Eqs. (6.6) to (6.8) that

$$M = \frac{1}{2}\int_0^1 \left\{ \left(\lim_{t\to+\infty} \frac{E[N(t)(\theta)]}{t} \right)_\alpha^L + \left(\lim_{t\to+\infty} \frac{E[N(t)(\theta)]}{t} \right)_\alpha^U \right\} \mathrm{d}\alpha.$$

The lemma is proved.

6.1 The Series System

Consider a series system composed of n components. Let X_i be the lifetime of component i and has a random fuzzy exponential distribution with parameter λ_i on the credibility space $(\Theta_i, \mathcal{P}(\Theta_i), \mathrm{Cr}_i)$, $i = 1, 2, \ldots, n$, and Y_i be the repair time of component i which has a random fuzzy exponential distribution with parameter μ_i on the credibility space $(\tau_i, \mathcal{P}(\tau_i), \mathrm{Cr}_i^{'})$, $i = 1, 2, \ldots, n$. We assume X_i and Y_i, $i = 1, 2, \ldots, n$ are independent. We also assume that the series system is in operating state at $t = 0$. If one of these components fails, the failed component is undergoing repair, all other components remain in "suspended animation." When the repair of the failed component is completed, the remaining components continue to work. At this instant, the series system is "as good as new," thus the moments the series system goes into work are regeneration points.

Theorem 6.1.1 *Let X_i and Y_i be the random fuzzy lifetime and the random fuzzy repair time of component i, $X_i \sim \mathcal{EXP}(\lambda_i)$ and $Y_i \sim \mathcal{EXP}(\mu_i)$, $i = 1, 2, \ldots, n$. Then the limiting availability of the repairable series system is*

$$A = E\left[\frac{1}{1 + \sum_{i=1}^{n} \frac{\lambda_i}{\mu_i}} \right].$$

Proof Let $A_{i,\alpha} = \{\theta_i \in \Theta_i | \mu\{\theta_i\} \geq \alpha\}$ and $B_{i,\alpha} = \{\vartheta_i \in \tau_i | \mu\{\vartheta_i\} \geq \alpha\}$, $i = 1, 2, \ldots, n$. Since the α-pessimistic values and the α-optimistic values of fuzzy variables $E[X_i(\theta_i)], E[Y_i(\vartheta_i)], \theta_i \in \Theta_i$, $\vartheta_i \in \tau_i$, $i = 1, 2, \ldots, n$ are continuous almost everywhere with respect to α, $\alpha \in (0, 1]$, then there at least exist the points $\theta_i^{'}, \theta_i^{''} \in A_{i,\alpha}$ and $\vartheta_i^{'}, \vartheta_i^{''} \in B_{i,\alpha}$, $i = 1, 2, \ldots, n$ such that

$$E[X_i(\theta_i^{'})] = E[X_i(\theta_i)]_\alpha^L,$$

$$E[X_i(\theta_i^{''})] = E[X_i(\theta_i)]_\alpha^U,$$

$$E[Y_i(\vartheta_i^{'})] = E[Y_i(\vartheta_i)]_\alpha^L,$$

$$E[Y_i(\vartheta_i^{''})] = E[Y_i(\vartheta_i)]_\alpha^U.$$

Then for $\forall\theta_{i,\alpha} \in A_{i,\alpha}$ and $\forall\vartheta_{i,\alpha} \in B_{i,\alpha}$, $i = 1, 2, \ldots, n$, we have

$$E[X_i(\theta_i')] \le E[X_i(\theta_{i,\alpha})] \le E[X_i(\theta_i'')] \tag{6.1.1}$$

and

$$E[Y_i(\vartheta_i')] \le E[Y_i(\vartheta_{i,\alpha})] \le E[Y_i(\vartheta_i'')]. \tag{6.1.2}$$

Since $X_i(\theta_i')$, $X_i(\theta_{i,\alpha})$, $X_i(\theta_i'')$ and $Y_i(\vartheta_i')$, $Y_i(\vartheta_{i,\alpha})$, $Y_i(\vartheta_i'')$ are random variables with exponential distributions, then by Eqs. (6.1.1) and (6.1.2), we can arrive at

$$\frac{1}{\lambda_i(\theta_i')} \le \frac{1}{\lambda_i(\theta_{i,\alpha})} \le \frac{1}{\lambda_i(\theta_i'')}$$

and

$$\frac{1}{\mu_i(\vartheta_i')} \le \frac{1}{\mu_i(\vartheta_{i,\alpha})} \le \frac{1}{\mu_i(\vartheta_i'')}.$$

Then we have

$$\lambda_i(\theta_i'') \le \lambda_i(\theta_{i,\alpha}) \le \lambda_i(\theta_i') \tag{6.1.3}$$

and

$$\mu_i(\vartheta_i'') \le \mu_i(\vartheta_{i,\alpha}) \le \mu_i(\vartheta_i'). \tag{6.1.4}$$

We can construct three repairable series systems:

(1) Series system 1: $X_i(\theta_i')$ and $Y_i(\vartheta_i'')$ are the lifetime and the repair time of component i, $i = 1, 2, \ldots, n$, respectively.
(2) Series system 2: $X_i(\theta_i'')$ and $Y_i(\vartheta_i')$ are the lifetime and the repair time of component i, $i = 1, 2, \ldots, n$, respectively.
(3) Series system 3: $X_i(\theta_{i,\alpha})$ and $Y_i(\vartheta_{i,\alpha})$ are the lifetime and the repair time of component i, $i = 1, 2, \ldots, n$, respectively.

It is easy to see that the system (1) and system (2) are two standard stochastic series systems. For any fixed $\theta_{i,\alpha} \in A_{i,\alpha}$ and $\vartheta_{i,\alpha} \in B_{i,\alpha}$, $i = 1, 2, \ldots, n$, system (3) is also a stochastic series system. The limiting availability of series systems (1), (2) and (3) are denoted by A_1, A_2 and A_3, respectively. From the result in classical reliability theory, we can arrive at

$$A_1 = \frac{1}{1 + \sum_{i=1}^{n} \frac{\lambda_i(\theta_i')}{\mu_i(\vartheta_i'')}},$$

$$A_2 = \frac{1}{1 + \sum_{i=1}^{n} \frac{\lambda_i(\theta_i^{''})}{\mu_i(\vartheta_i^{'})}}$$

and

$$A_3 = \frac{1}{1 + \sum_{i=1}^{n} \frac{\lambda_i(\theta_{i,\alpha})}{\mu_i(\vartheta_{i,\alpha})}}.$$

From Eqs. (6.1.3) and (6.1.4), we have

$$\frac{1}{1 + \sum_{i=1}^{n} \frac{\lambda_i(\theta_i^{'})}{\mu_i(\vartheta_i^{''})}} \leq \frac{1}{1 + \sum_{i=1}^{n} \frac{\lambda_i(\theta_{i,\alpha})}{\mu_i(\vartheta_{i,\alpha})}} \leq \frac{1}{1 + \sum_{i=1}^{n} \frac{\lambda_i(\theta_i^{''})}{\mu_i(\vartheta_i^{'})}}.$$

Since $\theta_{i,\alpha}$ and $\vartheta_{i,\alpha}$ are arbitrary points in $A_{i,\alpha}$ and $B_{i,\alpha}$, $i = 1, 2, \ldots, n$, we have

$$\left(\lim_{t \to +\infty} \Pr\{S(t)(\theta) = 1\}\right)_\alpha^L = \frac{1}{1 + \sum\limits_{i=1}^{n} \frac{\lambda_i(\theta_i^{'})}{\mu_i(\vartheta_i^{''})}} \tag{6.1.5}$$

and

$$\left(\lim_{t \to +\infty} \Pr\{S(t)(\theta) = 1\}\right)_\alpha^U = \frac{1}{1 + \sum\limits_{i=1}^{n} \frac{\lambda_i(\theta_i^{''})}{\mu_i(\vartheta_i^{'})}}. \tag{6.1.6}$$

By Eqs. (6.1.3) and (6.1.4), for $i = 1, 2, \ldots, n$, we have

$$\lambda_{i,\alpha}^L = \lambda_i(\theta_i^{''}), \tag{6.1.7}$$

$$\lambda_{i,\alpha}^U = \lambda_i(\theta_i^{'}), \tag{6.1.8}$$

$$\mu_{i,\alpha}^L = \mu_i(\vartheta_i^{''}), \tag{6.1.9}$$

$$\mu_{i,\alpha}^U = \mu_i(\vartheta_i^{'}). \tag{6.1.10}$$

By Lemma 6.1 and Eqs. (6.1.5)–(6.1.10), we can arrive at

$$A = \frac{1}{2}\left[\int_0^1 \left(\lim_{t \to +\infty} \Pr\{S(t)(\theta) = 1\}\right)_\alpha^L + \left(\lim_{t \to +\infty} \Pr\{S(t)(\theta) = 1\}\right)_\alpha^U\right] \mathrm{d}\alpha$$

$$
=\frac{1}{2}\int_0^1\left(\frac{1}{1+\sum_{i=1}^n \frac{\lambda_i(\theta_i^{'})}{\mu_i(\vartheta_i^{''})}\sum\limits_{i=1}^{n}}+\frac{1}{1+\sum_{i=1}^n \frac{\lambda_i(\theta_i^{''})}{\mu_i(\vartheta_i^{'})}}\right)\mathrm{d}\alpha
$$

$$
=\frac{1}{2}\int_0^1\left(\frac{1}{1+\sum_{i=1}^n \frac{\lambda_{i,\alpha}^U}{\mu_{i,\alpha}^L}}+\frac{1}{1+\sum_{i=1}^n \frac{\lambda_{i,\alpha}^L}{\mu_{i,\alpha}^U}}\right)\mathrm{d}\alpha
$$

$$
=\frac{1}{2}\int_0^1\left(\left[\frac{1}{1+\sum_{i=1}^n \frac{\lambda_i}{\mu_i}}\right]_\alpha^L+\left[\frac{1}{1+\sum_{i=1}^n \frac{\lambda_i}{\mu_i}}\right]_\alpha^U\right)\mathrm{d}\alpha=E\left[\frac{1}{1+\sum_{i=1}^n \frac{\lambda_i}{\mu_i}}\right].
$$

The theorem is proved.

Remark 6.1.1 If X_i and $Y_i, i=1,2,\ldots,n$ degenerate to random variables, then the result in Theorem 6.1.1 degenerates to the form

$$
A=\frac{1}{1+\sum_{i=1}^n \frac{\lambda_i}{\mu_i}},
$$

which is consistent with the result in stochastic case.

Remark 6.1.2 Under the conditions of Theorem 6.1.1, it follows that

$$
\begin{aligned}
\lim_{t\to+\infty}\mathrm{Ch}\{S(t)=0\}&=1-\lim_{t\to+\infty}\mathrm{Ch}\{S(t)=1\}\\
&=1-E\left[\frac{1}{1+\sum_{i=1}^n \frac{\lambda_i}{\mu_i}}\right].
\end{aligned}
$$

Theorem 6.1.2 *Let X_i and Y_i be the random fuzzy lifetime and the random fuzzy repair time of component i, $X_i \sim \mathcal{EXP}(\lambda_i)$ and $Y_i \sim \mathcal{EXP}(\mu_i)$, $i=1,2,\ldots,n$. Then the steady-state failure frequency of the repairable series system is*

$$
M=\frac{1}{2}\int_0^1\left\{\frac{\sum_{i=1}^n \lambda_{i,\alpha}^L}{1+\sum_{i=1}^n \frac{\lambda_{i,\alpha}^L}{\mu_{i,\alpha}^L}}+\frac{\sum_{i=1}^n \lambda_{i,\alpha}^U}{1+\sum_{i=1}^n \frac{\lambda_{i,\alpha}^U}{\mu_{i,\alpha}^U}}\right\}\mathrm{d}\alpha.
$$

Proof Based on $X_i(\theta_i^{'})$, $X_i(\theta_{i,\alpha})$, $X_i(\theta_i^{''})$ and $Y_i(\vartheta_i^{'})$, $Y_i(\vartheta_{i,\alpha})$, $Y_i(\vartheta_i^{''})$ in the proof of Theorem 6.1.1, we can also construct three repairable series systems:

(1) Series system 1: $X_i(\theta_i^{'})$ and $Y_i(\vartheta_i^{'})$ are the lifetime and the repair time of component i, $i=1,2,\ldots,n$, respectively.

(2) Series system 2: $X_i(\theta_i^{''})$ and $Y_i(\vartheta_i^{''})$ are the lifetime and the repair time of component i, $i=1,2,\ldots,n$, respectively.

(3) Series system 3: $X_i(\theta_{i,\alpha})$ and $Y_i(\vartheta_{i,\alpha})$ are the lifetime and the repair time of component i, $i = 1, 2, \ldots, n$, respectively.

It is easy to see that the system (1) and system (2) are two standard stochastic series systems. For any fixed $\theta_{i,\alpha} \in A_{i,\alpha}$ and $\vartheta_{i,\alpha} \in B_{i,\alpha}$, $i = 1, 2, \ldots, n$, the system (3) is also a stochastic series system. The steady-state failure frequency of series systems (1), (2) and (3) are denoted by M_1, M_2 and M_3, respectively. From the result in classical reliability theory, we can arrive at

$$M_1 = \frac{\sum_{i=1}^{n} \lambda_i(\theta_i^{'})}{1 + \sum_{i=1}^{n} \frac{\lambda_i(\theta_i^{'})}{\mu_i(\vartheta_i^{'})}},$$

$$M_2 = \frac{\sum_{i=1}^{n} \lambda_i(\theta_i^{''})}{1 + \sum_{i=1}^{n} \frac{\lambda_i(\theta_i^{''})}{\mu_i(\vartheta_i^{''})}}$$

and

$$M_3 = \frac{\sum_{i=1}^{n} \lambda_i(\theta_{i,\alpha})}{1 + \sum_{i=1}^{n} \frac{\lambda_i(\theta_{i,\alpha})}{\mu_i(\vartheta_{i,\alpha})}}.$$

Because the series system is a coherent system, we have

$$M_2 \le M_3 \le M_1.$$

Since $\theta_{i,\alpha}$ and $\vartheta_{i,\alpha}$ are arbitrary points in $A_{i,\alpha}$ and $B_{i,\alpha}$, $i = 1, 2, \ldots, n$, then we have

$$\left(\lim_{t\to+\infty} \frac{E[N(t)(\theta)]}{t}\right)_\alpha^L = M_2 = \frac{\sum_{i=1}^{n} \lambda_i(\theta_i^{''})}{1 + \sum_{i=1}^{n} \frac{\lambda_i(\theta_i^{''})}{\mu_i(\vartheta_i^{''})}} \tag{6.1.11}$$

and

$$\left(\lim_{t\to+\infty} \frac{E[N(t)(\theta)]}{t}\right)_\alpha^U = M_1 = \frac{\sum_{i=1}^{n} \lambda_i(\theta_i^{'})}{1 + \sum_{i=1}^{n} \frac{\lambda_i(\theta_i^{'})}{\mu_i(\vartheta_i^{'})}}. \tag{6.1.12}$$

By Lemma 6.2 and Eqs. (6.1.7)–(6.1.12), we can arrive at

$$M = \frac{1}{2}\int_0^1 \left\{\left(\lim_{t\to+\infty} \frac{E[N(t)(\theta)]}{t}\right)_\alpha^L + \left(\lim_{t\to+\infty} \frac{E[N(t)(\theta)]}{t}\right)_\alpha^U\right\} \mathrm{d}\alpha$$

$$= \frac{1}{2}\int_0^1 \left\{ \frac{\sum_{i=1}^n \lambda_i(\theta_i^{''})}{1+\sum_{i=1}^n \frac{\lambda_i(\theta_i^{''})}{\mu_i(\vartheta_i^{''})}} + \frac{\sum_{i=1}^n \lambda_i(\theta_i^{'})}{1+\sum_{i=1}^n \frac{\lambda_i(\theta_i^{'})}{\mu_i(\vartheta_i^{'})}} \right\} d\alpha$$

$$= \frac{1}{2}\int_0^1 \left\{ \frac{\sum_{i=1}^n \lambda_{i,\alpha}^L}{1+\sum_{i=1}^n \frac{\lambda_{i,\alpha}^L}{\mu_{i,\alpha}^L}} + \frac{\sum_{i=1}^n \lambda_{i,\alpha}^U}{1+\sum_{i=1}^n \frac{\lambda_{i,\alpha}^U}{\mu_{i,\alpha}^U}} \right\} d\alpha.$$

The theorem is proved.

Remark 6.1.3 If X_i and Y_i, $i = 1, 2, \ldots, n$ degenerate to random variables, then the result in Theorem 6.1.2 degenerates to the form

$$M = \frac{\sum_{i=1}^n \lambda_i}{1+\sum_{i=1}^n \frac{\lambda_i}{\mu_i}},$$

which is consistent with the result in stochastic case.

Example 6.1.1 Suppose that the series system composed by two components. The lifetime and the repair time of component i are X_i and Y_i, $i = 1, 2$, respectively. If $X_i \sim \mathcal{EXP}(\lambda_i)$ and $Y_i \sim \mathcal{EXP}(\mu_i)$, $i = 1, 2$, where $\lambda_1 = (2, 3, 4)$, $\lambda_2 = (2, 3, 4)$, $\mu_1 = (4, 5, 6)$ and $\mu_2 = (4, 5, 6)$. Then we can arrive at

$$\lambda_{1,\alpha}^L = \lambda_{2,\alpha}^L = 2+\alpha,\ \lambda_{1,\alpha}^U = \lambda_{2,\alpha}^U = 4-\alpha,$$
$$\mu_{1,\alpha}^L = \mu_{2,\alpha}^L = 4+\alpha,\ \mu_{1,\alpha}^U = \mu_{2,\alpha}^U = 6-\alpha.$$

By Theorem 6.1.1, the limiting availability of the series system is

$$\begin{aligned} A &= \frac{1}{2}\int_0^1 \left(\frac{1}{1+\sum_{i=1}^n \frac{\lambda_{i,\alpha}^U}{\mu_{i,\alpha}^L}} + \frac{1}{1+\sum_{i=1}^n \frac{\lambda_{i,\alpha}^L}{\mu_{i,\alpha}^U} \sum_{i=1}^n} \right) d\alpha \\ &= \frac{1}{2}\int_0^1 \left(\frac{1}{1+\frac{2(4-\alpha)}{4+\alpha}} + \frac{1}{1+\frac{2(2+\alpha)}{6-\alpha}} \right) d\alpha \\ &\approx 0.4586. \end{aligned}$$

By Theorem 6.1.2, the steady-state failure frequency of the series system is

$$M = \frac{1}{2}\int_0^1 \left\{ \frac{\sum_{i=1}^n \lambda_{i,\alpha}^L}{1+\sum_{i=1}^n \frac{\lambda_{i,\alpha}^L}{\mu_{i,\alpha}^L}} + \frac{\sum_{i=1}^n \lambda_{i,\alpha}^U}{1+\sum_{i=1}^n \frac{\lambda_{i,\alpha}^U}{\mu_{i,\alpha}^U}} \right\} d\alpha$$

$$= \frac{1}{2}\int_0^1 \left(\frac{2(2+\alpha)}{1+\frac{2(2+\alpha)}{4+\alpha}} + \frac{2(4-\alpha)}{1+\frac{2(4-\alpha)}{6-\alpha}} \right) \mathrm{d}\alpha$$

$$\approx 2.7231.$$

6.2 The Parallel System

Consider a repairable parallel system consisting of two nonidentical components and one repair facility. Let X_i and Y_i be the lifetime and the repair time of component i, $i = 1, 2$. We assume X_i has a random fuzzy exponential distribution with parameter λ_i defined on the credibility space $(\Theta_i, \mathcal{P}(\Theta_i), \mathrm{Cr}_i)$, $i = 1, 2$, and Y_i has a random fuzzy exponential distribution with parameter μ_i defined on the credibility space $(\tau_i, \mathcal{P}(\tau_i), \mathrm{Cr}_i^{'})$, $i = 1, 2$. We also assume X_i and Y_i, $i = 1, 2$ are independent. When the repair is completed, the parallel system goes into work instantly and the parallel system is assumed to be "as good as new".

Theorem 6.2.1 *The limiting availability of the repairable parallel system is*

$$A = \frac{1}{2}\int_0^1 \left[\frac{1}{\sum_{j=0}^2 m_j^{'} v_j^{'}} \left(\frac{v_0^{'}}{\lambda_{1,\alpha}^U + \lambda_{2,\alpha}^U} + \frac{v_1^{'}}{\lambda_{2,\alpha}^U + \mu_{1,\alpha}^L} + \frac{v_2^{'}}{\lambda_{1,\alpha}^U + \mu_{2,\alpha}^L} \right) \right.$$
$$\left. + \frac{1}{\sum_{j=0}^2 m_j^{''} v_j^{''}} \left(\frac{v_0^{''}}{\lambda_{1,\alpha}^L + \lambda_{2,\alpha}^L} + \frac{v_1^{''}}{\lambda_{2,\alpha}^L + \mu_{1,\alpha}^U} + \frac{v_2^{''}}{\lambda_{1,\alpha}^L + \mu_{2,\alpha}^U} \right) \right] \mathrm{d}\alpha,$$

in which

$$\begin{cases} m_0^{'} = \dfrac{1}{\lambda_{1,\alpha}^U + \lambda_{2,\alpha}^U} \\ m_1^{'} = \dfrac{1}{\mu_{1,\alpha}^L} \\ m_2^{'} = \dfrac{1}{\mu_{2,\alpha}^L}, \end{cases} \qquad \begin{cases} v_0^{'} = 1 - \dfrac{\lambda_{1,\alpha}^U \lambda_{2,\alpha}^U}{(\lambda_{2,\alpha}^U + \mu_{1,\alpha}^L)(\lambda_{1,\alpha}^U + \mu_{2,\alpha}^L)} \\ v_1^{'} = 1 - \dfrac{\lambda_{2,\alpha}^U \mu_{2,\alpha}^L}{(\lambda_{1,\alpha}^U + \lambda_{2,\alpha}^U)(\lambda_{1,\alpha}^U + \mu_{2,\alpha}^L)} \\ v_2^{'} = 1 - \dfrac{\lambda_{1,\alpha}^U \mu_{1,\alpha}^L}{(\lambda_{1,\alpha}^U + \lambda_{2,\alpha}^U)(\lambda_{2,\alpha}^U + \mu_{1,\alpha}^L)} \end{cases}$$

and

$$\begin{cases} m_0'' = \dfrac{1}{\lambda_{1,\alpha}^L + \lambda_{2,\alpha}^L} \\ m_1'' = \dfrac{1}{\mu_{1,\alpha}^U} \\ m_2'' = \dfrac{1}{\mu_{2,\alpha}^U}, \end{cases} \quad \begin{cases} v_0'' = 1 - \dfrac{\lambda_{1,\alpha}^L \lambda_{2,\alpha}^L}{(\lambda_{2,\alpha}^L + \mu_{1,\alpha}^U)(\lambda_{1,\alpha}^L + \mu_{2,\alpha}^U)} \\ v_1'' = 1 - \dfrac{\lambda_{2,\alpha}^L \mu_{2,\alpha}^U}{(\lambda_{1,\alpha}^L + \lambda_{2,\alpha}^L)(\lambda_{1,\alpha}^L + \mu_{2,\alpha}^L)} \\ v_2'' = 1 - \dfrac{\lambda_{1,\alpha}^L \mu_{1,\alpha}^U}{(\lambda_{1,\alpha}^L + \lambda_{2,\alpha}^L)(\lambda_{2,\alpha}^L + \mu_{1,\alpha}^U)}. \end{cases}$$

Proof Let $A_{i,\alpha} = \{\theta_i \in \Theta_i | \mu\{\theta_i\} \geq \alpha\}$ and $B_{i,\alpha} = \{\vartheta_i \in \tau_i | \mu\{\vartheta_i\} \geq \alpha\}$, $i = 1, 2$. The α-pessimistic values and α-optimistic values of fuzzy variables λ_i and μ_i are denoted by $\lambda_{i,\alpha}^L$, $\lambda_{i,\alpha}^U$, $\mu_{i,\alpha}^L$, $\mu_{i,\alpha}^U$, $i = 1, 2$, respectively. Then for any fixed $\theta_{i,\alpha} \in A_{i,\alpha}$ and $\vartheta_{i,\alpha} \in B_{i,\alpha}$, $i = 1, 2$, we have

$$\lambda_{i,\alpha}^L \leq \lambda_i(\theta_{i,\alpha}) \leq \lambda_{i,\alpha}^U \tag{6.2.1}$$

and

$$\mu_{i,\alpha}^L \leq \mu_i(\vartheta_{i,\alpha}) \leq \mu_{i,\alpha}^U. \tag{6.2.2}$$

Since X_i and Y_i have random fuzzy exponential distributions with fuzzy parameters λ_i and μ_i, $i = 1, 2$, respectively. When λ_i and μ_i, $i = 1, 2$, degenerate to crisp numbers, X_i and Y_i, $i = 1, 2$ degenerate to random variables with exponential distributions accordingly. We can construct three stochastic repairable parallel systems with two nonidentical components:

(1) Parallel system 1: let X_i' and Y_i'' be the lifetime and the repair time of component i, $X_i' \sim \exp(\lambda_{i,\alpha}^U)$, $Y_i'' \sim \exp(\mu_{i,\alpha}^L)$, $i = 1, 2$.
(2) Parallel system 2: let $X_i(\theta_{i,\alpha})$ and $Y_i(\vartheta_{i,\alpha})$ be the lifetime and the repair time of component i, $X_i(\theta_{i,\alpha}) \sim \exp(\lambda_i(\theta_{i,\alpha}))$, $Y_i(\vartheta_{i,\alpha}) \sim \exp(\mu_i(\vartheta_{i,\alpha}))$, $i = 1, 2$.
(3) Parallel system 3: let X_i'' and Y_i' be the lifetime and the repair time of component i, $X_i'' \sim \exp(\lambda_{i,\alpha}^L)$, $Y_i' \sim \exp(\mu_{i,\alpha}^U)$, $i = 1, 2$.

By Eqs. (6.2.1) and (6.2.2), we can see that

$$X_i' \leq_d X_i(\theta_{i,\alpha}) \leq_d X_i'' \quad \text{and} \quad Y_i' \leq_d Y_i(\vartheta_{i,\alpha}) \leq_d Y_i'', \; i = 1, 2.$$

The limiting availability of parallel systems (1), (2) and (3) is denoted by A_1, A_2 and A_3, respectively. Since the parallel system is a coherent system, we have

$$A_1 \leq A_2 \leq A_3.$$

Since $\theta_{i,\alpha}$ and $\vartheta_{i,\alpha}$ are arbitrary points in $A_{i,\alpha}$ and $B_{i,\alpha}$, $i = 1, 2$, respectively, from the result in stochastic case, we have

$$\left(\lim_{t\to+\infty}\Pr\{S(t)(\theta)=1\}\right)^L_\alpha$$
$$=A_1=\frac{1}{\sum_{j=0}^{2}m_j^{'}v_j^{'}}\left(\frac{v_0^{'}}{\lambda^U_{1,\alpha}+\lambda^U_{2,\alpha}}+\frac{v_1^{'}}{\lambda^U_{2,\alpha}+\mu^L_{1,\alpha}}+\frac{v_2^{'}}{\lambda^U_{1,\alpha}+\lambda^L_{2,\alpha}}\right) \quad (6.2.3)$$

and

$$\left(\lim_{t\to+\infty}\Pr\{S(t)(\theta)=1\}\right)^U_\alpha$$
$$=A_3=\frac{1}{\sum_{j=0}^{2}m_j^{''}v_j^{''}}\left(\frac{v_0^{''}}{\lambda^L_{1,\alpha}+\lambda^L_{2,\alpha}}+\frac{v_1^{''}}{\lambda^L_{2,\alpha}+\mu^U_{1,\alpha}}+\frac{v_2^{''}}{\lambda^L_{1,\alpha}+\lambda^U_{2,\alpha}}\right), \quad (6.2.4)$$

in which

$$\begin{cases} m_0^{'}=\dfrac{1}{\lambda^U_{1,\alpha}+\lambda^U_{2,\alpha}} \\ m_1^{'}=\dfrac{1}{\mu^L_{1,\alpha}} \\ m_2^{'}=\dfrac{1}{\mu^L_{2,\alpha}}, \end{cases} \quad \begin{cases} v_0^{'}=1-\dfrac{\lambda^U_{1,\alpha}\lambda^U_{2,\alpha}}{(\lambda^U_{2,\alpha}+\mu^L_{1,\alpha})(\lambda^U_{1,\alpha}+\mu^L_{2,\alpha})} \\ v_1^{'}=1-\dfrac{\lambda^U_{2,\alpha}\mu^L_{2,\alpha}}{(\lambda^U_{1,\alpha}+\lambda^U_{2,\alpha})(\lambda^U_{1,\alpha}+\mu^L_{2,\alpha})} \\ v_2^{'}=1-\dfrac{\lambda^U_{1,\alpha}\mu^L_{1,\alpha}}{(\lambda^U_{1,\alpha}+\lambda^U_{2,\alpha})(\lambda^U_{2,\alpha}+\mu^L_{1,\alpha})} \end{cases}$$

and

$$\begin{cases} m_0^{''}=\frac{1}{\lambda^L_{1,\alpha}+\lambda^L_{2,\alpha}} \\ m_1^{''}=\frac{1}{\mu^U_{1,\alpha}} \\ m_2^{''}=\frac{1}{\mu^U_{2,\alpha}}, \end{cases} \quad \begin{cases} v_0^{''}=1-\frac{\lambda^L_{1,\alpha}\lambda^L_{2,\alpha}}{(\lambda^L_{2,\alpha}+\mu^U_{1,\alpha})(\lambda^L_{1,\alpha}+\mu^U_{2,\alpha})} \\ v_1^{''}=1-\frac{\lambda^L_{2,\alpha}\mu^U_{2,\alpha}}{(\lambda^L_{1,\alpha}+\lambda^L_{2,\alpha})(\lambda^L_{1,\alpha}+\mu^L_{2,\alpha})} \\ v_2^{''}=1-\frac{\lambda^L_{1,\alpha}\mu^U_{1,\alpha}}{(\lambda^L_{1,\alpha}+\lambda^L_{2,\alpha})(\lambda^L_{2,\alpha}+\mu^U_{1,\alpha})}. \end{cases}$$

By Lemma 6.1, Eqs. (6.2.3) and (6.2.4), we can arrive at

$$A=\frac{1}{2}\int_0^1\left[\left(\lim_{t\to+\infty}\Pr\{S(t)(\theta)=1\}\right)^L_\alpha+\left(\lim_{t\to+\infty}\Pr\{S(t)(\theta)=1\}\right)^U_\alpha\right]\mathrm{d}\alpha$$
$$=\frac{1}{2}\int_0^1\left[\frac{1}{\sum_{j=0}^{2}m_j^{'}v_j^{'}}\left(\frac{v_0^{'}}{\lambda^U_{1,\alpha}+\lambda^U_{2,\alpha}}+\frac{v_1^{'}}{\lambda^U_{2,\alpha}+\mu^L_{1,\alpha}}+\frac{v_2^{'}}{\lambda^U_{1,\alpha}+\lambda^L_{2,\alpha}}\right)\right.$$
$$\left.+\frac{1}{\sum_{j=0}^{2}m_j^{''}v_j^{''}}\left(\frac{v_0^{''}}{\lambda^L_{1,\alpha}+\lambda^L_{2,\alpha}}+\frac{v_1^{''}}{\lambda^L_{2,\alpha}+\mu^U_{1,\alpha}}+\frac{v_2^{''}}{\lambda^L_{1,\alpha}+\lambda^U_{2,\alpha}}\right)\right]\mathrm{d}\alpha,$$

in which

$$\begin{cases} m_0^{'} = \frac{1}{\lambda_{1,\alpha}^U+\lambda_{2,\alpha}^U} \\ m_1^{'} = \frac{1}{\mu_{1,\alpha}^L} \\ m_2^{'} = \frac{1}{\mu_{2,\alpha}^L}, \end{cases} \quad \begin{cases} v_0^{'} = 1 - \frac{\lambda_{1,\alpha}^U\lambda_{2,\alpha}^U}{(\lambda_{2,\alpha}^U+\mu_{1,\alpha}^L)(\lambda_{1,\alpha}^U+\mu_{2,\alpha}^L)} \\ v_1^{'} = 1 - \frac{\lambda_{2,\alpha}^U\mu_{2,\alpha}^L}{(\lambda_{1,\alpha}^U+\lambda_{2,\alpha}^U)(\lambda_{1,\alpha}^U+\mu_{2,\alpha}^L)} \\ v_2^{'} = 1 - \frac{\lambda_{1,\alpha}^U\mu_{1,\alpha}^L}{(\lambda_{1,\alpha}^U+\lambda_{2,\alpha}^U)(\lambda_{2,\alpha}^U+\mu_{1,\alpha}^L)} \end{cases}$$

and

$$\begin{cases} m_0^{''} = \frac{1}{\lambda_{1,\alpha}^L+\lambda_{2,\alpha}^L} \\ m_1^{''} = \frac{1}{\mu_{1,\alpha}^U} \\ m_2^{''} = \frac{1}{\mu_{2,\alpha}^U}, \end{cases} \quad \begin{cases} v_0^{''} = 1 - \frac{\lambda_{1,\alpha}^L\lambda_{2,\alpha}^L}{(\lambda_{2,\alpha}^L+\mu_{1,\alpha}^U)(\lambda_{1,\alpha}^L+\mu_{2,\alpha}^U)} \\ v_1^{''} = 1 - \frac{\lambda_{2,\alpha}^L\mu_{2,\alpha}^U}{(\lambda_{1,\alpha}^L+\lambda_{2,\alpha}^L)(\lambda_{1,\alpha}^L+\mu_{2,\alpha}^L)} \\ v_2^{''} = 1 - \frac{\lambda_{1,\alpha}^L\mu_{1,\alpha}^U}{(\lambda_{1,\alpha}^L+\lambda_{2,\alpha}^L)(\lambda_{2,\alpha}^L+\mu_{1,\alpha}^U)}. \end{cases}$$

The theorem is proved.

Remark 6.2.1 If X_i and Y_i, $i = 1, 2$ degenerate to random variables, then the result in Theorem 6.2.1 degenerates to the form

$$A = \frac{1}{\sum_{j=0}^{2} m_j v_j}\left(\frac{v_0}{\lambda_1+\lambda_2} + \frac{v_1}{\lambda_2+\mu_1} + \frac{v_2}{\lambda_1+\mu_2}\right),$$

in which

$$\begin{cases} m_0 = \frac{1}{\lambda_1+\lambda_2} \\ m_1 = \frac{1}{\mu_1} \\ m_2 = \frac{1}{\mu_2}, \end{cases} \quad \begin{cases} v_0 = 1 - \frac{\lambda_1\lambda_2}{(\lambda_2+\mu_1)(\lambda_1+\mu_2)} \\ v_1 = 1 - \frac{\lambda_2\mu_2}{(\lambda_1+\lambda_2)(\lambda_1+\mu_2)} \\ v_2 = 1 - \frac{\lambda_1\mu_1}{(\lambda_1+\lambda_2)(\lambda_1+\mu_1)}, \end{cases}$$

which is consistent with the result in stochastic case.

Theorem 6.2.2 *The steady-state failure frequency of the repairable parallel system is*

$$M = \frac{1}{2}\int_0^1 \left[\frac{1}{\sum_{j=0}^{2} n_j^{'} u_j^{'}}\left(\frac{u_1^{'}\lambda_{2,\alpha}^U}{\lambda_{2,\alpha}^U + u_{1,\alpha}^U} + \frac{u_2^{'}\lambda_{1,\alpha}^U}{\lambda_{1,\alpha}^U + u_{2,\alpha}^U}\right) + \frac{1}{\sum_{j=0}^{2} n_j^{''} u_j^{''}}\left(\frac{u_1^{''}\lambda_{2,\alpha}^L}{\lambda_{2,\alpha}^L + u_{1,\alpha}^L} + \frac{u_2^{''}\lambda_{1,\alpha}^L}{\lambda_{1,\alpha}^L + u_{2,\alpha}^L}\right)\right] d\alpha,$$

in which

$$\begin{cases} n_0^{'} = \frac{1}{\lambda_{1,\alpha}^U+\lambda_{2,\alpha}^U} \\ n_1^{'} = \frac{1}{\mu_{1,\alpha}^U} \\ n_2^{'} = \frac{1}{\mu_{2,\alpha}^U}, \end{cases} \quad \begin{cases} u_0^{'} = 1 - \frac{\lambda_{1,\alpha}^U\lambda_{2,\alpha}^U}{(\lambda_{2,\alpha}^U+\mu_{1,\alpha}^U)(\lambda_{1,\alpha}^U+\mu_{2,\alpha}^U)} \\ u_1^{'} = 1 - \frac{\lambda_{2,\alpha}^U\mu_{2,\alpha}^U}{(\lambda_{1,\alpha}^U+\lambda_{2,\alpha}^U)(\lambda_{1,\alpha}^U+\mu_{2,\alpha}^U)} \\ u_2^{'} = 1 - \frac{\lambda_{1,\alpha}^U\mu_{1,\alpha}^U}{(\lambda_{1,\alpha}^U+\lambda_{2,\alpha}^U)(\lambda_{2,\alpha}^U+\mu_{1,\alpha}^U)} \end{cases}$$

and

$$\begin{cases} n_0^{''} = \frac{1}{\lambda_{1,\alpha}^L+\lambda_{2,\alpha}^L} \\ n_1^{''} = \frac{1}{\mu_{1,\alpha}^L} \\ n_2^{''} = \frac{1}{\mu_{2,\alpha}^L}, \end{cases} \quad \begin{cases} u_0^{''} = 1 - \frac{\lambda_{1,\alpha}^L\lambda_{2,\alpha}^L}{(\lambda_{2,\alpha}^L+\mu_{1,\alpha}^U)(\lambda_{1,\alpha}^L+\mu_{2,\alpha}^U)} \\ u_1^{''} = 1 - \frac{\lambda_{2,\alpha}^L\mu_{2,\alpha}^L}{(\lambda_{1,\alpha}^L+\lambda_{2,\alpha}^L)(\lambda_{1,\alpha}^L+\mu_{2,\alpha}^L)} \\ u_2^{''} = 1 - \frac{\lambda_{1,\alpha}^L\mu_{1,\alpha}^L}{(\lambda_{1,\alpha}^L+\lambda_{2,\alpha}^L)(\lambda_{2,\alpha}^L+\mu_{1,\alpha}^L)}. \end{cases}$$

Proof Based on $X_i^{'}$, $X_i(\theta_{i,\alpha})$, $X_i^{''}$ and $Y_i^{'}$, $Y_i(\vartheta_{i,\alpha})$, $Y_i^{''}$ in the proof of Theorem 6.2.1, we can construct three stochastic repairable parallel systems with two nonidentical components:

(1) Parallel system 1: let $X_i^{'}$ and $Y_i^{'}$ be the lifetime and the repair time of component i, $X_i^{'} \sim \exp(\lambda_{i,\alpha}^U)$, $Y_i^{'} \sim \exp(\mu_{i,\alpha}^U)$, $i = 1, 2$.
(2) Parallel system 2: let $X_i(\theta_{i,\alpha})$ and $Y_i(\vartheta_{i,\alpha})$ be the lifetime and the repair time of component i, $X_i(\theta_{i,\alpha}) \sim \exp(\lambda_i(\theta_{i,\alpha}))$, $Y_i(\vartheta_{i,\alpha}) \sim \exp(\mu_i(\vartheta_{i,\alpha}))$, $i = 1, 2$.
(3) Parallel system 3: let $X_i^{''}$ and $Y_i^{''}$ be the lifetime and the repair time of component i, $X_i^{''} \sim \exp(\lambda_{i,\alpha}^L)$, $Y_i^{''} \sim \exp(\mu_{i,\alpha}^L)$, $i = 1, 2$.

The steady-state failure frequency of parallel systems (1), (2) and (3) is denoted by M_1, M_2 and M_3, respectively. Since the parallel system is a coherent system, we have

$$M_3 \le M_2 \le M_1.$$

Because $\theta_{i,\alpha}$ and $\vartheta_{i,\alpha}$ are arbitrary points in $A_{i,\alpha}$ and $B_{i,\alpha}$, $i = 1, 2$, from the result in stochastic case, we have

$$\left(\lim_{t\to+\infty}\frac{E[N(t)(\theta)]}{t}\right)_\alpha^L = M_3 = \frac{1}{\sum_{j=0}^2 n_j^{''}u_j^{''}}\left(\frac{u_1^{''}\lambda_{2,\alpha}^L}{\lambda_{2,\alpha}^L+u_{1,\alpha}^L} + \frac{u_2^{''}\lambda_{1,\alpha}^L}{\lambda_{1,\alpha}^L+u_{2,\alpha}^L}\right) \tag{6.2.5}$$

and

$$\left(\lim_{t\to+\infty}\frac{E[N(t)(\theta)]}{t}\right)_\alpha^U = M_1 = \frac{1}{\sum_{j=0}^2 n_j^{'}u_j^{'}}\left(\frac{u_1^{'}\lambda_{2,\alpha}^U}{\lambda_{2,\alpha}^U+u_{1,\alpha}^U} + \frac{u_2^{'}\lambda_{1,\alpha}^U}{\lambda_{1,\alpha}^U+u_{2,\alpha}^U}\right), \tag{6.2.6}$$

in which

$$\begin{cases} n_0^{''} = \frac{1}{\lambda_{1,\alpha}^L+\lambda_{2,\alpha}^L} \\ n_1^{''} = \frac{1}{\mu_{1,\alpha}^L} \\ n_2^{''} = \frac{1}{\mu_{2,\alpha}^L}, \end{cases} \quad \begin{cases} u_0^{''} = 1 - \frac{\lambda_{1,\alpha}^L\lambda_{2,\alpha}^L}{(\lambda_{2,\alpha}^L+\mu_{1,\alpha}^U)(\lambda_{1,\alpha}^L+\mu_{2,\alpha}^U)} \\ u_1^{''} = 1 - \frac{\lambda_{2,\alpha}^L\mu_{2,\alpha}^L}{(\lambda_{1,\alpha}^L+\lambda_{2,\alpha}^L)(\lambda_{1,\alpha}^L+\mu_{2,\alpha}^L)} \\ u_2^{''} = 1 - \frac{\lambda_{1,\alpha}^L\mu_{1,\alpha}^L}{(\lambda_{1,\alpha}^L+\lambda_{2,\alpha}^L)(\lambda_{2,\alpha}^L+\mu_{1,\alpha}^L)} \end{cases}$$

and

$$\begin{cases} n_0' = \frac{1}{\lambda_{1,\alpha}^U + \lambda_{2,\alpha}^U} \\ n_1' = \frac{1}{\mu_{1,\alpha}^U} \\ n_2' = \frac{1}{\mu_{2,\alpha}^U}, \end{cases} \quad \begin{cases} u_0' = 1 - \frac{\lambda_{1,\alpha}^U \lambda_{2,\alpha}^U}{(\lambda_{2,\alpha}^U + \mu_{1,\alpha}^U)(\lambda_{1,\alpha}^U + \mu_{2,\alpha}^U)} \\ u_1' = 1 - \frac{\lambda_{2,\alpha}^U \mu_{2,\alpha}^U}{(\lambda_{1,\alpha}^U + \lambda_{2,\alpha}^U)(\lambda_{1,\alpha}^U + \mu_{2,\alpha}^U)} \\ u_2' = 1 - \frac{\lambda_{1,\alpha}^U \mu_{1,\alpha}^U}{(\lambda_{1,\alpha}^U + \lambda_{2,\alpha}^U)(\lambda_{2,\alpha}^U + \mu_{1,\alpha}^U)}. \end{cases}$$

By Lemma 6.2, Eqs. (6.2.5) and (6.2.6), we can arrive at

$$M = \frac{1}{2}\int_0^1 \left\{ \left(\lim_{t\to+\infty} \frac{E[N(t)(\theta)]}{t} \right)_\alpha^L + \left(\lim_{t\to+\infty} \frac{E[N(t)(\theta)]}{t} \right)_\alpha^U \right\} d\alpha$$

$$= \frac{1}{2}\int_0^1 \left[\frac{1}{\sum_{j=0}^2 n_j' u_j'} \left(\frac{u_1' \lambda_{2,\alpha}^U}{\lambda_{2,\alpha}^U + u_{1,\alpha}^U} + \frac{u_2' \lambda_{1,\alpha}^U}{\lambda_{1,\alpha}^U + u_{2,\alpha}^U} \right) \right.$$

$$\left. + \frac{1}{\sum_{j=0}^2 n_j'' u_j''} \left(\frac{u_1'' \lambda_{2,\alpha}^L}{\lambda_{2,\alpha}^L + u_{1,\alpha}^L} + \frac{u_2'' \lambda_{1,\alpha}^L}{\lambda_{1,\alpha}^L + u_{2,\alpha}^L} \right) \right] d\alpha,$$

in which

$$\begin{cases} n_0' = \frac{1}{\lambda_{1,\alpha}^U + \lambda_{2,\alpha}^U} \\ n_1' = \frac{1}{\mu_{1,\alpha}^U} \\ n_2' = \frac{1}{\mu_{2,\alpha}^U}, \end{cases} \quad \begin{cases} u_0' = 1 - \frac{\lambda_{1,\alpha}^U \lambda_{2,\alpha}^U}{(\lambda_{2,\alpha}^U + \mu_{1,\alpha}^U)(\lambda_{1,\alpha}^U + \mu_{2,\alpha}^U)} \\ u_1' = 1 - \frac{\lambda_{2,\alpha}^U \mu_{2,\alpha}^U}{(\lambda_{1,\alpha}^U + \lambda_{2,\alpha}^U)(\lambda_{1,\alpha}^U + \mu_{2,\alpha}^U)} \\ u_2' = 1 - \frac{\lambda_{1,\alpha}^U \mu_{1,\alpha}^U}{(\lambda_{1,\alpha}^U + \lambda_{2,\alpha}^U)(\lambda_{2,\alpha}^U + \mu_{1,\alpha}^U)} \end{cases}$$

and

$$\begin{cases} n_0'' = \frac{1}{\lambda_{1,\alpha}^L + \lambda_{2,\alpha}^L} \\ n_1'' = \frac{1}{\mu_{1,\alpha}^L} \\ n_2'' = \frac{1}{\mu_{2,\alpha}^L}, \end{cases} \quad \begin{cases} u_0'' = 1 - \frac{\lambda_{1,\alpha}^L \lambda_{2,\alpha}^L}{(\lambda_{2,\alpha}^L + \mu_{1,\alpha}^U)(\lambda_{1,\alpha}^L + \mu_{2,\alpha}^U)} \\ u_1'' = 1 - \frac{\lambda_{2,\alpha}^L \mu_{2,\alpha}^L}{(\lambda_{1,\alpha}^L + \lambda_{2,\alpha}^L)(\lambda_{1,\alpha}^L + \mu_{2,\alpha}^L)} \\ u_2'' = 1 - \frac{\lambda_{1,\alpha}^L \mu_{1,\alpha}^L}{(\lambda_{1,\alpha}^L + \lambda_{2,\alpha}^L)(\lambda_{2,\alpha}^L + \mu_{1,\alpha}^L)}. \end{cases}$$

The theorem is proved.

Remark 6.2.2 If X_i and Y_i, $i = 1, 2$ degenerate to random variables, then the result in Theorem 6.2.2 degenerates to the form

$$M = \frac{1}{\sum_{j=0}^2 n_j u_j} \left(\frac{u_1 \lambda_2}{\lambda_2 + \mu_1} + \frac{u_2 \lambda_1}{\lambda_1 + \mu_2} \right),$$

in which

$$\begin{cases} n_0 = \frac{1}{\lambda_1+\lambda_2} \\ n_1 = \frac{1}{\mu_1} \\ n_2 = \frac{1}{\mu_2}, \end{cases} \quad \begin{cases} u_0 = 1 - \frac{\lambda_1\lambda_2}{(\lambda_2+\mu_1)(\lambda_1+\mu_2)} \\ u_1 = 1 - \frac{\lambda_2\mu_2}{(\lambda_1+\lambda_2)(\lambda_1+\mu_2)} \\ u_2 = 1 - \frac{\lambda_1\mu_1}{(\lambda_1+\lambda_2)(\lambda_1+\mu_1)}, \end{cases}$$

which is consistent with the result in stochastic case.

Example 6.2.1 Suppose that the repairable parallel system composed by two nonidentical components. The lifetime and the repair time of component i are denoted by X_i and $Y_i, i = 1, 2$, respectively. If $X_i \sim \mathcal{EXP}(\lambda_i)$ and $Y_i \sim \mathcal{EXP}(\mu_i), i = 1, 2$, where $\lambda_1 = (0.5, 1.5, 2.5), \lambda_2 = (1, 2, 3), \mu_1 = (2, 3, 4)$ and $\mu_2 = (3, 4, 5)$. Then we can arrive at

$$\begin{cases} \lambda_{1,\alpha}^L = 0.5 + \alpha \\ \lambda_{1,\alpha}^U = 2.5 - \alpha, \end{cases} \begin{cases} \lambda_{2,\alpha}^L = 1 + \alpha \\ \lambda_{2,\alpha}^U = 3 - \alpha, \end{cases} \begin{cases} \mu_{1,\alpha}^L = 2 + \alpha \\ \mu_{1,\alpha}^U = 4 - \alpha, \end{cases} \begin{cases} \mu_{2,\alpha}^L = 3 + \alpha \\ \mu_{2,\alpha}^U = 5 - \alpha. \end{cases}$$

To obtain the limiting availability, we should first calculate the following formula by Theorem 6.2.1,

$$\begin{cases} m_0^{'} = \frac{1}{\lambda_{1,\alpha}^U+\lambda_{2,\alpha}^U} = \frac{1}{5.5-2\alpha} \\ m_1^{'} = \frac{1}{\mu_{1,\alpha}^L} = \frac{1}{2+\alpha} \\ m_2^{'} = \frac{1}{\mu_{2,\alpha}^L} = \frac{1}{3+\alpha}, \end{cases} \quad \begin{cases} m_0^{''} = \frac{1}{\lambda_{1,\alpha}^L+\lambda_{2,\alpha}^L} = \frac{1}{1.5+2\alpha} \\ m_1^{''} = \frac{1}{\mu_{1,\alpha}^U} = \frac{1}{4-\alpha} \\ m_2^{''} = \frac{1}{\mu_{2,\alpha}^U} = \frac{1}{5-\alpha}, \end{cases}$$

$$\begin{cases} v_0^{'} = 1 - \dfrac{\lambda_{1,\alpha}^U\lambda_{2,\alpha}^U}{(\lambda_{2,\alpha}^U + \mu_{1,\alpha}^L)(\lambda_{1,\alpha}^U + \mu_{2,\alpha}^L)} = 1 - \dfrac{(2.5-\alpha)(3-\alpha)}{27.5} \\ v_1^{'} = 1 - \dfrac{\lambda_{2,\alpha}^U\mu_{2,\alpha}^L}{(\lambda_{1,\alpha}^U + \lambda_{2,\alpha}^U)(\lambda_{1,\alpha}^U + \mu_{2,\alpha}^L)} = 1 - \dfrac{(3-\alpha)(3+\alpha)}{5.5(5.5-2\alpha)} \\ v_2^{'} = 1 - \dfrac{\lambda_{1,\alpha}^U\mu_{1,\alpha}^L}{(\lambda_{1,\alpha}^U + \lambda_{2,\alpha}^U)(\lambda_{2,\alpha}^U + \mu_{1,\alpha}^L)} = 1 - \dfrac{(2.5-\alpha)(2+\alpha)}{5(5.5-2\alpha)}, \end{cases}$$

$$\begin{cases} v_0^{''} = 1 - \dfrac{\lambda_{1,\alpha}^L\lambda_{2,\alpha}^L}{(\lambda_{2,\alpha}^L + \mu_{1,\alpha}^U)(\lambda_{1,\alpha}^L + \mu_{2,\alpha}^U)} = 1 - \dfrac{(0.5+\alpha)(1+\alpha)}{27.5} \\ v_1^{''} = 1 - \dfrac{\lambda_{2,\alpha}^L\mu_{2,\alpha}^U}{(\lambda_{1,\alpha}^L + \lambda_{2,\alpha}^L)(\lambda_{1,\alpha}^L + \mu_{2,\alpha}^L)} = 1 - \dfrac{(1+\alpha)(5-\alpha)}{5.5(1.5+2\alpha)} \\ v_2^{''} = 1 - \dfrac{\lambda_{1,\alpha}^L\mu_{1,\alpha}^U}{(\lambda_{1,\alpha}^L + \lambda_{2,\alpha}^L)(\lambda_{2,\alpha}^L + \mu_{1,\alpha}^U)} = 1 - \dfrac{(0.5+\alpha)(4-\alpha)}{5(1.5+2\alpha)}. \end{cases}$$

The limiting availability of the repairable parallel system is

$$A=\frac{1}{2}\int_0^1\left[\frac{1}{\sum_{j=0}^{2}m_j^{'}v_j^{'}}\left(\frac{v_0^{'}}{\lambda_{1,\alpha}^U+\lambda_{2,\alpha}^U}+\frac{v_1^{'}}{\lambda_{2,\alpha}^U+\mu_{1,\alpha}^L}+\frac{v_2^{'}}{\lambda_{1,\alpha}^U+\mu_{2,\alpha}^L}\right)\right.$$
$$\left.+\frac{1}{\sum_{j=0}^{2}m_j^{''}v_j^{''}}\left(\frac{v_0^{''}}{\lambda_{1,\alpha}^L+\lambda_{2,\alpha}^L}+\frac{v_1^{''}}{\lambda_{2,\alpha}^L+\mu_{1,\alpha}^U}+\frac{v_2^{''}}{\lambda_{1,\alpha}^L+\mu_{2,\alpha}^U}\right)\right]\mathrm{d}\alpha$$
$$\approx 0.808326.$$

To obtain steady-state failure frequency, we should first calculate the following formula,

$$\begin{cases} n_0^{'}=\frac{1}{\lambda_{1,\alpha}^U+\lambda_{2,\alpha}^U}=\frac{1}{5.5-2\alpha}\\ n_1^{'}=\frac{1}{\mu_{1,\alpha}^U}=\frac{1}{4-\alpha}\\ n_2^{'}=\frac{1}{\mu_{2,\alpha}^U}=\frac{1}{5-\alpha},\end{cases}\qquad \begin{cases} n_0^{''}=\frac{1}{\lambda_{1,\alpha}^L+\lambda_{2,\alpha}^L}=\frac{1}{1.5+2\alpha}\\ n_1^{''}=\frac{1}{\mu_{1,\alpha}^L}=\frac{1}{2+\alpha}\\ n_2^{''}=\frac{1}{\mu_{2,\alpha}^L}=\frac{1}{3+\alpha},\end{cases}$$

$$\begin{cases} u_0^{'}=1-\frac{\lambda_{1,\alpha}^U\lambda_{2,\alpha}^U}{(\lambda_{2,\alpha}^U+\mu_{1,\alpha}^U)(\lambda_{1,\alpha}^U+\mu_{2,\alpha}^U)}=1-\frac{(2.5-\alpha)(3-\alpha)}{(7-2\alpha)(7.5-2\alpha)}\\ u_1^{'}=1-\frac{\lambda_{2,\alpha}^U\mu_{2,\alpha}^U}{(\lambda_{1,\alpha}^U+\lambda_{2,\alpha}^U)(\lambda_{1,\alpha}^U+\mu_{2,\alpha}^U)}=1-\frac{(3-\alpha)(5-\alpha)}{(5.5-2\alpha)(7.5-2\alpha)}\\ u_2^{'}=1-\frac{\lambda_{1,\alpha}^U\mu_{1,\alpha}^U}{(\lambda_{1,\alpha}^U+\lambda_{2,\alpha}^U)(\lambda_{2,\alpha}^U+\mu_{1,\alpha}^U)}=1-\frac{(2.5-\alpha)(4-\alpha)}{(5.5-2\alpha)(7-2\alpha)},\end{cases}$$

$$\begin{cases} u_0^{''}=1-\frac{\lambda_{1,\alpha}^L\lambda_{2,\alpha}^L}{(\lambda_{2,\alpha}^L+\mu_{1,\alpha}^L)(\lambda_{1,\alpha}^L+\mu_{2,\alpha}^L)}=1-\frac{(0.5+\alpha)(1+\alpha)}{(3+2\alpha)(3.5+2\alpha)}\\ u_1^{''}=1-\frac{\lambda_{2,\alpha}^L\mu_{2,\alpha}^U}{(\lambda_{1,\alpha}^L+\lambda_{2,\alpha}^L)(\lambda_{1,\alpha}^L+\mu_{2,\alpha}^L)}=1-\frac{(1+\alpha)(3+\alpha)}{(1.5+2\alpha)(3.5+2\alpha)}\\ u_2^{''}=1-\frac{\lambda_{1,\alpha}^L\mu_{1,\alpha}^L}{(\lambda_{1,\alpha}^L+\lambda_{2,\alpha}^L)(\lambda_{2,\alpha}^L+\mu_{1,\alpha}^L)}=1-\frac{(0.5+\alpha)(2+\alpha)}{(1.5+2\alpha)(3+2\alpha)}.\end{cases}$$

Then the steady-state failure frequency is

$$M=\frac{1}{2}\int_0^1\left[\frac{1}{\sum_{j=0}^{2}n_j^{'}u_j^{'}}\left(\frac{u_1^{'}\lambda_{2,\alpha}^U}{\lambda_{2,\alpha}^U+u_{1,\alpha}^U}+\frac{u_2^{'}\lambda_{1,\alpha}^U}{\lambda_{1,\alpha}^U+u_{2,\alpha}^U}\right)\right.$$
$$\left.+\frac{1}{\sum_{j=0}^{2}n_j^{''}u_j^{''}}\left(\frac{u_1^{''}\lambda_{2,\alpha}^L}{\lambda_{2,\alpha}^L+u_{1,\alpha}^L}+\frac{u_2^{''}\lambda_{1,\alpha}^L}{\lambda_{1,\alpha}^L+u_{2,\alpha}^L}\right)\right]\mathrm{d}\alpha$$
$$\approx 0.683212.$$

6.3 Cold Standby Systems

In this section, the repairable cold standby systems with two identical components and two nonidentical components are considered, respectively. Initially, one component is in operating and the other one is in cold standby. The standby component is active instantaneously when the operating component fails, and then the failed component is repaired immediately. We assume the conversion switch is completely reliable and the transition is instantaneous. The component works as a new one after repair. If the operating component fails during the repair process of another component, the cold standby system fails.

6.3.1 The Two Identical Components Case

Consider a cold standby system consisting of two identical components. Let X be the lifetime of the component in operating and has a random fuzzy exponential distribution with fuzzy parameter λ defined on the credibility space $(\Theta, \mathcal{P}(\Theta), \mathrm{Cr})$, and Y be the repair time of the failed component and has a random fuzzy exponential distribution with fuzzy parameter μ defined on the credibility space $(\tau, \mathcal{P}(\tau), \mathrm{Cr}^{'})$. We also assume X and Y are mutually independent.

Theorem 6.3.1 *The limiting availability of the repairable cold standby system is*

$$A = E\left[\frac{2}{2 + \frac{\lambda^2}{\mu(\mu+\lambda)}}\right].$$

Proof Let $P_\alpha = \{\theta \in \Theta | \mu(\theta) \geq \alpha\}$, $Q_\alpha = \{\vartheta \in \tau | \mu(\vartheta) \geq \alpha\}$. Since the α-pessimistic values and α-optimistic values of fuzzy variables $E[X(\theta)]$, $\theta \in \Theta$ and $E[Y(\vartheta)]$, $\vartheta \in \tau$ are continuous almost everywhere with respect to α, $\alpha \in (0, 1]$, then there at least exist points $\theta^1, \theta^2 \in P_\alpha$, $\vartheta^1, \vartheta^2 \in Q_\alpha$ such that

$$E[X(\theta^1)] = E[X(\theta)]_\alpha^L, \quad E[X(\theta^2)] = E[X(\theta)]_\alpha^U$$

and

$$E[Y(\vartheta^1)] = E[Y(\vartheta)]_\alpha^L, \quad E[Y(\vartheta^2)] = E[Y(\vartheta)]_\alpha^U.$$

Then for $\forall \theta \in P_\alpha$ and $\forall \vartheta \in Q_\alpha$, we have

$$E[X(\theta^1)] \leq E[X(\theta)] \leq E[X(\theta^2)] \tag{6.3.1}$$

and

$$E[Y(\vartheta^1)] \le E[Y(\vartheta)] \le E[Y(\vartheta^2)]. \tag{6.3.2}$$

Obviously, $X(\theta^1)$, $X(\theta)$, $X(\theta^2)$ and $Y(\vartheta^1)$, $Y(\vartheta)$, $Y(\vartheta^2)$ are exponentially distributed random variables. It is easy to calculate their expected values. Therefore, Eqs. (6.3.1) and (6.3.2) can be expressed by

$$\frac{1}{\lambda(\theta^1)} \le \frac{1}{\lambda(\theta)} \le \frac{1}{\lambda(\theta^2)}$$

and

$$\frac{1}{\mu(\vartheta^1)} \le \frac{1}{\mu(\vartheta)} \le \frac{1}{\mu(\vartheta^2)}.$$

That is

$$\lambda(\theta^2) \le \lambda(\theta) \le \lambda(\theta^1) \tag{6.3.3}$$

and

$$\mu(\vartheta^2) \le \mu(\vartheta) \le \mu(\vartheta^1). \tag{6.3.4}$$

Based on random variables $X(\theta^1)$, $X(\theta)$, $X(\theta^2)$ and $Y(\vartheta^1)$, $Y(\vartheta)$, $Y(\vartheta^2)$, we can construct three stochastic repairable cold standby systems:

(1) Cold standby system 1: the lifetime and the repair time of each component are $X(\theta^1)$ and $Y(\vartheta^2)$, respectively.
(2) Cold standby system 2: the lifetime and the repair time of each component are $X(\theta^2)$ and $Y(\vartheta^1)$, respectively.
(3) Cold standby system 3: the lifetime and the repair time of each component are $X(\theta)$ and $Y(\vartheta)$, respectively.

The limiting availability of cold standby systems (1), (2) and (3) is denoted by A_1, A_2 and A_3, respectively. From the result in traditional reliability theory, we have

$$A_1 = \frac{2}{2 + \frac{\lambda^2(\theta^1)}{\mu(\vartheta^2)(\mu(\vartheta^2)+\lambda(\theta^1))}},$$

$$A_2 = \frac{2}{2 + \frac{\lambda^2(\theta^2)}{\mu(\vartheta^1)(\mu(\vartheta^1)+\lambda(\theta^2))}}$$

and

$$A_3 = \frac{2}{2 + \frac{\lambda^2(\theta)}{\mu(\vartheta)(\mu(\vartheta)+\lambda(\theta))}}.$$

By Eqs. (6.3.3) and (6.3.4), we can arrive at

$$A_1 \leq A_3 \leq A_2.$$

Since θ and ϑ are arbitrary points in P_α and Q_α, from the result in stochastic case, we have

$$\left(\lim_{t\to+\infty} \Pr\{S(t)(\theta)=1\}\right)_\alpha^L = A_1 = \frac{2}{2+\frac{\lambda^2(\theta^1)}{\mu(\vartheta^2)(\mu(\vartheta^2)+\lambda(\theta^1))}} \tag{6.3.5}$$

and

$$\left(\lim_{t\to+\infty} \Pr\{S(t)(\theta)=1\}\right)_\alpha^U = A_2 = \frac{2}{2+\frac{\lambda^2(\theta^2)}{\mu(\vartheta^1)(\mu(\vartheta^1)+\lambda(\theta^2))}}. \tag{6.3.6}$$

By Eqs. (6.3.3) and (6.3.4), we can arrive at

$$\lambda_\alpha^L = \lambda(\theta^2), \tag{6.3.7}$$

$$\lambda_\alpha^U = \lambda(\theta^1), \tag{6.3.8}$$

$$\mu_\alpha^L = \mu(\vartheta^2), \tag{6.3.9}$$

$$\mu_\alpha^U = \mu(\vartheta^1). \tag{6.3.10}$$

By Lemma 6.1 and Eqs. (6.3.5)–(6.3.10), we can arrive at

$$\begin{aligned}
A &= \frac{1}{2}\int_0^1 \left[\left(\lim_{t\to+\infty} \Pr\{S(t)(\theta)=1\}\right)_\alpha^L + \left(\lim_{t\to+\infty} \Pr\{S(t)(\theta)=1\}\right)_\alpha^U\right] \mathrm{d}\alpha \\
&= \frac{1}{2}\int_0^1 \left[\frac{2}{2+\frac{\lambda^2(\theta^1)}{\mu(\vartheta^2)(\mu(\vartheta^2)+\lambda(\theta^1))}} + \frac{2}{2+\frac{\lambda^2(\theta^2)}{\mu(\vartheta^1)(\mu(\vartheta^1)+\lambda(\theta^2))}}\right] \mathrm{d}\alpha \\
&= \frac{1}{2}\int_0^1 \left[\frac{2}{2+\frac{(\lambda_\alpha^U)^2}{\mu_\alpha^L(\mu_\alpha^L+\lambda_\alpha^U)}} + \frac{2}{2+\frac{(\lambda_\alpha^L)^2}{\mu_\alpha^U(\mu_\alpha^U+\lambda_\alpha^L)}}\right] \mathrm{d}\alpha \\
&= \frac{1}{2}\int_0^1 \left\{\left[\frac{2}{2+\frac{\lambda^2}{\mu(\mu+\lambda)}}\right]_\alpha^L + \left[\frac{2}{2+\frac{\lambda^2}{\mu(\mu+\lambda)}}\right]_\alpha^U\right\} \mathrm{d}\alpha \\
&= E\left[\frac{2}{2+\frac{\lambda^2}{\mu(\mu+\lambda)}}\right].
\end{aligned}$$

The theorem is proved.

Remark 6.3.1 If X and Y degenerate to random variables, then the result in Theorem 6.3.1 degenerates to the form

$$A = \frac{2}{2 + \frac{\lambda^2}{\mu(\mu+\lambda)}},$$

which is consistent with the result in stochastic case.

Theorem 6.3.2 *The steady-state failure frequency of the repairable cold standby system is*

$$M = \frac{1}{2}\int_0^1 \left[\frac{2\mu_\alpha^L(\lambda_\alpha^L)^2}{2(\mu_\alpha^L)^2 + 2\mu_\alpha^L\lambda_\alpha^L + (\lambda_\alpha^L)^2} + \frac{2\mu_\alpha^U(\lambda_\alpha^U)^2}{2(\mu_\alpha^U)^2 + 2\mu_\alpha^U\lambda_\alpha^U + (\lambda_\alpha^U)^2}\right] d\alpha.$$

Proof Based on $X(\theta^1)$, $X(\theta)$, $X(\theta^2)$ and $Y(\vartheta^1)$, $Y(\vartheta)$,$Y(\vartheta^2)$, we can construct three stochastic repairable cold standby systems:

(1) Cold standby system 1: the lifetime and the repair time of each component are $X(\theta^1)$ and $Y(\vartheta^1)$, respectively.
(2) Cold standby system 2: the lifetime and the repair time of each component are $X(\theta^2)$ and $Y(\vartheta^2)$, respectively.
(3) Cold standby system 3: the lifetime and the repair time of each component are $X(\theta)$ and $Y(\vartheta)$, respectively.

The steady-state failure frequency of cold standby systems (1), (2) and (3) is denoted by M_1, M_2 and M_3, respectively. Since the cold standby repairable system is a coherent system, then $M_2 \le M_3 \le M_1$. Because θ and ϑ are arbitrary points in P_α and Q_α, respectively, from the result in traditional reliability theory, we have

$$\left(\lim_{t\to+\infty}\frac{E[N(t)(\theta)]}{t}\right)_\alpha^L = M_2 = \frac{2\mu(\vartheta^2)\lambda^2(\theta^2)}{2\mu^2(\vartheta^2) + 2\mu(\vartheta^2)\lambda(\theta^2) + \lambda^2(\theta^2)} \tag{6.3.11}$$

and

$$\left(\lim_{t\to+\infty}\frac{E[N(t)(\theta)]}{t}\right)_\alpha^U = M_1 = \frac{2\mu(\vartheta^1)\lambda^2(\theta^1)}{2\mu^2(\vartheta^1) + 2\mu(\vartheta^1)\lambda(\theta^1) + \lambda^2(\theta^1)}. \tag{6.3.12}$$

By Lemma 6.2 and Eqs. (6.3.7)–(6.3.12), we can arrive at

$$\begin{aligned} M &= \frac{1}{2}\int_0^1 \left\{\left(\lim_{t\to+\infty}\frac{E[N(t)(\theta)]}{t}\right)_\alpha^L + \left(\lim_{t\to+\infty}\frac{E[N(t)(\theta)]}{t}\right)_\alpha^U\right\} d\alpha \\ &= \frac{1}{2}\int_0^1 \left[\frac{2\mu(\vartheta^1)\lambda^2(\theta^1)}{2\mu^2(\vartheta^1) + 2\mu(\vartheta^1)\lambda(\theta^1) + \lambda^2(\theta^1)}\right. \end{aligned}$$

$$
+ \frac{2\mu(\vartheta^2)\lambda^2(\theta^2)}{2\mu^2(\vartheta^2) + 2\mu(\vartheta^2)\lambda(\theta^2) + \lambda^2(\theta^2)}\Bigg] \mathrm{d}\alpha
$$

$$
= \frac{1}{2}\int_0^1 \left[\frac{2\mu_\alpha^L(\lambda_\alpha^L)^2}{2(\mu_\alpha^L)^2 + 2\mu_\alpha^L\lambda_\alpha^L + (\lambda_\alpha^L)^2} + \frac{2\mu_\alpha^U(\lambda_\alpha^U)^2}{2(\mu_\alpha^U)^2 + 2\mu_\alpha^U\lambda_\alpha^U + (\lambda_\alpha^U)^2}\right] \mathrm{d}\alpha.
$$

The theorem is proved.

Remark 6.3.2 If X and Y degenerate to random variables, then the result in Theorem 6.3.2 degenerates to the form

$$
M = \frac{2\mu\lambda^2}{2\mu^2 + 2\mu\lambda + \lambda^2},
$$

which is consistent with the result in stochastic case.

Example 6.3.1 Suppose the cold standby system composed by two identical components. Let X and Y be the lifetime and the repair time of each component, respectively. If $X \sim \mathcal{EXP}(\lambda)$ and $Y \sim \mathcal{EXP}(\mu)$, where $\lambda = (2, 3, 4)$ and $\mu = (4, 5, 6)$. We should first calculate the following formula

$$
\lambda_\alpha^L = 2 + \alpha, \quad \lambda_\alpha^U = 4 - \alpha, \quad \mu_\alpha^L = 4 + \alpha, \quad \mu_\alpha^U = 6 - \alpha.
$$

By Theorem 6.3.1, the limiting availability of the cold standby system is

$$
\begin{aligned}
A &= \frac{1}{2}\int_0^1 \left[\frac{2}{2 + \frac{(\lambda_\alpha^U)^2}{\mu_\alpha^L(\mu_\alpha^L + \lambda_\alpha^U)}} + \frac{2}{2 + \frac{(\lambda_\alpha^L)^2}{\mu_\alpha^U(\mu_\alpha^U + \lambda_\alpha^L)}}\right] \mathrm{d}\alpha \\
&= \frac{1}{2}\int_0^1 \left[\frac{2}{2 + \frac{(4-\alpha)^2}{(4+\alpha)(4+\alpha+4-\alpha)}} + \frac{2}{2 + \frac{(2+\alpha)^2}{(6-\alpha)(6-\alpha+2+\alpha)}}\right] \mathrm{d}\alpha \\
&= 4\ln\frac{5}{4} \approx 0.8926.
\end{aligned}
$$

By Theorem 6.3.2, the steady-state failure frequency of the cold standby system is

$$
\begin{aligned}
M &= \frac{1}{2}\int_0^1 \left[\frac{2\mu_\alpha^L(\lambda_\alpha^L)^2}{2(\mu_\alpha^L)^2 + 2\mu_\alpha^L\lambda_\alpha^L + (\lambda_\alpha^L)^2} + \frac{2\mu_\alpha^U(\lambda_\alpha^U)^2}{2(\mu_\alpha^U)^2 + 2\mu_\alpha^U\lambda_\alpha^U + (\lambda_\alpha^U)^2}\right] \mathrm{d}\alpha \\
&= \frac{1}{2}\int_0^1 \Bigg[\frac{2(4+\alpha)(2+\alpha)^2}{2(4+\alpha)^2 + 2(4+\alpha)(2+\alpha) + (2+\alpha)^2}
\end{aligned}
$$

$$+ \frac{2(6-\alpha)(4-\alpha)^2}{2(6-\alpha)^2+2(6-\alpha)(4-\alpha)+(4-\alpha)^2}\Bigg]\mathrm{d}\alpha$$
$$= 1.0119.$$

6.3.2 The Two Nonidentical Components Case

Consider a cold standby system consisting of two nonidentical components. We assume X_i is the lifetime of component i and has a random fuzzy exponential distribution with fuzzy parameter λ_i on the credibility space $(\Theta_i, \mathcal{P}(\Theta_i), \mathrm{Cr}_i)$, and Y_i is the repair time of component i and has a random fuzzy exponential distribution with fuzzy parameter μ_i on the credibility space $(\tau_i, \mathcal{P}(\tau_i), \mathrm{Cr}_i')$, $i = 1, 2$. We also assume X_i and Y_i, $i = 1, 2$ are independent.

Theorem 6.3.3 *The limiting availability of the repairable cold standby system is*

$$A = E\left[\frac{\frac{1}{\lambda_1}+\frac{1}{\lambda_2}}{\frac{1}{\lambda_1}+\frac{1}{\lambda_2}+\frac{\lambda_1}{\mu_2(\lambda_1+\mu_2)}+\frac{\lambda_2}{\mu_1(\lambda_2+\mu_1)}}\right].$$

Proof Let $P_{i,\alpha} = \{\theta_i \in \Theta_i | \mu(\theta_i) \geq \alpha\}$, $Q_{i,\alpha} = \{\vartheta \in \tau_i | \mu(\vartheta_i) \geq \alpha\}$, $i = 1, 2$. Since the α-pessimistic values and α-optimistic values of fuzzy variables $E[X_i(\theta_i)]$, $E[Y_i(\vartheta_i)]$, $\theta_i \in \Theta_i$, $\vartheta_i \in \tau_i$, $i = 1, 2$ are continuous almost everywhere with respect to α, $\alpha \in (0, 1]$, then there at least exist points $\theta_i^1, \theta_i^2 \in P_{i,\alpha}$, $\vartheta_i^1, \vartheta_i^2 \in Q_{i,\alpha}$, $i = 1, 2$ such that

$$E[X_i(\theta_i^1)] = E[X_i(\theta_i)]_\alpha^L, \quad E[X_i(\theta_i^2)] = E[X_i(\theta_i)]_\alpha^U$$

and

$$E[Y_i(\vartheta_i^1)] = E[Y_i(\vartheta_i)]_\alpha^L, \quad E[Y_i(\vartheta_i^2)] = E[Y_i(\vartheta_i)]_\alpha^U.$$

Then for $\forall\theta_i \in P_{i,\alpha}$, $\forall\vartheta_i \in Q_{i,\alpha}$, $i = 1, 2$, we have

$$E[X_i(\theta_i^1)] \leq E[X_i(\theta_i)] \leq E[X_i(\theta_i^2)] \tag{6.3.13}$$

and

$$E[Y_i(\vartheta_i^1)] \leq E[Y_i(\vartheta_i)] \leq E[Y_i(\vartheta_i^2)]. \tag{6.3.14}$$

It is easy to see that $X_i(\theta_i^1)$, $X_i(\theta_i)$, $X_i(\theta_i^2)$ and $Y_i(\vartheta_i^1)$, $Y_i(\vartheta_i)$, $Y_i(\vartheta_i^2)$, $i = 1, 2$ are exponentially distributed random variables. Then we can calculate their expected values. By Eqs. (6.3.13) and (6.3.14), we have

$$\frac{1}{\lambda_i(\theta_i^1)} \le \frac{1}{\lambda_i(\theta_i)} \le \frac{1}{\lambda_i(\theta_i^2)}$$

and

$$\frac{1}{\mu_i(\vartheta_i^1)} \le \frac{1}{\mu_i(\vartheta_i)} \le \frac{1}{\mu_i(\vartheta_i^2)}.$$

Therefore, for $i = 1, 2$, we have

$$\lambda_i(\theta_i^2) \le \lambda_i(\theta_i) \le \lambda_i(\theta_i^1) \tag{6.3.15}$$

and

$$\mu_i(\vartheta_i^2) \le \mu_i(\vartheta_i) \le \mu_i(\vartheta_i^1). \tag{6.3.16}$$

Based on $X_i(\theta_i^1)$, $X_i(\theta_i)$, $X_i(\theta_i^2)$ and $Y_i(\vartheta_i^1)$, $Y_i(\vartheta_i)$, $Y_i(\vartheta_i^2)$, $i = 1, 2$, we can construct three stochastic repairable cold standby systems:

(1) Cold standby system 1: the lifetime and the repair time of component i are $X_i(\theta_i^1)$ and $Y_i(\vartheta_i^2)$, $i = 1, 2$, respectively.
(2) Cold standby system 2: the lifetime and the repair time of component i are $X_i(\theta_i^2)$ and $Y_i(\vartheta_i^1)$, $i = 1, 2$, respectively.
(3) Cold standby system 3: the lifetime and the repair time of component i are $X_i(\theta_i)$ and $Y_i(\vartheta_i)$, $i = 1, 2$, respectively.

The limiting availability of cold standby systems (1), (2) and (3) is denoted by A_1, A_2 and A_3, respectively. From the result in traditional reliability theory, we can arrive at

$$A_1 = \frac{\frac{1}{\lambda_1(\theta_1^1)} + \frac{1}{\lambda_2(\theta_2^1)}}{\frac{1}{\lambda_1(\theta_1^1)} + \frac{1}{\lambda_2(\theta_2^1)} + \frac{\lambda_1(\theta_1^1)}{\mu_2(\vartheta_2^2)(\lambda_1(\theta_1^1)+\mu_2(\vartheta_2^2))} + \frac{\lambda_2(\theta_2^1)}{\mu_1(\vartheta_1^2)(\lambda_2(\theta_2^1)+\mu_1(\vartheta_1^2))}},$$

$$A_2 = \frac{\frac{1}{\lambda_1(\theta_1^2)} + \frac{1}{\lambda_2(\theta_2^2)}}{\frac{1}{\lambda_1(\theta_1^2)} + \frac{1}{\lambda_2(\theta_2^2)} + \frac{\lambda_1(\theta_1^1)}{\mu_2(\vartheta_2^1)(\lambda_1(\theta_1^2)+\mu_2(\vartheta_2^1))} + \frac{\lambda_2(\theta_2^2)}{\mu_1(\vartheta_1^1)(\lambda_2(\theta_2^2)+\mu_1(\vartheta_1^1))}}$$

and

$$A_3 = \frac{\frac{1}{\lambda_1(\theta_1)} + \frac{1}{\lambda_2(\theta_2)}}{\frac{1}{\lambda_1(\theta_1)} + \frac{1}{\lambda_2(\theta_2)} + \frac{\lambda_1(\theta_1^1)}{\mu_2(\vartheta_2)(\lambda_1(\theta_1)+\mu_2(\vartheta_2))} + \frac{\lambda_2(\theta_2)}{\mu_1(\vartheta_1)(\lambda_2(\theta_2)+\mu_1(\vartheta_1))}}.$$

By Eqs. (6.3.15) and (6.3.16), we can arrive at

$$A_1 \le A_3 \le A_2.$$

Since θ_i and ϑ_i are arbitrary points in $P_{i,\alpha}$ and $Q_{i,\alpha}$, $i = 1, 2$, from the result in traditional reliability theory, we have

$$\left(\lim_{t\to+\infty} \Pr\{S(t)(\theta) = 1\}\right)_\alpha^L$$
$$= A_1 = \frac{\frac{1}{\lambda_1(\theta_1^1)} + \frac{1}{\lambda_2(\theta_2^1)}}{\frac{1}{\lambda_1(\theta_1^1)} + \frac{1}{\lambda_2(\theta_2^1)} + \frac{\lambda_1(\theta_1^1)}{\mu_2(\vartheta_2^2)(\lambda_1(\theta_1^1)+\mu_2(\vartheta_2^2))} + \frac{\lambda_2(\theta_2^1)}{\mu_1(\vartheta_1^2)(\lambda_2(\theta_2^1)+\mu_1(\vartheta_1^2))}} \tag{6.3.17}$$

and

$$\left(\lim_{t\to+\infty} \Pr\{S(t)(\theta) = 1\}\right)_\alpha^U$$
$$= A_2 = \frac{\frac{1}{\lambda_1(\theta_1^2)} + \frac{1}{\lambda_2(\theta_2^2)}}{\frac{1}{\lambda_1(\theta_1^2)} + \frac{1}{\lambda_2(\theta_2^2)} + \frac{\lambda_1(\theta_1^1)}{\mu_2(\vartheta_2^1)(\lambda_1(\theta_1^2)+\mu_2(\vartheta_2^1))} + \frac{\lambda_2(\theta_2^2)}{\mu_1(\vartheta_1^1)(\lambda_2(\theta_2^2)+\mu_1(\vartheta_1^1))}}. \tag{6.3.18}$$

By Eqs. (6.3.15) and (6.3.16), for $i = 1, 2$, we have

$$\lambda_{i,\alpha}^L = \lambda_i(\theta_i^2), \tag{6.3.19}$$

$$\lambda_{i,\alpha}^U = \lambda(\theta_i^1), \tag{6.3.20}$$

$$\mu_{i,\alpha}^L = \mu(\vartheta_i^2), \tag{6.3.21}$$

$$\mu_{i,\alpha}^U = \mu(\vartheta_i^1). \tag{6.3.22}$$

By Lemma 6.1, Eqs. (6.3.17)–(6.3.22), we can arrive at

$$A = \frac{1}{2}\int_0^1 \left[\left(\lim_{t\to+\infty} \Pr\{S(t)(\theta) = 1\}\right)_\alpha^L + \left(\lim_{t\to+\infty} \Pr\{S(t)(\theta) = 1\}\right)_\alpha^U\right] \mathrm{d}\alpha$$
$$= \frac{1}{2}\int_0^1 \left[\frac{\frac{1}{\lambda_1(\theta_1^1)} + \frac{1}{\lambda_2(\theta_2^1)}}{\frac{1}{\lambda_1(\theta_1^1)} + \frac{1}{\lambda_2(\theta_2^1)} + \frac{\lambda_1(\theta_1^1)}{\mu_2(\vartheta_2^2)(\lambda_1(\theta_1^1)+\mu_2(\vartheta_2^2))} + \frac{\lambda_2(\theta_2^1)}{\mu_1(\vartheta_1^2)(\lambda_2(\theta_2^1)+\mu_1(\vartheta_1^2))}}\right.$$
$$\left. + \frac{\frac{1}{\lambda_1(\theta_1^2)} + \frac{1}{\lambda_2(\theta_2^2)}}{\frac{1}{\lambda_1(\theta_1^2)} + \frac{1}{\lambda_2(\theta_2^2)} + \frac{\lambda_1(\theta_1^1)}{\mu_2(\vartheta_2^1)(\lambda_1(\theta_1^2)+\mu_2(\vartheta_2^1))} + \frac{\lambda_2(\theta_2^2)}{\mu_1(\vartheta_1^1)(\lambda_2(\theta_2^2)+\mu_1(\vartheta_1^1))}}\right] \mathrm{d}\alpha$$
$$= \frac{1}{2}\int_0^1 \left[\frac{\frac{1}{\lambda_{1,\alpha}^U} + \frac{1}{\lambda_{2,\alpha}^U}}{\frac{1}{\lambda_{1,\alpha}^U} + \frac{1}{\lambda_{2,\alpha}^U} + \frac{\lambda_{1,\alpha}^U}{\mu_{2,\alpha}^L(\lambda_{1,\alpha}^U+\mu_{2,\alpha}^L)} + \frac{\lambda_{2,\alpha}^U}{\mu_{1,\alpha}^L(\lambda_{2,\alpha}^U+\mu_{1,\alpha}^L)}}\right.$$

$$+\frac{\frac{1}{\lambda^L_{1,\alpha}}+\frac{1}{\lambda^L_{2,\alpha}}}{\frac{1}{\lambda^L_{1,\alpha}}+\frac{1}{\lambda^L_{2,\alpha}}+\frac{\lambda^L_{1,\alpha}}{\mu^U_{2,\alpha}(\lambda^L_{1,\alpha}+\mu^U_{2,\alpha})}+\frac{\lambda^L_{2,\alpha}}{\mu^U_{1,\alpha}(\lambda^L_{2,\alpha}+\mu^U_{1,\alpha})}}\Bigg]d\alpha$$

$$=\frac{1}{2}\int_0^1\left\{\left[\frac{\frac{1}{\lambda_1}+\frac{1}{\lambda_2}}{\frac{1}{\lambda_1}+\frac{1}{\lambda_2}+\frac{\lambda_1}{\mu_2(\lambda_1+\mu_2)}+\frac{\lambda_2}{\mu_1(\lambda_2+\mu_1)}}\right]^L_\alpha\right.$$

$$\left.+\left[\frac{\frac{1}{\lambda_1}+\frac{1}{\lambda_2}}{\frac{1}{\lambda_1}+\frac{1}{\lambda_2}+\frac{\lambda_1}{\mu_2(\lambda_1+\mu_2)}+\frac{\lambda_2}{\mu_1(\lambda_2+\mu_1)}}\right]^U_\alpha\right\}d\alpha$$

$$=E\left[\frac{\frac{1}{\lambda_1}+\frac{1}{\lambda_2}}{\frac{1}{\lambda_1}+\frac{1}{\lambda_2}+\frac{\lambda_1}{\mu_2(\lambda_1+\mu_2)}+\frac{\lambda_2}{\mu_1(\lambda_2+\mu_1)}}\right].$$

The theorem is proved.

Remark 6.3.3 If X_i and Y_i, $i=1,2$ degenerate to random variables, then the result in Theorem 6.3.3 degenerates to the form

$$A=\frac{\frac{1}{\lambda_1}+\frac{1}{\lambda_2}}{\frac{1}{\lambda_1}+\frac{1}{\lambda_2}+\frac{\lambda_1}{\mu_2(\lambda_1+\mu_2)}+\frac{\lambda_2}{\mu_1(\lambda_2+\mu_1)}},$$

which is consistent with the result in stochastic case.

Theorem 6.3.4 *The steady-state failure frequency of the repairable cold standby system is*

$$M=\frac{1}{2}\int_0^1\left[\frac{\frac{\lambda^U_{2,\alpha}}{\mu^U_{1,\alpha}+\lambda^U_{2,\alpha}}+\frac{\lambda^U_{1,\alpha}}{\mu^U_{2,\alpha}+\lambda^U_{1,\alpha}}}{\frac{1}{\lambda^U_{1,\alpha}}+\frac{1}{\lambda^U_{2,\alpha}}+\frac{1}{\mu^U_{1,\alpha}}+\frac{1}{\mu^U_{2,\alpha}}-\frac{1}{\mu^U_{1,\alpha}+\lambda^U_{2,\alpha}}-\frac{1}{\mu^U_{2,\alpha}+\lambda^U_{1,\alpha}}}\right.$$

$$\left.+\frac{\frac{\lambda^L_{2,\alpha}}{\mu^L_{1,\alpha}+\lambda^L_{2,\alpha}}+\frac{\lambda^L_{1,\alpha}}{\mu^L_{2,\alpha}+\lambda^L_{1,\alpha}}}{\frac{1}{\lambda^L_{1,\alpha}}+\frac{1}{\lambda^L_{2,\alpha}}+\frac{1}{\mu^L_{1,\alpha}}+\frac{1}{\mu^L_{2,\alpha}}-\frac{1}{\mu^L_{1,\alpha}+\lambda^L_{2,\alpha}}-\frac{1}{\mu^L_{2,\alpha}+\lambda^L_{1,\alpha}}}\right]d\alpha.$$

Proof Based on $X_i(\theta_i^1)$, $X_i(\theta_i)$, $X_i(\theta_i^2)$ and $Y_i(\vartheta_i^1)$, $Y_i(\vartheta_i)$, $Y_i(\vartheta_i^2)$, $i=1,2$, we can construct three stochastic repairable cold standby systems:

(1) Cold standby system 1: the lifetime and the repair time of component i are $X_i(\theta_i^1)$ and $Y_i(\vartheta_i^1)$, $i=1,2$, respectively.
(2) Cold standby system 2: the lifetime and the repair time of component i are $X_i(\theta_i^2)$ and $Y_i(\vartheta_i^2)$, $i=1,2$, respectively.
(3) Cold standby system 3: the lifetime and the repair time of component i are $X_i(\theta_i)$ and $Y_i(\vartheta_i)$, $i=1,2$, respectively.

The steady-state failure frequency of cold standby systems (1), (2) and (3) is denoted by M_1, M_2 and M_3, respectively. Since the cold standby system is a coherent system, we have

$$M_2 \le M_3 \le M_1.$$

Since θ_i and ϑ_i are arbitrary points in $P_{i,\alpha}$ and $Q_{i,\alpha}$, $i = 1, 2$, from the result in traditional reliability theory, we have

$$\left(\lim_{t\to+\infty}\frac{E[N(t)(\theta)]}{t}\right)_\alpha^L = M_2 = \frac{\frac{\lambda_2(\theta_2^2)}{\mu_1(\vartheta_1^2)+\lambda_2(\theta_2^2)}+\frac{\lambda_1(\theta_1^2)}{\mu_2(\vartheta_2^2)+\lambda_1(\theta_1^2)}}{\frac{1}{\lambda_1(\theta_1^2)}+\frac{1}{\lambda_2(\theta_2^2)}+\frac{1}{\mu_1(\vartheta_1^2)}+\frac{1}{\mu_2(\vartheta_2^2)}-\frac{1}{\mu_1(\vartheta_1^2)+\lambda_2(\theta_2^2)}-\frac{1}{\mu_2(\vartheta_2^2)+\lambda_1(\theta_1^2)}} \tag{6.3.23}$$

and

$$\left(\lim_{t\to+\infty}\frac{E[N(t)(\theta)]}{t}\right)_\alpha^U = M_1 = \frac{\frac{\lambda_2(\theta_2^1)}{\mu_1(\vartheta_1^1)+\lambda_2(\theta_2^1)}+\frac{\lambda_1(\theta_1^1)}{\mu_2(\vartheta_2^1)+\lambda_1(\theta_1^1)}}{\frac{1}{\lambda_1(\theta_1^1)}+\frac{1}{\lambda_2(\theta_2^1)}+\frac{1}{\mu_1(\vartheta_1^1)}+\frac{1}{\mu_2(\vartheta_2^1)}-\frac{1}{\mu_1(\vartheta_1^1)+\lambda_2(\theta_2^1)}-\frac{1}{\mu_2(\vartheta_2^1)+\lambda_1(\theta_1^1)}}. \tag{6.3.24}$$

By Lemma 6.2, Eqs. (6.3.19)–(6.3.24), we can arrive at

$$\begin{aligned}
M &= \frac{1}{2}\int_0^1\left\{\left(\lim_{t\to+\infty}\frac{E[N(t)(\theta)]}{t}\right)_\alpha^L+\left(\lim_{t\to+\infty}\frac{E[N(t)(\theta)]}{t}\right)_\alpha^U\right\}\mathrm{d}\alpha\\
&= \frac{1}{2}\int_0^1\left[\frac{\frac{\lambda_2(\theta_2^2)}{\mu_1(\vartheta_1^2)+\lambda_2(\theta_2^2)}+\frac{\lambda_1(\theta_1^2)}{\mu_2(\vartheta_2^2)+\lambda_1(\theta_1^2)}}{\frac{1}{\lambda_1(\theta_1^2)}+\frac{1}{\lambda_2(\theta_2^2)}+\frac{1}{\mu_1(\vartheta_1^2)}+\frac{1}{\mu_2(\vartheta_2^2)}-\frac{1}{\mu_1(\vartheta_1^2)+\lambda_2(\theta_2^2)}-\frac{1}{\mu_2(\vartheta_2^2)+\lambda_1(\theta_1^2)}}\right.\\
&\quad\left.+\frac{\frac{\lambda_2(\theta_2^1)}{\mu_1(\vartheta_1^1)+\lambda_2(\theta_2^1)}+\frac{\lambda_1(\theta_1^1)}{\mu_2(\vartheta_2^1)+\lambda_1(\theta_1^1)}}{\frac{1}{\lambda_1(\theta_1^1)}+\frac{1}{\lambda_2(\theta_2^1)}+\frac{1}{\mu_1(\vartheta_1^1)}+\frac{1}{\mu_2(\vartheta_2^1)}-\frac{1}{\mu_1(\vartheta_1^1)+\lambda_2(\theta_2^1)}-\frac{1}{\mu_2(\vartheta_2^1)+\lambda_1(\theta_1^1)}}\right]\mathrm{d}\alpha\\
&= \frac{1}{2}\int_0^1\left[\frac{\frac{\lambda_{2,\alpha}^U}{\mu_{1,\alpha}^U+\lambda_{2,\alpha}^U}+\frac{\lambda_{1,\alpha}^U}{\mu_{2,\alpha}^U+\lambda_{1,\alpha}^U}}{\frac{1}{\lambda_{1,\alpha}^U}+\frac{1}{\lambda_{2,\alpha}^U}+\frac{1}{\mu_{1,\alpha}^U}+\frac{1}{\mu_{2,\alpha}^U}-\frac{1}{\mu_{1,\alpha}^U+\lambda_{2,\alpha}^U}-\frac{1}{\mu_{2,\alpha}^U+\lambda_{1,\alpha}^U}}\right.\\
&\quad\left.+\frac{\frac{\lambda_{2,\alpha}^L}{\mu_{1,\alpha}^L+\lambda_{2,\alpha}^L}+\frac{\lambda_{1,\alpha}^L}{\mu_{2,\alpha}^L+\lambda_{1,\alpha}^L}}{\frac{1}{\lambda_{1,\alpha}^L}+\frac{1}{\lambda_{2,\alpha}^L}+\frac{1}{\mu_{1,\alpha}^L}+\frac{1}{\mu_{2,\alpha}^L}-\frac{1}{\mu_{1,\alpha}^L+\lambda_{2,\alpha}^L}-\frac{1}{\mu_{2,\alpha}^L+\lambda_{1,\alpha}^L}}\right]\mathrm{d}\alpha.
\end{aligned}$$

The theorem is proved.

Remark 6.3.3 If X_i and Y_i, $i = 1, 2$ degenerate to random variables, then the result in Theorem 6.3.4 degenerates to the form

$$M = \frac{\frac{\lambda_2}{\mu_1+\lambda_2} + \frac{\lambda_1}{\mu_2+\lambda_1}}{\frac{1}{\lambda_1} + \frac{1}{\lambda_2} + \frac{1}{\mu_1} + \frac{1}{\mu_2} - \frac{1}{\mu_1+\lambda_2} - \frac{1}{\mu_2+\lambda_1}},$$

which is consistent with the result in stochastic case.

Example 6.3.2 Suppose the cold standby system composed by two nonidentical components. Let X_i and Y_i be the lifetime and the repair time of component $i, i = 1, 2$. If $X_i \sim \mathcal{EXP}(\lambda_i)$ and $Y_i \sim \mathcal{EXP}(\mu_i), i = 1, 2$, where $\lambda_1 = (1, 2, 3), \lambda_2 = (2, 3, 4)$, $\mu_1 = (3, 4, 5)$ and $\mu_2 = (4, 5, 6)$. We can arrive at

$$\lambda^L_{1,\alpha} = 1 + \alpha,\ \lambda^U_{1,\alpha} = 3 - \alpha,\ \lambda^L_{2,\alpha} = 2 + \alpha,\ \lambda^U_{2,\alpha} = 4 - \alpha,$$
$$\mu^L_{1,\alpha} = 3 + \alpha,\ \mu^U_{1,\alpha} = 5 - \alpha,\ \mu^L_{2,\alpha} = 4 + \alpha,\ \mu^U_{2,\alpha} = 6 - \alpha.$$

It follows from Theorem 6.3.3 that

$$\begin{aligned} A &= \frac{1}{2}\int_0^1 \left[\frac{\frac{1}{\lambda^U_{1,\alpha}} + \frac{1}{\lambda^U_{2,\alpha}}}{\frac{1}{\lambda^U_{1,\alpha}} + \frac{1}{\lambda^U_{2,\alpha}} + \frac{\lambda^U_{1,\alpha}}{\mu^L_{2,\alpha}(\lambda^U_{1,\alpha}+\mu^L_{2,\alpha})} + \frac{\lambda^U_{2,\alpha}}{\mu^L_{1,\alpha}(\lambda^U_{2,\alpha}+\mu^L_{1,\alpha})}} \right. \\ &\quad \left. + \frac{\frac{1}{\lambda^L_{1,\alpha}} + \frac{1}{\lambda^L_{2,\alpha}}}{\frac{1}{\lambda^L_{1,\alpha}} + \frac{1}{\lambda^L_{2,\alpha}} + \frac{\lambda^L_{1,\alpha}}{\mu^U_{2,\alpha}(\lambda^L_{1,\alpha}+\mu^U_{2,\alpha})} + \frac{\lambda^L_{2,\alpha}}{\mu^U_{1,\alpha}(\lambda^L_{2,\alpha}+\mu^U_{1,\alpha})}} \right] \mathrm{d}\alpha \\ &= \frac{1}{2}\int_0^1 \left[\frac{\frac{1}{3-\alpha} + \frac{1}{4-\alpha}}{\frac{1}{3-\alpha} + \frac{1}{4-\alpha} + \frac{3-\alpha}{(4+\alpha)(3-\alpha+4+\alpha)} + \frac{4-\alpha}{(3+\alpha)(4-\alpha+3+\alpha)}} \right. \\ &\quad \left. + \frac{\frac{1}{1+\alpha} + \frac{1}{2+\alpha}}{\frac{1}{1+\alpha} + \frac{1}{2+\alpha} + \frac{1+\alpha}{(6-\alpha)(1+\alpha+6-\alpha)} + \frac{2+\alpha}{(5-\alpha)(2+\alpha+5-\alpha)}} \right] \mathrm{d}\alpha \\ &\approx 0.9052. \end{aligned}$$

It follows from Theorem 6.3.4 that

$$\begin{aligned} M &= \frac{1}{2}\int_0^1 \left[\frac{\frac{\lambda^U_{2,\alpha}}{\mu^U_{1,\alpha}+\lambda^U_{2,\alpha}} + \frac{\lambda^U_{1,\alpha}}{\mu^U_{2,\alpha}+\lambda^U_{1,\alpha}}}{\frac{1}{\lambda^U_{1,\alpha}} + \frac{1}{\lambda^U_{2,\alpha}} + \frac{1}{\mu^U_{1,\alpha}} + \frac{1}{\mu^U_{2,\alpha}} - \frac{1}{\mu^U_{1,\alpha}+\lambda^U_{2,\alpha}} - \frac{1}{\mu^U_{2,\alpha}+\lambda^U_{1,\alpha}}} \right. \\ &\quad \left. + \frac{\frac{\lambda^L_{2,\alpha}}{\mu^L_{1,\alpha}+\lambda^L_{2,\alpha}} + \frac{\lambda^L_{1,\alpha}}{\mu^L_{2,\alpha}+\lambda^L_{1,\alpha}}}{\frac{1}{\lambda^L_{1,\alpha}} + \frac{1}{\lambda^L_{2,\alpha}} + \frac{1}{\mu^L_{1,\alpha}} + \frac{1}{\mu^L_{2,\alpha}} - \frac{1}{\mu^L_{1,\alpha}+\lambda^L_{2,\alpha}} - \frac{1}{\mu^L_{2,\alpha}+\lambda^L_{1,\alpha}}} \right] \mathrm{d}\alpha \end{aligned}$$

$$= \frac{1}{2}\int_0^1 \left[\frac{\frac{(4-\alpha)}{(5-\alpha)+(4-\alpha)} + \frac{(3-\alpha)}{(6-\alpha)+(3-\alpha)}}{\frac{1}{(3-\alpha)} + \frac{1}{(4-\alpha)} + \frac{1}{(5-\alpha)} + \frac{1}{(6-\alpha)} - \frac{1}{(5-\alpha)+(4-\alpha)} - \frac{1}{(6-\alpha)+(3-\alpha)}} \right.$$

$$\left. + \frac{\frac{(2+\alpha)}{(3+\alpha)+(2+\alpha)} + \frac{(1+\alpha)}{(4+\alpha)+(1+\alpha)}}{\frac{1}{(1+\alpha)} + \frac{1}{(2+\alpha)} + \frac{1}{(3+\alpha)} + \frac{1}{(4+\alpha)} - \frac{1}{(3+\alpha)+(2+\alpha)} - \frac{1}{(4+\alpha)+(1+\alpha)}} \right] d\alpha$$

$$= 0.7148.$$

6.4 The Repairable Coherent System

Consider a repairable coherent system composed by n components and enough repairmen. If a component fails, the whole system goes into repair state instantly and the components in operating also can fail in this case. When repair is completed, the system is considered to be "as good as new". Let X_i and Y_i be the lifetime and the repair time of component $i, i = 1, 2, \ldots, n$. We assume that X_i and Y_i are random fuzzy variables defined on the credibility spaces $(\Theta_i, \mathcal{P}(\Theta_i), \text{Cr}_i)$ and $(\tau_i, \mathcal{P}(\tau_i), \text{Cr}_i')$, $i = 1, 2, \ldots, n$, respectively. We also assume X_i and $Y_i, i = 1, 2, \ldots, n$ are mutually independent.

Theorem 6.4.1 *The limiting availability of the random fuzzy repairable coherent system is*

$$A = \frac{1}{2}\int_0^1 \left(h(A_1^{(1)}, A_2^{(1)}, \ldots, A_n^{(1)}) + h(A_1^{(3)}, A_2^{(3)}, \ldots, A_n^{(3)}) \right) d\alpha,$$

where $h(\cdot)$ is the reliability function of the coherent system,

$$A_i^{(1)} = \frac{E[X_i(\theta_i)]_\alpha^L}{E[X_i(\theta_i)]_\alpha^L + E[Y_i(\vartheta_i)]_\alpha^U} \text{ and } A_i^{(3)} = \frac{E[X_i(\theta_i)]_\alpha^U}{E[X_i(\theta_i)]_\alpha^U + E[Y_i(\vartheta_i)]_\alpha^L}$$

for $\theta_i \in \Theta_i, \vartheta_i \in \tau_i, i = 1, 2, \ldots, n$.

Proof Let $A_{i,\alpha} = \{\theta_i \in \Theta_i | \mu\{\theta_i\} \geq \alpha\}$ and $B_{i,\alpha} = \{\vartheta_i \in \tau_i | \mu\{\vartheta_i\} \geq \alpha\}$, $i = 1, 2, \ldots, n$. Since the α-pessimistic values and the α-optimistic values of fuzzy variables $E[X_i(\theta_i)]$, $E[Y_i(\vartheta_i)]$, $\theta_i \in \Theta_i, \vartheta_i \in \tau_i, i = 1, 2, \ldots, n$ are continuous almost everywhere with respect to α, $\alpha \in (0, 1]$, then there at least exist points $\theta_i^1, \theta_i^2 \in A_{i,\alpha}$ and $\vartheta_i^1, \vartheta_i^2 \in B_{i,\alpha}$, $i = 1, 2, \ldots, n$ such that

$$E[X_i(\theta_i^1)] = E[X_i(\theta_i)]_\alpha^L, \tag{6.4.1}$$

$$E[X_i(\theta_i^2)] = E[X_i(\theta_i)]_\alpha^U, \tag{6.4.2}$$

$$E[Y_i(\vartheta_i^1)] = E[Y_i(\vartheta_i)]_\alpha^L, \tag{6.4.3}$$

$$E[Y_i(\vartheta_i^2)] = E[Y_i(\vartheta_i)]_\alpha^U. \tag{6.4.4}$$

For $\forall\theta_{i,\alpha} \in A_{i,\alpha}, \forall\vartheta_{i,\alpha} \in B_{i,\alpha}, i = 1, 2, \ldots, n$, we have

$$E[X_i(\theta_i^1)] \leq E[X_i(\theta_{i,\alpha})] \leq E[X_i(\theta_i^2)] \tag{6.4.5}$$

and

$$E[Y_i(\vartheta_i^1)] \leq E[Y_i(\vartheta_{i,\alpha})] \leq E[Y_i(\vartheta_i^2)]. \tag{6.4.6}$$

By Theorem 1.1.21, for $i = 1, 2, \ldots, n$, we have

$$X_i(\theta_i^1) \leq_d X_i(\theta_{i,\alpha}) \leq_d X_i(\theta_i^2)$$

and

$$Y_i(\vartheta_i^1) \leq_d Y_i(\vartheta_{i,\alpha}) \leq_d Y_i(\vartheta_i^2).$$

Based on $X_i(\theta_i^1)$, $X_i(\theta_{i,\alpha})$, $X_i(\theta_i^2)$ and $Y_i(\vartheta_i^1)$, $Y_i(\vartheta_{i,\alpha})$, $Y_i(\vartheta_i^2)$, we can construct three stochastic repairable coherent systems with random lifetimes and repair times:

(1) Coherent system 1: the lifetime and the repair time of component i are $X_i(\theta_i^1)$ and $Y_i(\vartheta_i^2)$, $i = 1, 2, \ldots, n$, respectively.
(2) Coherent system 2: the lifetime and the repair time of component i are $X_i(\theta_{i,\alpha})$ and $Y_i(\vartheta_{i,\alpha})$, $i = 1, 2, \ldots, n$, respectively.
(3) Coherent system 3: the lifetime and the repair time of component i are $X_i(\theta_i^2)$ and $Y_i(\vartheta_i^1)$, $i = 1, 2, \ldots, n$, respectively.

The limiting availability of coherent systems (1), (2) and (3) is denoted by $A^{(1)}$, $A^{(2)}$ and $A^{(3)}$, respectively. Let $h(\cdot)$ be the reliability function of the coherent system, $A_i^{(1)}$, $A_i^{(2)}$ and $A_i^{(3)}$ be the limiting availability of component $i, i = 1, 2, \ldots, n$ in coherent systems (1), (2) and (3), respectively. By the result in classical reliability theory, we can arrive at

$$\begin{cases} A^{(1)} = h(A_1^{(1)}, A_2^{(1)}, \ldots, A_n^{(1)}) \\ A_i^{(1)} = \dfrac{E[X_i(\theta_i^1)]}{E[X_i(\theta_i^1)] + E[Y_i(\vartheta_i^2)]}, i = 1, 2, \ldots, n, \end{cases} \tag{6.4.7}$$

$$\begin{cases} A^{(2)} = h(A_1^{(2)}, A_2^{(2)}, \ldots, A_n^{(2)}) \\ A_i^{(2)} = \dfrac{E[X_i(\theta_{i,\alpha})]}{E[X_i(\theta_{i,\alpha})] + E[Y_i(\vartheta_{i,\alpha})]}, i = 1, 2, \ldots, n \end{cases} \tag{6.4.8}$$

and

$$\begin{cases} A^{(3)} = h(A_1^{(3)}, A_2^{(3)}, \ldots, A_n^{(3)}) \\ A_i^{(3)} = \dfrac{E[X_i(\theta_i^2)]}{E[X_i(\theta_i^2)] + E[Y_i(\vartheta_i^1)]}, \ i = 1, 2, \ldots, n. \end{cases} \tag{6.4.9}$$

By Eqs. (6.4.5)–(6.4.9), we can arrive at

$$A_i^{(1)} \le A_i^{(2)} \le A_i^{(3)}, i = 1, 2, \ldots, n.$$

Since $\theta_{i,\alpha}$ and $\vartheta_{i,\alpha}$ are arbitrary points in $A_{i,\alpha}$ and $B_{i,\alpha}$, $i = 1, 2, \ldots, n$, we have

$$\lim_{t \to +\infty} \mathrm{Pr}_\alpha^L\{S(t)(\theta) = 1\} = A^{(1)} = h(A_1^{(1)}, A_2^{(1)}, \ldots, A_n^{(1)}) \tag{6.4.10}$$

and

$$\lim_{t \to +\infty} \mathrm{Pr}_\alpha^U\{S(t)(\theta) = 1\} = A^{(3)} = h(A_1^{(3)}, A_2^{(3)}, \ldots, A_n^{(3)}), \tag{6.4.11}$$

in which

$$A_i^{(1)} = \frac{E[X_i(\theta_i^1)]}{E[X_i(\theta_i^1)] + E[Y_i(\vartheta_i^2)]}, \ A_i^{(3)} = \frac{E[X_i(\theta_i^2)]}{E[X_i(\theta_i^2)] + E[Y_i(\vartheta_i^1)]}, \ i = 1, 2, \ldots, n.$$

By Lemma 6.1, Eqs. (6.4.1)–(6.4.4), (6.4.10) and (6.4.11), we can arrive at

$$A = \frac{1}{2} \int_0^1 \Big(h(A_1^{(1)}, A_2^{(1)}, \ldots, A_n^{(1)}) + h(A_1^{(3)}, A_2^{(3)}, \ldots, A_n^{(3)}) \Big) \mathrm{d}\alpha,$$

in which

$$A_i^{(1)} = \frac{E[X_i(\theta_i)]_\alpha^L}{E[X_i(\theta_i)]_\alpha^L + E[Y_i(\vartheta_i)]_\alpha^U},$$
$$A_i^{(3)} = \frac{E[X_i(\theta_i)]_\alpha^U}{E[X_i(\theta_i)]_\alpha^U + E[Y_i(\vartheta_i)]_\alpha^L}, i = 1, 2, \ldots, n.$$

The theorem is proved.

Remark 6.4.1 If X_i and Y_i, $i = 1, 2, \ldots, n$ degenerate to random variables, then the result in Theorem 6.4.1 degenerates to the form

$$\begin{cases} A = h(A_1, A_2, \ldots, A_n) \\ A_i = \dfrac{E[X_i]}{E[X_i] + E[Y_i]}, i = 1, 2, \ldots, n, \end{cases}$$

which is consistent with the result in stochastic case.

Theorem 6.4.2 *The steady-state failure frequency of the repairable coherent system is*

$$M = \frac{1}{2}\sum_{i=1}^{n}\int_0^1 \left\{ \frac{h(1_i, A^{(1)}) - h(0_i, A^{(1)})}{E[X_i(\theta_i)]_\alpha^L + E[Y_i(\vartheta_i)]_\alpha^L} + \frac{h(1_i, A^{(3)}) - h(0_i, A^{(3)})}{E[X_i(\theta_i)]_\alpha^U + E[Y_i(\vartheta_i)]_\alpha^U} \right\} \mathrm{d}\alpha,$$

where $\theta_i \in \Theta_i$, $\vartheta_i \in \tau_i$, $h(\cdot)$ *is the reliability function of the coherent system,*

$$\mathrm{A}^{(1)} = \left(\frac{E[X_1(\theta_1)]_\alpha^L}{E[X_1(\theta_1)]_\alpha^L + E[Y_1(\vartheta_1)]_\alpha^L}, \ldots, \frac{E[X_n(\theta_n)]_\alpha^L}{E[X_n(\theta_n)]_\alpha^L + E[Y_n(\vartheta_n)]_\alpha^L} \right)$$

and

$$\mathrm{A}^{(3)} = \left(\frac{E[X_1(\theta_1)]_\alpha^U}{E[X_1(\theta_1)]_\alpha^U + E[Y_1(\vartheta_1)]_\alpha^U}, \ldots, \frac{E[X_n(\theta_n)]_\alpha^U}{E[X_n(\theta_n)]_\alpha^U + E[Y_n(\vartheta_n)]_\alpha^U} \right).$$

Proof Based on the random variables $X_i(\theta_i^1)$, $X_i(\theta_{i,\alpha})$, $X_i(\theta_i^2)$ and $Y_i(\vartheta_i^1)$, $Y_i(\vartheta_{i,\alpha})$, $Y_i(\vartheta_i^2)$, $i = 1, 2, \ldots, n$ in the proof of Theorem 6.4.1, we can construct three stochastic repairable coherent systems:

(1) Coherent system 1: the lifetime and the repair time of component i are $X_i(\theta_i^1)$ and $Y_i(\vartheta_i^1)$, $i = 1, 2, \ldots, n$, respectively.
(2) Coherent system 2: the lifetime and the repair time of component i are $X_i(\theta_{i,\alpha})$ and $Y_i(\vartheta_{i,\alpha})$, $i = 1, 2, \ldots, n$, respectively.
(3) Coherent system 3: the lifetime and the repair time of component i are $X_i(\theta_i^2)$ and $Y_i(\vartheta_i^2)$, $i = 1, 2, \ldots, n$, respectively.

The steady-state failure frequency of coherent systems (1), (2) and (3) is denoted by $M^{(1)}$, $M^{(2)}$ and $M^{(3)}$, respectively. From the result in stochastic case, we can arrive at

$$\begin{cases} M^{(1)} = \sum_{i=1}^{n} \dfrac{h(1_i, A^{(1)}) - h(0_i, A^{(1)})}{E[X_i(\theta_i^1)] + E[Y_i(\vartheta_i^1)]} \\ A^{(1)} = \left(\dfrac{E[X_1(\theta_1^1)]}{E[X_1(\theta_1^1)] + E[Y_1(\vartheta_1^1)]}, \ldots, \dfrac{E[X_n(\theta_n^1)]}{E[X_n(\theta_n^1)] + E[Y_n(\vartheta_n^1)]} \right), \end{cases} \tag{6.4.12}$$

$$\begin{cases} M^{(2)} = \sum_{i=1}^{n} \frac{h(1_i, A^{(2)}) - h(0_i, A^{(2)})}{E[X_i(\theta_i)] + E[Y_i(\vartheta_i)]} \\ A^{(2)} = \left(\frac{E[X_1(\theta_{1,\alpha})]}{E[X_1(\theta_{1,\alpha})] + E[Y_1(\vartheta_{1,\alpha})]}, \ldots, \frac{E[X_n(\theta_{n,\alpha})]}{E[X_n(\theta_{n,\alpha})] + E[Y_n(\vartheta_{n,\alpha})]} \right) \end{cases} \tag{6.4.13}$$

and

$$\begin{cases} M^{(3)} = \sum_{i=1}^{n} \frac{h(1_i, A^{(3)}) - h(0_i, A^{(3)})}{E[X_i(\theta_i^2)] + E[Y_i(\vartheta_i^2)]} \\ A^{(3)} = \left(\frac{E[X_1(\theta_1^2)]}{E[X_1(\theta_1^2)] + E[Y_1(\vartheta_1^2)]}, \ldots, \frac{E[X_n(\theta_n^2)]}{E[X_n(\theta_n^2)] + E[Y_n(\vartheta_n^2)]} \right). \end{cases} \tag{6.4.14}$$

It is easy to see that

$$M^{(3)} \le M^{(2)} \le M^{(1)}.$$

Since $\theta_{i,\alpha}$ and $\vartheta_{i,\alpha}$ are arbitrary points in $A_{i,\alpha}$ and $B_{i,\alpha}$, $i = 1, 2, \ldots, n$, we have

$$\begin{cases} \lim_{t \to +\infty} \frac{E[N(t)(\theta)]_\alpha^L}{t} = M^{(3)} = \sum_{i=1}^{n} \frac{h(1_i, A^{(3)}) - h(0_i, A^{(3)})}{E[X_i(\theta_i^2)] + E[Y_i(\vartheta_i^2)]} \\ A^{(3)} = \left(\frac{E[X_1(\theta_1^2)]}{E[X_1(\theta_1^2)] + E[Y_1(\vartheta_1^2)]}, \ldots, \frac{E[X_n(\theta_n^2)]}{E[X_n(\theta_n^2)] + E[Y_n(\vartheta_n^2)]} \right) \end{cases} \tag{6.4.15}$$

and

$$\begin{cases} \lim_{t \to +\infty} \frac{E[N(t)(\theta)]_\alpha^U}{t} = M^{(1)} = \sum_{i=1}^{n} \frac{h(1_i, A^{(1)}) - h(0_i, A^{(1)})}{E[X_i(\theta_i^1)] + E[Y_i(\vartheta_i^1)]} \\ A^{(1)} = \left(\frac{E[X_1(\theta_1^1)]}{E[X_1(\theta_1^1)] + E[Y_1(\vartheta_1^1)]}, \ldots, \frac{E[X_n(\theta_n^1)]}{E[X_n(\theta_n^1)] + E[Y_n(\vartheta_n^1)]} \right). \end{cases} \tag{6.4.16}$$

By Lemma 6.2, Eqs. (6.4.1)–(6.4.4), (6.4.15) and (6.4.16), we can arrive at

$$\begin{aligned} M &= \frac{1}{2} \int_0^1 \left\{ \sum_{i=1}^{n} \frac{h(1_i, A^{(3)}) - h(0_i, A^{(3)})}{E[X_i(\theta_i)]_\alpha^U + E[Y_i(\vartheta_i)]_\alpha^U} + \sum_{i=1}^{n} \frac{h(1_i, A^{(1)}) - h(0_i, A^{(1)})}{E[X_i(\theta_i)]_\alpha^L + E[Y_i(\vartheta_i)]_\alpha^L} \right\} d\alpha \\ &= \frac{1}{2} \sum_{i=1}^{n} \int_0^1 \left\{ \frac{h(1_i, A^{(1)}) - h(0_i, A^{(1)})}{E[X_i(\theta_i)]_\alpha^L + E[Y_i(\vartheta_i)]_\alpha^L} + \frac{h(1_i, A^{(3)}) - h(0_i, A^{(3)})}{E[X_i(\theta_i)]_\alpha^U + E[Y_i(\vartheta_i)]_\alpha^U} \right\} d\alpha, \end{aligned}$$

in which

$$A^{(1)} = \left(\frac{E[X_1(\theta_1)]_\alpha^L}{E[X_1(\theta_1)]_\alpha^L + E[Y_1(\vartheta_1)]_\alpha^L}, \ldots, \frac{E[X_n(\theta_n)]_\alpha^L}{E[X_n(\theta_n)]_\alpha^L + E[Y_n(\vartheta_n)]_\alpha^L} \right)$$

and

$$A^{(3)} = \left(\frac{E[X_1(\theta_1)]_\alpha^U}{E[X_1(\theta_1)]_\alpha^U + E[Y_1(\vartheta_1)]_\alpha^U}, \ldots, \frac{E[X_n(\theta_n)]_\alpha^U}{E[X_n(\theta_n)]_\alpha^U + E[Y_n(\vartheta_n)]_\alpha^U} \right).$$

The theorem is proved.

Remark 6.4.2 If X_i and Y_i, $i = 1, 2, \ldots, n$ degenerate to random variables, then the result in Theorem 6.4.2 degenerates to the form

$$M = \sum_{i=1}^{n} \frac{h(1_i, A) - h(0_i, A)}{E[X_i] + E[Y_i]},$$

in which

$$A = \left(\frac{E[X_1]}{E[X_1] + E[Y_1]}, \ldots, \frac{E[X_n]}{E[X_n] + E[Y_n]} \right),$$

which is consistent with the result in stochastic case.

Example 6.4.1 The structure of a repairable series–parallel system is shown in Fig. 6.1. The lifetimes and the repair times of components 1, 2 and 3 are denoted by X_1, X_2, X_3 and Y_1, Y_2, Y_3, respectively. We assumed that $X_i \sim \mathcal{EXP}(\lambda_i)$ and $Y_i \sim \mathcal{EXP}(\mu_i), i = 1, 2, 3$, where $\lambda_1 = (1, 2, 3)$, $\lambda_2 = (0.5, 1.5, 2.5)$, $\lambda_3 = (1.5, 2.5, 3.5)$, $\mu_1 = (3, 4, 5)$, $\mu_2 = (2.5, 3.5, 4.5)$ and $\mu_3 = (3.5, 4.5, 5.5)$. We can arrive at

$$\begin{aligned}
&\lambda_{1,\alpha}^L = 1 + \alpha, \quad \lambda_{2,\alpha}^L = 0.5 + \alpha, \quad \lambda_{3,\alpha}^L = 1.5 + \alpha, \\
&\lambda_{1,\alpha}^U = 3 - \alpha, \quad \lambda_{2,\alpha}^U = 2.5 - \alpha, \quad \lambda_{3,\alpha}^U = 3.5 - \alpha, \\
&\mu_{1,\alpha}^L = 3 + \alpha, \quad \mu_{2,\alpha}^L = 2.5 + \alpha, \quad \mu_{3,\alpha}^L = 3.5 + \alpha, \\
&\mu_{1,\alpha}^U = 5 - \alpha, \quad \mu_{2,\alpha}^U = 4.5 - \alpha, \quad \mu_{3,\alpha}^U = 5.5 - \alpha.
\end{aligned}$$

Then we have

$$\begin{aligned}
&E[X_1(\theta)]_\alpha^L = \tfrac{1}{3-\alpha}, \quad E[X_2(\theta)]_\alpha^L = \tfrac{1}{2.5-\alpha}, \quad E[X_3(\theta)]_\alpha^L = \tfrac{1}{3.5-\alpha}, \\
&E[X_1(\theta)]_\alpha^U = \tfrac{1}{1+\alpha}, \quad E[X_2(\theta)]_\alpha^U = \tfrac{1}{0.5+\alpha}, \quad E[X_3(\theta)]_\alpha^U = \tfrac{1}{1.5+\alpha}, \\
&E[Y_1(\vartheta)]_\alpha^L = \tfrac{1}{5-\alpha}, \quad E[Y_2(\vartheta)]_\alpha^L = \tfrac{1}{4.5-\alpha}, \quad E[Y_3(\vartheta)]_\alpha^L = \tfrac{1}{5.5-\alpha}, \\
&E[Y_1(\vartheta)]_\alpha^U = \tfrac{1}{3+\alpha}, \quad E[Y_2(\vartheta)]_\alpha^U = \tfrac{1}{2.5+\alpha}, \quad E[Y_3(\vartheta)]_\alpha^U = \tfrac{1}{3.5+\alpha}.
\end{aligned}$$

To obtain the limiting availability, we should first calculate the following formula

$$A_1^{(1)} = \frac{1}{6}(3 + \alpha), \quad A_2^{(1)} = \frac{1}{5}(2.5 + \alpha), \quad A_3^{(1)} = \frac{1}{7}(3.5 + \alpha)$$

and

$$A_1^{(3)} = \frac{1}{6}(5 - \alpha), \quad A_2^{(3)} = \frac{1}{5}(4.5 - \alpha), \quad A_3^{(3)} = \frac{1}{7}(5.5 - \alpha)$$

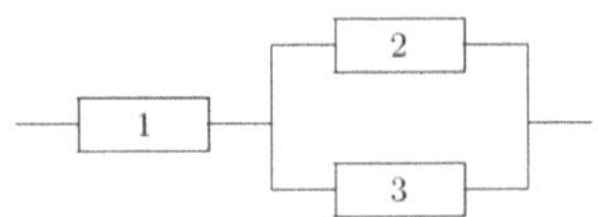

Fig. 6.1 Repairable series–parallel system

By Theorem 6.4.1, the limiting availability of the series–parallel system is

$$A = \frac{1}{2}\int_0^1 \Big((h(A_1^{(1)}, A_2^{(1)}, A_3^{(1)}) + h(A_1^{(3)}, A_2^{(3)}, A_3^{(3)})\Big)\mathrm{d}\alpha$$
$$= \frac{1}{2}\int_0^1 \left\{\frac{1}{6}(3+\alpha)\left[1-\left(1-\frac{1}{5}(2.5+\alpha)\right)\left(1-\frac{1}{7}(3.5+\alpha)\right)\right]\right.$$
$$\left. + \frac{1}{6}(5-\alpha)\left[1-\left(1-\frac{1}{5}(4.5+\alpha)\right)\left(1-\frac{1}{7}(5.5+\alpha)\right)\right]\right\}\mathrm{d}\alpha \approx 0.5952.$$

To obtain the steady-state failure frequency, we can arrive at

$$\mathrm{A}^{(1)} = \left(\frac{5-\alpha}{8-2\alpha}, \frac{4.5-\alpha}{7-2\alpha}, \frac{5.5-\alpha}{9-2\alpha}\right) \text{and } A^{(3)} = \left(\frac{3+\alpha}{4+2\alpha}, \frac{2.5+\alpha}{3+2\alpha}, \frac{3.5+\alpha}{5+2\alpha}\right).$$

By Theorem 6.4.2, the steady-state failure frequency of the series–parallel system is

$$M = \frac{1}{2}\int_0^1 \left[\frac{1-(1-\frac{4.5-\alpha}{7-2\alpha})(1-\frac{5.5-\alpha}{9-2\alpha})}{\frac{1}{3-\alpha}+\frac{1}{5-\alpha}} + \frac{\frac{5-\alpha}{8-2\alpha}(1-\frac{5.5-\alpha}{9-2\alpha})}{\frac{1}{2.5-\alpha}+\frac{1}{4.5-\alpha}}\right.$$
$$+ \frac{\frac{5-\alpha}{8-2\alpha}(1-\frac{4.5-\alpha}{7-2\alpha})}{\frac{1}{3.5-\alpha}+\frac{1}{5.5-\alpha}} + \frac{1-(1-\frac{2.5-\alpha}{3+2\alpha})(1-\frac{3.5+\alpha}{5+2\alpha})}{\frac{1}{1+\alpha}+\frac{1}{3+\alpha}}$$
$$\left. + \frac{\frac{3+\alpha}{4+2\alpha}(1-\frac{3.5+\alpha}{5+2\alpha})}{\frac{1}{0.5+\alpha}+\frac{1}{2.5-\alpha}} + \frac{\frac{3+\alpha}{4+2\alpha}(1-\frac{2.5+\alpha}{3+2\alpha})}{\frac{1}{1.5+\alpha}+\frac{1}{3.5+\alpha}}\right]\mathrm{d}\alpha \approx 1.6999.$$

Example 6.4.2 Consider a lighting system composed by three lamps. The entire lighting system is functioning when at least two lamps are functioning. The lighting system is just a 2.out-of-3 system, and the structure is shown in Fig. 6.2. The random fuzzy lifetimes and the repair times of lamp 1, lamp 2 and lamp 3 are denoted by X_1, X_2, X_3 and Y_1, Y_2, Y_3, respectively. If $X_i \sim \mathcal{EXP}(\lambda_i)$, $Y_i \sim \mathcal{EXP}(\mu_i)$, $i = 1, 2, 3$, in which $\lambda_1 = (1, 2, 3)$, $\lambda_2 = (0.5, 1.5, 2.5)$, $\lambda_3 = (1.5, 2.5, 3.5)$, $\mu_1 = (3, 4, 5)$, $\mu_2 = (2.5, 3.5, 4.5)$ and $\mu_3 = (3.5, 4.5, 5.5)$.

By Theorem 6.4.1, the limiting availability of the lighting system is

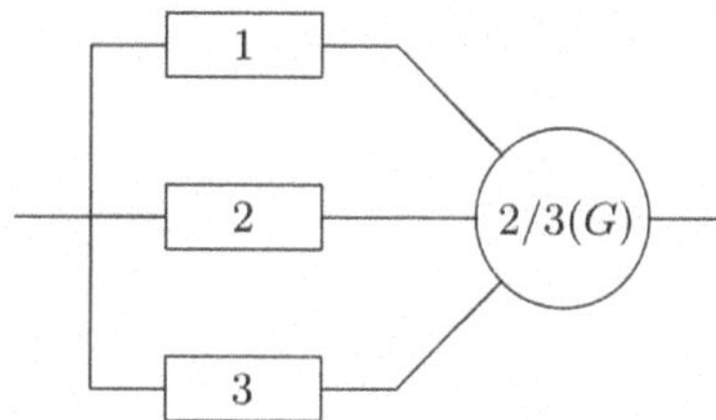

Fig. 6.2 2-out-of-3 system

$$A = \frac{1}{2}\int_0^1 \Big((h(A_1^{(1)}, A_2^{(1)}, A_3^{(1)}) + h(A_1^{(3)}, A_2^{(3)}, A_3^{(3)})\Big)\mathrm{d}\alpha$$

$$= \frac{1}{2}\int_0^1 \Big(A_1^{(1)}A_2^{(1)} + A_1^{(1)}A_3^{(1)} + A_2^{(1)}A_3^{(1)} - 2A_1^{(1)}A_2^{(1)}A_3^{(1)}$$

$$+ A_1^{(3)}A_2^{(3)} + A_1^{(3)}A_3^{(3)} + A_2^{(3)}A_3^{(3)} - 2A_1^{(3)}A_2^{(3)}A_3^{(3)}\Big)\mathrm{d}\alpha$$

$$= \frac{1}{2}\int_0^1 \Big[\frac{1}{30}(3+\alpha)(2.5+\alpha) + \frac{1}{42}(3+\alpha)(3.5+\alpha) + \frac{1}{35}(2.5+\alpha)(3.5+\alpha)$$

$$- \frac{1}{105}(3+\alpha)(2.5+\alpha)(3.5+\alpha) + \frac{1}{30}(5-\alpha)(4.5-\alpha) + \frac{1}{42}(5-\alpha)(5.5-\alpha)$$

$$+ \frac{1}{35}(4.5-\alpha)(5.5-\alpha) - \frac{1}{105}(5-\alpha)(4.5-\alpha)(5.5-\alpha)\Big]\mathrm{d}\alpha$$

$$= \frac{103}{140} \approx 0.7357.$$

By Theorem 6.4.2, the steady-state failure frequency of the lighting system is

$$M = \frac{1}{2}\int_0^1 \Big[\Big(\frac{4.5-\alpha}{7-2\alpha} + \frac{5.5-\alpha}{9-2\alpha} - 2\cdot\frac{4.5-\alpha}{7-2\alpha}\cdot\frac{5.5-\alpha}{9-2\alpha}\Big)\Big/\Big(\frac{1}{3-\alpha} + \frac{1}{5-\alpha}\Big)$$

$$+ \Big(\frac{5-\alpha}{8-2\alpha} + \frac{5.5-\alpha}{9-2\alpha} - 2\cdot\frac{5-\alpha}{8-2\alpha}\cdot\frac{5.5-\alpha}{9-2\alpha}\Big)\Big/\Big(\frac{1}{2.5-\alpha} + \frac{1}{4.5-\alpha}\Big)$$

$$+ \Big(\frac{5-\alpha}{8-2\alpha} + \frac{4.5-\alpha}{7-2\alpha} - 2\cdot\frac{5-\alpha}{8-2\alpha}\cdot\frac{4.5-\alpha}{7-2\alpha}\Big)\Big/\Big(\frac{1}{3.5-\alpha} + \frac{1}{5.5-\alpha}\Big)$$

$$+ \Big(\frac{2.5+\alpha}{3+2\alpha} + \frac{3.5+\alpha}{5+2\alpha} - 2\cdot\frac{2.5+\alpha}{3+2\alpha}\cdot\frac{3.5+\alpha}{5+2\alpha}\Big)\Big/\Big(\frac{1}{1+\alpha} + \frac{1}{3+\alpha}\Big)$$

$$+ \Big(\frac{3+\alpha}{4+2\alpha} + \frac{3.5+\alpha}{5+2\alpha} - 2\cdot\frac{3+\alpha}{4+2\alpha}\cdot\frac{3.5+\alpha}{5+2\alpha}\Big)\Big/\Big(\frac{1}{0.5+\alpha} + \frac{1}{2.5+\alpha}\Big)$$

$$+ \Big(\frac{3+\alpha}{4+2\alpha} + \frac{2.5+\alpha}{3+2\alpha} - 2\cdot\frac{3+\alpha}{4+2\alpha}\cdot\frac{2.5+\alpha}{3+2\alpha}\Big)\Big/\Big(\frac{1}{1.5+\alpha} + \frac{1}{3.5+\alpha}\Big)\Big]\mathrm{d}\alpha$$

$$\approx 1.7454.$$

Bibliography

1. R. von Mises, Wahrscheinlichkeitsrechnung und ihre Anwendung in der Statistik und Theoretischen Physik. Leipzig and Wien, Franz Deuticke (1931)
2. A. Kolmogorov, Grundbegriffe der Wahrscheinlichkeitsrechnung. Julius Springer, Berlin, 1933
3. B. Liu, *Theory and Practice of Uncertain Programming* (Physica-Verlag, Heidelberg, 2002).
4. B. Liu, *Uncertainty Theory: An Introduction to Its Axiomatic Foundations* (Springer, Berlin, 2004).
5. B. Liu, A survey of credibility theory. Fuzzy Optim. Decis. Making **5**(4), 387–408 (2006)
6. B. Liu, Y. Liu, Expected value of fuzzy variable and fuzzy expected value models. IEEE Trans. Fuzzy Syst. **10**(4), 445–450 (2002)
7. Y. Liu, B. Liu, Random fuzzy programming with chance measures defined by fuzzy integrals. Math. Comput. Modell. **36**(4–5), 509–524 (2002)
8. Y. Liu, B. Liu, Fuzzy random variables: a scalar expected value. Fuzzy Optim. Decis. Making **2**(2), 143–160 (2003)
9. Y. Liu, B. Liu, Expected value operator of random fuzzy variable and random fuzzy expected value models. Int. J. Uncertainty Fuzziness Knowl. Based Syst. **11**(2), 195–215 (2003)
10. Y. Liu, X. Li, G. Yang, Reliability analysis of random fuzzy unrepairable cold standby systems with imperfect conversion switches. Inf. Int. Interdisc. J. **14**(2), 283–296 (2011)
11. Y. Liu, X. Li, Reliability analysis of random fuzzy unrepairable warm standby systems, in *Proceedings of the Fourth International Forum on Decision Sciences. Uncertainty and Operations Research* (Springer, Berlin, 2017), pp. 461–476
12. Y. Liu, X. Li, Z. Du, Reliability analysis of a random fuzzy repairable parallel system with two non-identical components. J. Intell. Fuzzy Syst. **27**(6), 2775–2784 (2014)
13. Y. Liu, X. Li, J. Li, Reliability analysis of random fuzzy unrepairable systems. Discrete Dyn. Nat. Soc. Article ID 625985 (2014), p. 15
14. Y. Liu, X. Li, L. Wang, Random fuzzy unrepairable warm standby systems. Fuzzy Inf. Eng. Oper. Res. Manage. Adv. Intell. Syst. Comput. **211**, 491–500 (2014)
15. Y. Liu, X. Li, G. Yang, Reliability analysis of random fuzzy repairable series system. Fuzzy Inf. Eng. Adv. Soft Comput. **78**(1/2), 281–296 (2010)
16. Y. Liu, X. Li, Y. Yu, D. Wang, Reliability analysis of random fuzzy repairable cold standby systems with two units, in 13th *International Conference on Natural Computation, Fuzzy Systems and Knowledge Discovery* (2018), pp. 1349–1356
17. Y. Liu, X. Li, Y. Zhang, Random fuzzy repairable coherent systems with independent components. Int. J. Uncertainty Fuzziness Knowl. Based Syst. **24**(6), 859–872 (2016)
18. Y. Liu, W. Tang, R. Zhao, Reliability and mean time to failure of unrepairable systems with fuzzy random lifetimes. IEEE Trans. Fuzzy Syst. **15**(5), 1009–1026 (2007)
19. Y. Liu, W. Tang, X. Li, Random fuzzy shock models and bivariate random fuzzy exponential distribution. Appl. Math. Model. **35**(5), 2408–2418 (2011)

Y. Liu, *Reliability Theory Based on Uncertain Lifetimes*,
https://doi.org/10.1007/978-981-16-0995-4

20. Y. Liu, H. Zhu, Reliability analysis of fuzzy unrepairable systems. Inf. Int. Interdisc. J. **15**(10), 3935–3944 (2012)
21. L. Zadeh, Fuzzy sets. Inf. Control **8**(3), 338–353 (1965)
22. L. Zadeh, Fuzzy sets as a basis for a theory of possibility. Fuzzy Sets Syst. **1**(1), 3–28 (1978)
23. R. Zhao, W. Tang, H. Yun, Random fuzzy renewal process. Eur. J. Oper. Res. **169**(1), 189–201 (2006)
24. J. Cao, K. Cheng, *Introduction to Reliability Mathematics*, Revised. (Higher Education Press, Beijing, 2012).

The manufacturer's authorised representative in the EU is Springer Nature Customer Service Centre GmbH, Europaplatz 3, 69115 Heidelberg, Germany. If you have any concerns regarding our products, please contact ProductSafety@springernature.com

Printed and bound by CPI Group (UK) Ltd, Croydon, CR0 4YY
15/07/2026
02167632-0003